NUTRIENTS IN
NATURAL WATERS

ENVIRONMENTAL SCIENCE AND TECHNOLOGY

A Wiley-Interscience Series of Texts and Monographs

Edited by ROBERT L. METCALF, *University of Illinois*
JAMES H. PITTS, Jr., *University of California*

PRINCIPLES AND PRACTICES OF INCINERATION
Richard C. Corey

AN INTRODUCTION TO EXPERIMENTAL AEROBIOLOGY
Robert L. Dimmick

AIR POLLUTION CONTROL, Part I
Werner Strauss

AIR POLLUTION CONTROL, Part II
Werner Strauss

APPLIED STREAM SANITATION
Clarence J. Velz

ENVIRONMENT, POWER, AND SOCIETY
Howard T. Odum

PHYSICOCHEMICAL PROCESSES FOR WATER QUALITY CONTROL
Walter J. Weber, Jr.

ENVIRONMENTAL SANITATION
Joseph A. Salvato, Jr.

NUTRIENTS IN NATURAL WATERS
Herbert E. Allen and James R. Kramer, Editors

NUTRIENTS IN NATURAL WATERS

EDITED BY

HERBERT E. ALLEN

University of Michigan
Ann Arbor, Michigan

JAMES R. KRAMER

McMaster University
Hamilton, Ontario, Canada

A WILEY-INTERSCIENCE PUBLICATION

A DIVISION OF JOHN WILEY & SONS

NEW YORK • LONDON • SYDNEY • TORONTO

Library of Congress Cataloging in Publication Data

Allen, Herbert Ellis, 1939–
Nutrients in natural waters.

(Environmental science and technology)
1. Limnology. 2. Eutrophication. 3. Water—Pollution. 4. Water—Analysis. I. Kramer, James Richard, 1931– joint author. II. Title.
QH96.A48 574.92′9 72–3786

ISBN O–471–02328–0

Printed in the United States of America

10 9 8 7 6 5 4 3 2 1

SERIES PREFACE

Environmental Science and Technology

The Environmental Science and Technology Series of Monographs, Textbooks, and Advances is devoted to the study of the quality of the environment and to the technology of its conservation. Environmental science therefore relates to the chemical, physical, and biological changes in the environment through contamination or modification, to the physical nature and biological behavior of air, water, soil, food, and waste as they are affected by man's agricultural, industrial, and social activities, and to the application of science and technology to the control and improvement of environmental quality.

The deterioration of environmental quality, which began when man first collected into villages and utilized fire, has existed as a serious problem under the ever-increasing impacts of exponentially increasing population and of industrializing society, environmental contamination of air, water, soil, and food has become a threat to the continued existence of many plant and animal communities of the ecosystem and may ultimately threaten the very survival of the human race.

It seems clear that if we are to preserve for future generations some semblance of the biological order of the world of the past and hope to improve on the deteriorating standards of urban public health environmental science and technology must quickly come to play a dominant role in designing our social and industrial structure for tomorrow. Scientifically rigorous criteria of environmental quality must be developed. Based in part on these criteria, realistic standards must be established and our technological progress must be tailored to meet them. It is obvious that civilization will continue to require increasing amounts of fuel, transportation, industrial chemicals, fertilizers, pesticides, and countless other products and that it will continue to produce waste products of all descriptions. What is urgently needed is a total systems approach to modern civilization through which the pooled talents of scientists and engineers, in cooperation with social scientists and the medical profession, can be focused on the development of order and equilibrium to the presently disparate segments of the human environment. Most of the skills and tools that are needed are already in existence. Surely a technology that has created such manifold environmental problems is also capable of solving them. It is our hope that this Series in Environmental Sciences and Technology will not only serve to make this challenge more explicit to the established professional but that it also will help to stimulate the student toward the career opportunities in this vital area.

Robert L. Metcalf
James N. Pitts, Jr.

PREFACE

The serious aquatic chemistry problems include nutrient concentrations in natural waters, methods to control these levels, and the effects of nutrient enrichment. As our rivers and lakes become utilized by larger numbers of persons for both recreation and waste disposal, the problems mount. Adequate solution of our nutrient problems requires familiarity, on the part of the scientists, engineers, and government officials doing research and setting policy on nutrient matters, with all aspects of the chemistry, geochemistry, and biochemistry of nutrients as these fields relate to natural waters. Moreover, those involved must have enough additional knowledge to be able to cross boundaries set by professional and educational traditions.

This book has been prepared to fill the need just defined. The latest thinking and research on nutrients have been brought together by experts in the chemical and biological sciences and in engineering. The authors represent educational institutions, government, and industry.

The book features a broad discussion of the topic. In addition to chapters on the three principal nutrients (nitrogen, phosphorus, and carbon), chapters on nutrient geochemistry and nutrient regeneration in hypolimnetic waters are included. Both chemical and bioassay analyses of nutrients are discussed by experts in their respective fields. Computer models of nutrient biochemical processes are described with the use of many illustrations, so that even those unfamiliar with computer techniques can appreciate their utility.

Evidence for the rate of increase of nutrient loading is given by geochemical studies of sediments. Atmospheric contributions of chemicals to the aquatic system are dealt with specific reference to Lake Michigan. Chapter XI explains the testing program to which a new detergent is subjected, as well as the requirements of a detergent. The chapters on biological and physical-chemical waste treatment will serve not only to shed light on our present capabilities for nutrient removal but also to indicate future directions for research into this most important area. Finally, in the last two chapters, the nutrient control policies of both the United States and Canada are explained.

With the exception of Chapter II, the chapters in this book were adapted by the respective authors from papers presented in the symposium "Nutrients in Natural Waters" at the 161st National Meeting of the American Chemical Society in Los Angeles, California, March 28-29, 1971. The symposium was sponsored by the Division of Water, Air and Waste Chemistry and was organized by Herbert E. Allen and James R. Kramer.

Completion of the book would have been impossible without the help of Deena Allen, Rebecka Andress, Cathey Bernhard, Maida de Stein, Susan Grossman, and Janet Smith. We would also like to acknowledge the assistance of the Department of Environmental and Industrial Health, University of Michigan; the University of Michigan Sea Grant Program; and the Department of Geology, McMaster University.

HERBERT E. ALLEN
JAMES R. KRAMER

Ann Arbor, Michigan
Hamilton, Ontario
August 1972

TABLE OF CONTENTS

NUTRIENTS IN NATURAL WATERS

Chapter I

NITROGEN: SOURCES AND TRANSFORMATIONS IN NATURAL WATERS

Patrick L. Brezonik

Department of Environmental Engineering,
University of Florida, Gainesville, Florida

Of the major nutrient cycles in natural waters, the nitrogen cycle is perhaps the most interesting, the most complex and the least understood from a quantitative point of view. The geocycle of nitrogen is largely a biochemical phenomenon; in natural waters it is nearly wholly so. Thus the nitrogen cycle, like the carbon and phosphorus cycles, is inextricably related to aquatic organic productivity. Although many elements and compounds are required for biosynthesis, nitrogen and phosphorus have long been considered to be the principal limiting nutrients for primary production; evidence lately obtained suggests that carbon may also limit production in some situations. The great recent concern over cultural eutrophication has stimulated much new research in the following areas: the chemistry and biochemistry of nutrients in aquatic systems, the quantification of the sources and sinks of nutrients, and the dynamics of nutrient uptake and release. In this chapter we discuss these subjects with respect to the cycle of nitrogen in natural waters.

GENERAL ASPECTS OF THE NITROGEN CYCLE

Nitrogen occurs in the biosphere in a variety of forms

ranging in oxidation state from +5 to -3. Inorganic nitrogen is present primarily as highly oxidized nitrite and nitrate, as reduced ammonia, and as molecular nitrogen. A variety of intermediate gaseous oxides of nitrogen are important in atmospheric chemistry but not in natural waters. Naturally occurring organic nitrogen consists primarily of amino and amide (proteinaceous) nitrogen, along with some heterocyclic compounds such as purines and pyrimidines. Nitrogen compounds are present as cellular constituents, as nonliving particulate matter, as soluble organic compounds, and as inorganic ions in solution. All these forms are interrelated by a series of reactions known collectively as the "nitrogen cycle," which portrays the flow of nitrogen from inorganic forms in soil, air, and water into living systems and then back again into inorganic forms. This cyclic phenomenon can be illustrated in various ways. In, Figure 1, which presents a simplified reaction sequence of the interconversions between organic nitrogen and the main inorganic forms, the principal reactions are labeled. Figure 2 illustrates the nitrogen cycle as it may occur in an idealized stratified lake. Except for the process of ammonia exchange with sediments, it is apparent that all reactions are biologically mediated.

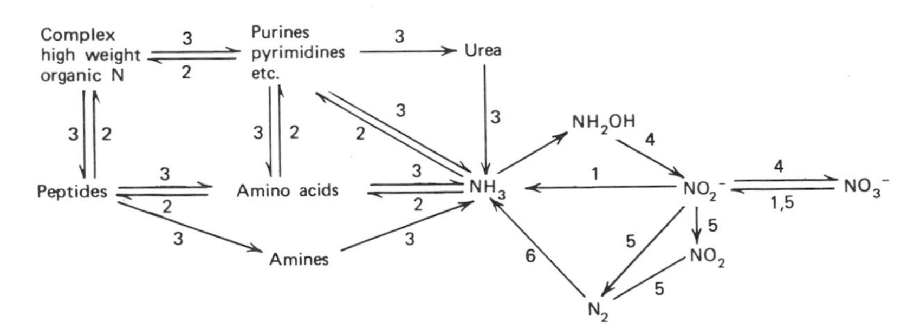

Fig. 1. Simplified nitrogen cycle showing main molecular transformations: 1. nitrate assimilation, 2. ammonia assimilation, 3. ammonification, 4. nitrification, 5. denitrification, 6. nitrogen fixation

By far the greatest influx of inorganic nitrogen into organisms results from ammonia and nitrate assimilation. These reactions predominate in surface waters and are mediated primarily by phytoplankton and macrophytes. Nitrate tends to be the predominant inorganic nitrogen form in surface waters, but there is considerable evidence

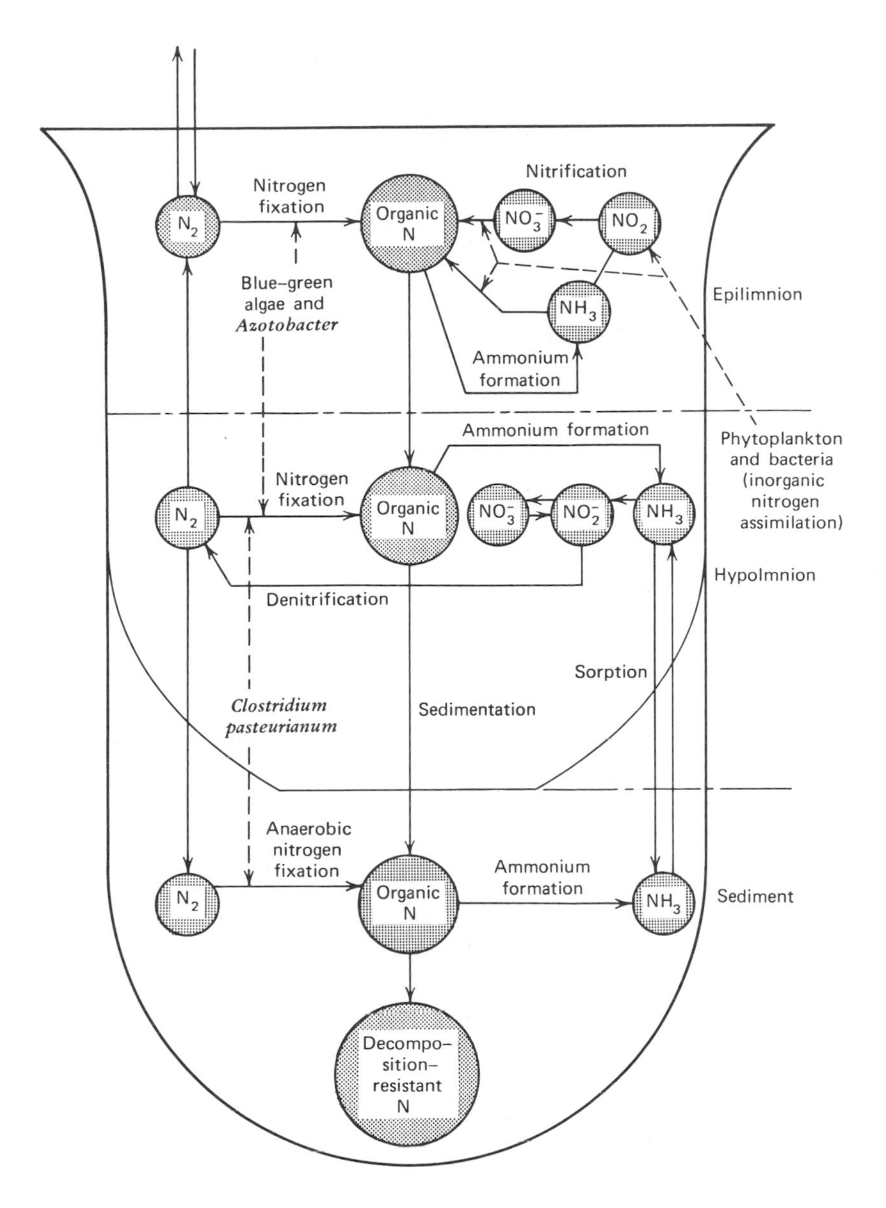

Fig. 2. Nitrogen cycle reactions in an idealized stratified lake. Note that both aerobic and anaerobic transformations are shown in the hypolimnion, although in a real lake they would not occur simultaneously. Adapted from Kuznetsov (1).

(discussed in a later section) that ammonia is the preferred form for planktonic assimilation, since it is already at the reduction level of organic nitrogen. Nitrate can be used by most plants [Strickland (2) and others cite some flagellates as exceptions].

Organisms using nitrate as their nitrogen source must reduce it to the level of ammonia before incorporating it into organic forms, and this process requires a reduction system including the enzyme nitrate reductase. This inducible enzyme is present in algal cells only when nitrate is being used as the nitrogen source (3), which suggests a mechanism for determining the form of nitrogen an algal population is using.

The reverse of assimilation is ammonification, whereby organic nitrogen is returned to the inorganic nitrogen pool as ammonia. This is a complicated process involving several mechanisms. Early workers considered only bacterial decomposition of soluble organic nitrogen and organic detritus as important (4), but more recent studies have revealed significant excretion of ammonia and amino acids by zooplankton feeding on phytoplankton and detritus. Johannes (5) suggested this as the dominant mechanism of ammonification in surface waters and reviewed previous work indicating that net zooplankton release ammounts of dissolved nitrogen and phosphorus equal to their total body content of these nutrients in 20 to 200 hr. A second mechanism for ammonification is direct autolysis after cell death, which may account for 30 to 50% of the nutrients released from plant and animal material (5,6).

Current information and theories suggest that bacterial decomposition of organic matter accounts for a minor portion of nutrient regeneration in marine surface waters and in shallow lakes (although this is probably not true in systems receiving large organic inputs from sewage outfalls). In soils and bottom sediments and in anoxic hypolimnia of lakes, nutrient regeneration by bacteria and fungi is the dominant process. Ammonification is obviously important in renewing a limited supply of nitrogen for assimilation and growth of primary producers.

In the process of nitrification, ammonia is oxidized to nitrite and nitrate by a select group of aerobic autotrophic bacteria which obtain their energy by nitrogen oxidation and their cellular carbon by reduction of carbon dioxide. It has also been reported that a variety of heterotrophic bacteria, actinomycetes, and fungi (7) are capable of nitrification, generally at much slower rates; but the role of these organisms in aquatic nitrification is imperfectly understood. The significance of nitrification in the nitrogen cycle lies in the conversion of labile ammonia (which tends to be lost from solution by sorption onto sediments and by volatilization at high pH) to a more stable form (nitrate). On the other hand, denitrification reactions reduce nitrate to molecular nitrogen, and nitrification is important in producing the reactants for this nitrogen sink.

In the process of denitrification, nitrate serves as a terminal electron acceptor by facultative and anerobic bacteria in the absence of oxygen. Nitrite is formed as the first intermediate in the process; nitrous oxide can sometimes be formed, along with molecular nitrogen, although it is not an essential intermediate (7). Since the principal end product (N_2) is a nitrogen form not utilizable by most organisms, this reaction acts as a nitrogen sink and can assume importance in considerations of the nitrogen balances of lakes and other aquatic systems subject to intermittent anoxia, (e.g., soils) (7).

Denitrification may account in part for the difficulty encountered in obtaining nitrogen balances in waste treatment plants (8). The reaction may occur in anoxic microzones of activated sludge floc or in anoxic settling basins. The occurrence of denitrification has been demonstrated in oxygen-depleted layers of the tropical Pacific (9,10), and the quantitative significance of the reaction in the nitrogen budget of a stratified eutrophic lake has been estimated (11). Nitrification followed by intentional denitrification has been proposed as a means of removing nitrogen from biological waste treatment plants (12,13).

Probably more attention has been devoted to nitrogen fixation than to all the other nitrogen cycle reactions combined. This process is important on several levels. Geochemically, nitrogen fixation is essential in maintaining a nitrogen balance in the biosphere, which would otherwise become depleted as a consequence of denitrification over a time scale of millions of years (14). The reaction is important in agriculture in maintaining or increasing soil fertility, and in natural waters nitrogen fixation acts as a source of nitrogen and permits continued organic production when the supply of fixed nitrogen becomes depleted.

A variety of organisms are capable of nitrogen fixation, including a number of blue-green algae, apparently all photosynthetic bacteria, various aerobic bacteria (e.g., _Azotobacter_), anaerobic bacteria (e.g., _Clostridium_), many facultative bacteria [but only under anoxic conditions (15)], legume root nodules, and nonleguminous root-nodulated plants such as _Podocarpus_ and the alder tree (*Alnus* sp.). Most studies of nitrogen fixation in natural waters have emphasized the role of filamentous, heterocystous blue-green algae such as *Anabaena*, *Gloeotrichia*, and *Nostoc*. The agents and occurrence of nitrogen fixation in the biosphere have been extensively reviewed (16-18).

Nitrogen fixation is generally considered to be an adaptive process used by organisms only when the supply of fixed nitrogen is depleted. Nitrogen-fixing algae usually bloom in lakes only after nutrients have been depleted by blooms of other algae (i.e., late summer in temperate lakes). However, contrary to earlier opinion, small to moderate concentrations of ammonia do not necessarily inhibit fixation, although synthesis of the enzyme nitrogenase is repressed at high levels. Stewart (18) suggests that the levels of combined nitrogen in most natural ecosystems are insufficient to inhibit fixation immediately or even to persist long enough for existing nitrogenase to be diluted out. Low levels of combined nitrogen may actually be advantageous to nitrogen-fixing plants by enabling more efficient and healthy growth than could be achieved on N_2 alone.

Reduction of N_2 to the level of ammonia requires energy and a source of reduced hydrogen; both may be obtained from photosynthetic production or from oxidation of organic carbon. Nitrate reduction to ammonia similarly requires energy obtainable by oxidizing organic substrates. The net reactions in each case are exergonic, but they are nevertheless energy sinks, since the cells apparently have no means of trapping and storing the energy (as ATP) given off by the reactions. To this extent, use of nitrate and of molecular nitrogen rather than ammonia for assimilation is wasteful of energy.

TRANSPORT OF NITROGEN INTO LAKES

The concentration and forms of nitrogen in a lake at a given time are a product of input rates, the interconversion reactions occurring within the lake, and the rates of loss by way of outflow, denitrification, and sediment deposition. Lakes act in general as nutrient traps, implying an accumulation of nitrogen in bottom sediments as a lake gradually fills in. Both natural and cultural nitrogen sources can be significant in lacustrine nitrogen budgets, but recent interest in quantifying nitrogen sources stems from the problems induced by excessive cultural additions. The natural nutrient input to a lake is a function of such factors as drainage basin geochemistry (i.e., the potential for nutrient leaching from soils and substratum), drainage basin size, hydrology, and precipitation patterns (19). Superimposed on these are a multitude of human or cultural variables, which can be expressed in terms of drainage basin land use patterns and population characteristics. Quantitative information on lacustrine nitrogen budgets and the significance of individual sources is sparse, but concern over cultural eutrophication has stimulated much needed measurements of the nitrogen (and phosphorus) contributions from various sources, along with development of elementary nutrient models.

Table 1 lists the possible nitrogen sources and sinks that must be considered in calculating the nitrogen budget for a lake. The classical approach to evaluation of a lake's nutrient budget is actual measurement of nutrient concentrations and flows for each source over a reasonable time span (e.g., one year). Obviously this requires a large time and manpower expenditure, especially for large lakes, and some diffuse sources (e.g., groundwater seepage) may not be amenable to direct or accurate evaluation. Because of these difficulties, it is probable that complete measurements of all sources and sinks have not been accomplished for any single lake, although approximate nutrient budgets have been established for a few American and European lakes (19-21).

An alternative to actual measurement is development of a simple simulation model of nutrient transport based on knowledge of the lake's drainage basin size, population, hydrology, and land use patterns, as well as literature values for nutrient contributions from the various sources (22,23). Actual measurements of nutrient contributions are undoubtedly more accurate than literature estimates, but as the literature on nutrient export rates from various land uses and other sources becomes more complete, the accuracy of literature-based nutrient budgets will increase. Obviously the accuracy of such a nutrient budget depends both on the accuracy of the original measurements and on their applicability to the system under consideration. However, considering the expense and difficulties of direct measurement, literature-based models seem to be the only reasonable approach for many lakes receiving cultural stress. Thus refinement of nutrient contribution values from the various types of sources should receive high priority.

Table 2 summarizes the nitrogen and phosphorus contributions from various sources and lists references that have evaluated or reviewed the significance of the sources. Nutrient contributions from undisturbed forest land are

Table 1. Sources and Sinks for the Nitrogen Budget of a Lake

Sources	Sinks
Airborne	Effluent loss
Rainwater	
Aerosols and dust	Groundwater recharge
Leaves and miscellaneous debris	
	Fish harvest
Surface	
Agricultural (cropland) and drainage	Weed harvest
Animal waste runoff	Insect emergence
Marsh drainage	
Runoff from uncultivated and forest land	Volatilization (of NH_3)
Urban storm water runoff	Evaporation (aerosol formation from surface foam)
Domestic waste effluents	
Industrial waste effluent	
Wastes from boating activities	
	Denitrification
Underground	
Natural groundwater	Sediment deposition of detritus
Subsurface agricultural and urban drainage	
Subsurface drainage from septic tanks near lake shore	Sorption of ammonia onto sediments
In situ	
Nitrogen fixation	
Sediment leaching	

low, but fertilization and clear-cutting tend to increase nutrient export (26,54). For example, Cole and Gessel (26) found that nitrogen output through percolation water from a Douglas fir forest increased from 0.54 kg/ha to 0.69 and 1.04 kg/ha when plots were fertilized with urea and ammonium sulfate, respectively. Clear-cutting increased nitrogen export to 0.96 kg/ha. Clear-cutting also tends to increase runoff and sediment transport. When regrowth was prevented with herbicides after leveling a hardwood forest in one study (54), a large alteration in the system's nitrogen cycle resulted, as evidenced by greatly increased nitrate levels (>10 mg/liter) in stream water and by increased nitrogen outflow from less than 2 kg/(ha)(yr) to more than 60. Thus properly managed and

Table 2. Nitrogen and Phosphorus Contents of Various Nutrient Sources for Lakes[a]

Source	Nitrogen	Phosphorus	References
Natural			
Forest runoff	1.3-5.0	0.084-0.18	24
	1.5-3.4	0.83-0.86	25
Forest percolation water	0.54	0.034	26
Swamp and marsh runoff	?	?	21,22
Meadowland runoff	?	?	21
Precipitation on	0.58	0.044	19
lake surface	0.18-0.98	0.015-0.060	21,27-29
Aquatic birds	0.48-0.95	0.09-0.18	30-32
Leaves and pollen	?	?	21,33-36
Cultural			
Domestic sewage	39.4	0.80	21,37-39
Agricultural areas			
Citrus	22.4	0.18	40
Pasture	8.5	0.18	41
	1-5	0.15-0.75	21
Cropland	5-120	0.22-1.0	21,42-44
Farm animals, feedlots	?	?	21,45-48
Urban runoff	8.8	1.1	27,49-51
Septic tanks	?	?	23,52
Marsh and landfill drainage	?	?	53

[a]All units in kg/(ha)(yr) except rainfall [g/(m^2)(lake area)(yr)], birds [kg/(duck)(yr)], and domestic sewage (annual kilograms per capita).

undisturbed forests retain their accumulated nutrients efficiently, but management and harvesting practices can have a significant effect on nutrient export from them.

Undisturbed meadowland is probably comparable to forestland in retention of nutrients, but intensively used pastureland is a somewhat greater source. In many areas pasture land is fertilized to increase the yield of grass, and this fertilizer, coupled with animal wastes, undoubtedly increase nutrient export rates. Swamp and marshland have frequently been cited as being of major importance both in trapping nutrients before they reach the lake (55) and in contributing to the nutrient budget of a lake (56), but no quantitative information is available on this source (22). Pollen was recently shown to be an insignificant source of nitrogen for Lake Tahoe (35), but it may be more significant (by a few percentage points or so) in small alpine or subalpine lakes with limited nutrient inflow and greater shoreline-to-volume ratio. We know that leaves exert a significant effect on water quality in streams (36); however, quantitative assessment of their role in lacustrine nutrient budgets would be difficult to obtain.

Nutrient contributions of water fowl are not well known but may be significant in some cases, especially where large rookeries and migratory bird preserves occur on small water bodies. It was recently estimated, for example, that a community of about 50 mallard ducks maintained for research purposes on the shore of Lake Mize (a small, deep lake in north-central Florida) had increased the nitrogen and phosphorus loading rates beyond those suggested as critical for eutrophication (57). Duck contributions were calculated to represent 50 and 75% of the nitrogen and phosphorus loading rates of 8.1 and 0.73 $g/(m^2)(yr)$, respectively.

Waste production of small and medium-sized birds (ducks, chickens) has been assessed in relation to pollution from commercial breeding, production, and processing plants (30,31). The role of wild aquatic birds in the nutrient regime of a water body may be more that of cycling agents than of direct sources (i.e., much of their food may be taken from the lake itself--fish--or from the immediate watershed). However, such activities can still stimulate increased production and apparent eutrophication by accelerating nutrient cycling and degrading a normally more stable (organic) nutrient reservoir.

The nitrogen content of rainfall has been widely studied and reviewed (21,27-29), and it is apparent that nitrogen levels are, unfortunately, highly variable. Both nitrate and ammonia occur in significant amounts, and ammonia is more significant in temperate regions than in the tropics (28). Total nitrogen concentrations frequently approach 1 mg N/liter, but the great temporal and areal variability (Tables 3 and 4) makes mean values rather meaningless for simple predictive models. Since both natural and cultural sources are responsible for ntirogen in rainfall, the correlation of high-rainfall ammonia with alkaline soils, and low-rainfall ammonia with acid soils suggests that sorption of ammonia onto soil clays is an inportant factor in the hydrospheric nitrogen cycle (29). The large cultural sources of atmos-

Table 3. Temporal Variations of Nutrients in Rain Showers at Gainesville, Florida, in Summer, 1969

Date and weather	Time, min after start of rainfall	Concentrations, mg/liter				
		Total P	Ortho P	Total Organic Nitrogen	NH_3-N	NO_3^--N
August 4,	Thundershower after weekend of dry weather					
	0-5	0.058	0.056	1.85	0.92	0.08
	5-10	0.032	0.032	1.00	0.84	0.08
	10-20	0.015	0.010	0.83	0.09	0
	20-30	0.005	0.005	0.10	0.09	0
	30-60	0.005	0.005	0.15	0.10	0
September 3,	Moderate and steady rain following day with about 2 in. of rain					
	0-5	0.008	0.008	0.15	0.10	0.08
	5-10	0.005	0.005	0.10	0.10	0.005
	10-20	0	0	0	0.08	0
	20-30	0	0	0	0	0
	30-60	0	0	0.10	0	0

pheric nitrogen (ammonia from fertilizer, nitrogen oxides from auto emissions) indicates a possible correlation between high rainfall nitrogen and areas of industrial or cultural activity, but there is conflicting evidence in this regard (21,29). Careful analysis of historical records to discern possible long-term trends toward increased rainfall nitrogen resulting from increased fertilizer use or from other cultural sources has not been reported.

Agricultural sources of nitrogen are difficult to summarize because of large local variations in soil-retention capacities, in irrigation and fertilization practices, and in differences associated with climate and type of crops grown. Waste output for the common farm animals (cattle, hogs, sheep, horses) has been ascertained with reasonable accuracy (21,46), and data on total animal population are readily available in areal units at least as small as counties. However, the amount of animal waste that enters and is transported by the source waters for a lake depends largely on local circumstances, and few quantitative figures are available. Ammonia volatilized

Table 4. Variations in Nutrient Content and Volume of Rainfall in Gainesville, Florida, Area, August 29 to September 1, 1969

Sampling station[a]	Volume collected, ml[b]	Concentrations, mg/liter			
		Total P	Ortho P	NH_3-N	NO_3^--N
1	250	0.033	0.006	0.35	0.02
2	135	0.65[c]	0.043	0.10	0.09
3	290	0.125	0.009	0.20	0.11
4	20	---	0.018	0.05	0.07
5	130	0.076	0.011	0.14	0.07
6	180	0.061	0.018	0.02	0.07
7	130	0.045	0.004	0.05	0.05
8	290	0.32[c]	0.021	0.10	0.05
9	170	0.022	0.002	0.15	0.06

[a]Sampling stations were scattered over a 15-mile^2 area of the city except for station 8, located 10 miles west of the city limits.
[b]Proportional to inches of rainfall, since all collecting funnels were the same diameter.
[c]Sample contained large amount of particulates, evidently from dry fallout.

from cattle feed lots has been shown to contribute significantly to the nitrogen budgets of downwind lakes (47). For example, it was estimated that this source adds about 0.6 mg NH_3-N/(liter)(yr) to Seely Lake, Colorado, a small lake 2 km downwind from a 90,000-unit cattle feed lot. Groundwater in areas of cattle feedlots is frequently contaminated with high nitrate levels, but present data are insufficient to allow prediction of nitrogen fluxes for individual situations.

Nitrogen contributions from sewered populations are well known on a per capita basis, and the total nitrogen concentration in raw and treated domestic sewage is probably sufficiently constant to permit evaluation of this source, for nutrient budget purposes, from readily available sewage flow and population records. Nonsewered human waste contributions, as from septic tank drainage, are far more difficult to evaluate. Septic tanks are undoubtedly important factors in the cultural eutrophication of many recreational lakes. However, almost no data can be cited to quantify this source. Urban runoff composition has been studied in detail in a few places (27, 49-51), and storm drainage flows are more widely known, but the general applicability of existing information outside the original study areas is unknown.

In summary, a large volume of data is available on the various external nitrogen sources for lakes. The significance of a few point sources (e.g., domestic sewage) is well known, and such diffuse sources as rainfall can be estimated with fair accuracy, but other important sources require much further evaluation.

From land use and population data and literature values for nutrient export coefficients, Shannon (23) recently computed nitrogen and phosphorus budgets for 55 lakes in north central Florida. On an areal basis, nitrogen-loading rates ranged from 1.2 g/(m^2)(yr) for a small ultraoligotrophic lake (Swan Lake, Putnam County) to more than 90 g/(m^2)(yr) for Lake Alice, a shallow, 45-acre hypereutrophic lake, which receives 1.5 to 2.0 million gal/day secondary effluent from the University of Florida campus treatment plant. Expressed on a volumetric loading basis, nitrogen extremes are 0.18 to 106 g/(m^3)(yr).

Loading rates in general correlate with the trophic states of the lakes. In order to study the statistical relationships between trophic conditions and watershed characteristics, a quantitative index of trophic state (TSI) was developed. The index is based on values for seven trophic state indicators (including primary production, Secchi disc, and nitrogen and phosphorus levels) for the 55 lakes, using the multivariate statistical technique of principal component analyses (23,58). The index was derived so that increasing TSI values correspond to increasing eutrophic conditions. It is felt that the index values, which agree with qualitative rankings of trophic conditions in the 55 lakes, serve to quantify at least approximately the concept of trophic state.

Stepwise regression analysis of the trophic state index versus the nitrogen and phosphorus loading rates (expressed on both areal and volumetric bases) were performed (59) using an additive model [TSI = $f(N + P)$] and a multiplicative model, [TSI = $f(N \times P)$]; the results appear in Table 5. For the multiplicative model, the data were transformed logarithmically before regression analysis. The regressions are all significant at the 99% confidence level, and the independent variables are arranged in the regression equations in order of their partial correlation with the dependent variable. Thus phosphorus loading is the more important variable (in a statistical sense) in three of the four regression equations.

Interestingly, the simple additive model explains more of the total variation in TSI than the theoretically

Table 5. Regression Analysis of Trophic State Index (TSI) Versus Nitrogen and Phosphorus Loading Rates for 55 Florida Lakes (59)

Model	Loading rate units	Equation[a]	F Ratio[b]	Multiple correlation coefficient	Percentage variance explained by equation
Additive					
1.	per unit lake surface area	$TSI = 0.62(N_{SL}) + 10.06(P_{SL})$	46.44	0.804	64.5
2.	per unit lake volume	$TSI = 26.1(P_{VL}) + 0.90(N_{VL})$	43.20	0.793	62.9
Multiplicative					
3.	per unit lake surface area	$TSI = 0.84(P_{SL})^{0.48}(N_{SL})^{0.20}$	14.08	0.600	36.0
4.	per unit lake volume	$TSI = 1.08(P_{VL})^{0.42}(N_{VL})^{0.04}$	15.64	0.620	38.5

[a]Abbreviations: TSI = trophic state index (dimensionless); N_{SL} and P_{SL} = nitrogen and phosphorus surface loading rates, $g/(m^2)(yr)$; N_{VL} and P_{VL} = nitrogen and phosphorus volumetric loading rates, $g/(m^3)(yr)$.

[b]All significant at the 99% confidence level.

more attractive multiplicative model. Since both nitrogen and phosphorus can limit productivity (which is highly correlated with trophic state), an analytical model of production as a function of nutrients and other environmental variables would include N and P as multiplicative factors (60). Thus trophic state would not seem to be a simple additive function of the rates of N and P loading, rather, it should be dependent on both in a second-order type of expression. We must bear in mind, however, that regression analysis is inherently empirical; its value lies in its predictive abilities rather than in any analytical or determinative potential. If the additive regression equations fit the data better, they are obviously preferable for use in prediction. The relationship between TSI and nitrogen and phosphorus loading is fair, accounting for about 63% of the variance in TSI. However, a multiple regression of TSI versus watershed factors themselves (e.g., fertilized, forested, urban areas, sewered population) accounted for about 80% of the variance in TSI (23).

Of great interest in control of cultural eutrophication is the development for nitrogen and phosphorus of critical loading rates above which eutrophic conditions can be expected to ensue. Sawyer (37) was the first to propose quantitative guidelines of this sort. Based on data from Wisconsin lakes, he suggested that 0.015 mg/liter of inorganic phosphorus and 0.3 mg/liter of inorganic nitrogen at the spring maximum are critical levels; above these, algal blooms can normally be expected. In the absence of any other studies, these values have been widely quoted and applied to many types of lakes in diverse geologic and climatic situations. Recently Vollenweider (21) analyzed the available data on nutrient loading rates and corresponding trophic conditions and proposed permissible and critical loading rates for nitrogen and phosphorus as a function of lake mean depth (Table 6).

Critical loading rates for nitrogen and phosphorus can be estimated for Florida lakes from the regression equations in Table 5. Since the normal nitrogen-to-phosphorus ratio in algal cells is about 15:1 by atoms (61), this ratio was applied in determining critical loading rates. Comparison of TSI values and trophic conditions in the 55 lakes (23,58) indicates that eutrophic conditions can be expected when TSI values are greater than 7.0 and that some problems occur when TSI values are greater than about 4.0. Inserting these TSI values into equation 1 of Table 5 and constraining the N:P ratio to 15:1 yields the following permissible and dangerous rates:

	g/(m²)(yr) N	P
Permissible loading up to	1.9	0.28
Dangerous loading in excess of	3.4	0.49

Although somewhat larger than Vollenweider's permissible and dangerous rates, these values are in the same general range.

Table 6. Permissible and Dangerous Loading Rates for Nitrogen and Phosphorus from Vollenweider (21)

Mean depth (up to),	Permissible loading (up to), g/(m²)(yr)		Dangerous loading (in excess of), g/(m²)(yr)	
m	N	P	N	P
5	1.0	0.07	2.0	0.13
10	1.5	0.10	3.0	0.20
50	4.0	0.25	8.0	0.50

In general, deep lakes can assimilate more nutrients without adverse effects than shallow ones can. The question of critical loading rates as a function of lake mean depth can be approached by plotting loading rates versus mean depth and delimiting the areas in which eutrophic and oligotrophic lakes predominate. This has been done in Figure 3, where lake coordinates with respect to mean depth and N loading are denoted by symbols referring to the lake's trophic state (as determined by qualitative inspection of its trophic conditions and by application of the TSI concept). Lines delineating eutrophic and oligotrophic loading regions were drawn by inspection of trophic state trends. A similar analysis based on data from somewhat deeper lakes was performed by Vollenweider (21), and his critical lines delineating eutrophic and oligotrophic loading regions are included in Figure 3 for comparison. Some differences occur between the two analyses, but the results are of the same magnitude. Florida lakes appear to be capable of assimilating somewhat greater quantities of nutrients before becoming mesotrophic or eutrophic than Vollenweider's analysis would suggest.

The general agreement between Vollenweider's analysis

and our own and the agreement between lake trophic state indices and nutrient loading rates substantiate the validity of the overall approach, although there is much room for refinement of the data and techniques.

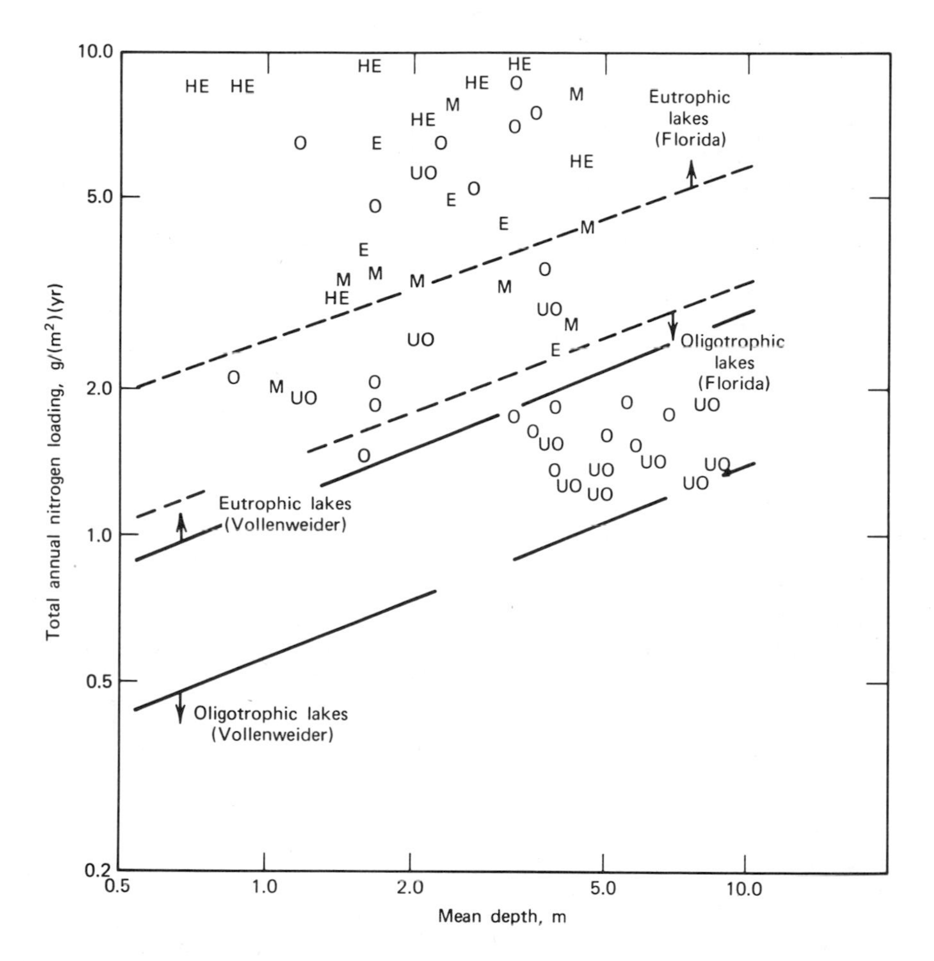

Fig. 3. Annual nitrogen loading versus mean depth for 55 Florida lakes. Each datum represents a lake having the trophic state denoted by the following symbols: UO, ultraoligotrophic; O, oligotrophic; M, mesotrophic; E, eutrophic; HE, hypereutrophic. From Ref. 23.

INTERNAL NITROGEN SOURCES AND SINKS

Aside from the external nitrogen sources discussed previously, several mechanisms within a lake can effect an increase or decrease in the total nitrogen content of the lake. Nitrogen fixation is an in situ source, whereas denitrification is a sink. Sediments must act as nitrogen sinks over the geological life span of a lake, but over shorter periods of time they may act as sources to the overlying water. Nitrogen is lost to sediments by deposition of particulate matter (detritus, silt), and by sorption of ammonia onto clays that are in the sediments or being deposited onto them. Nitrogen can be released from sediments by activities of burrowing animals, decomposition of organic nitrogen to ammonia and its diffusion into the overlying water, and desorption of ammonia from clays and other sorbents in the sediment.

For many years it was thought that blue-green algae in eutrophic surface waters were the only significant nitrogen fixers in the aquatic environment. Their role was recognized by Hutchinson as early as 1941 (62), and the first quantitative in situ measurements of nitrogen fixation rates were reported by Dugdale et al. in 1959 (63). Since then nitrogen fixation rates have been measured with ^{15}N techniques in a variety of lake and marine waters (64-68). High rates have been correlated with blooms of blue-green algae, primarily *Anabaena*, *Gloeotrichia*, and *Aphanizomenon* in lakes, and *Trichodesmium* in subtropical marine waters.

The expense and difficulties associated with ^{15}N tracer techniques have precluded routine and detailed investigations of nitrogen fixation in aquatic environments until recently, when a simple, indirect method of assessing nitrogen fixation rates was developed (69). This method utilizes the property of the nitrogen-fixing enzyme complex to reduce acetylene to ethylene, which can be sensitively and easily determined by gas chromatography.

An extensive survey of nitrogen fixation in Florida lakes and an intensive study of fixation in two eutrophic lakes over a one-year period was recently completed by the author. The survey involved 55 lakes with widely varying trophic and chemical characteristics in north and central Florida, and the waters were sampled periodically for over a year in conjunction with a large-scale regional limnology and eutrophication study (19,70). Detailed results will be presented elsewhere; for the present discussion it is pertinent to consider the frequency of nitrogen fixation in these 55 lakes, assuming them to be representative of Florida lakes as a whole.

All 55 lakes were sampled four times and 19 of the lakes were sampled seven times over one year. Out of a total of 272 lake samplings (not including separate samples taken in depth profiles on some of the lakes), 42 samples from 15 different lakes gave positive fixation rates using the acetylene reduction method (69). The infrequent sampling routine probably missed fixation in four or five more lakes that seem to be likely environments for sporadic nitrogen fixation. (This assumption is based on the trophic condition of the lakes and the nature of their plankton.) Low (undetectable) fixation may be still more widespread or frequent in lakes giving occasional high rates, but no attempt was made to concentrate the algae by filtration or centrifugation before incubation. Fixation was detected on only one occasion in five of the 15 lakes and in five other lakes it was found twice. However, in the remaining five lakes it is apparently a common occurrence; rates in these lakes are summarized in Table 7. Rates expressed as equivalent N fixed in μmoles $NH_3/(m^3)(hr)$ ranged from just detectable (about 0.5) to a high of 175, but most rates were low (less than 10).

Table 7. Nitrogen-Fixation Rates in Selected Florida Lakes[a]

Date	Bivins Arm	Lake Dora	Lake Hawthorne	Newnan's Lake	Unnamed #20
1969					
June	4.8	1.4	9.6	0	0
August	73.0	20.0	2.8	0	0
October	0	1.9	0.7	1.4	6.8
December	0	9.5	3.5	27.4	3.4
1970					
February	11.8	4.1	10.8	9.5	0
April	1.2	6.3	36.2	2.2	2.8
June	31.4	15.0	0	0	1.9

[a]All rates in micromoles of nitrogen per meter^3 per hour derived from micromoles of ethylene produced per meter^3 per hour using theoretical factor of 1.5 moles of ethylene produced per mole of ammonia fixed.

Yet even low rates can contribute significant amounts of nitrogen to a lake's nitrogen budget if the rates continue over a long period. For example, fixation was found in Lake Dora on all seven sampling dates and in Lake Hawthorne on six out of seven sampling dates. Mean fixation rates in the two lakes are 8.3 and 9.2 μmoles $N/(m^3)(hr)$, respectively. Assuming that this mean rate occurs over a year and that fixation occurs on the average for 8 hr a day, fixation would contribute about 0.34 and 0.37 mg N/liter to Lakes Dora and Hawthorne, respectively, on an annual basis.

The seasonal pattern of nitrogen fixation in temperate lakes has been well established (64,65), and fixation is primarily a late-summer phenomenon. In subarctic lakes, significant fixation is apparently limited to parts of the short ice-free season (66). In tropical and subtropical lakes, seasonal cycles are less pronounced and fixation is possible year-round. Detailed studies of nitrogen fixation in two highly eutrophic lakes near Gainesville, Florida, partially bear this out and further illustrate the highly dynamic and rapidly changing character of the nutrient and biological conditions in such lakes (71).

Figure 4 shows the concentrations of major nitrogen forms and rates of nitrogen fixation in Bivins Arm and Newnan's Lake from June, 1969, to June, 1970. In spite of geographical proximity and similar enriched conditions, these two lakes exhibit patterns of fixation that are quite dissimilar. Newnan's Lake is a highly colored, soft water lake, 2430 ha in area and 1.5 m in mean depth; it supports blooms of blue-green algae throughout the year, but an especially dense bloom of _Aphanizomenon_ usually develops in midwinter or early spring. In winter of 1969-1970, the bloom occurred in two pulses. The first pulse began in mid-December and died out in early January, apparently because of adverse weather (an abrupt cold wave and an extended period of heavy rain). The population stayed low in January and gradually increased during February to high levels through most of March.

Except for one high rate of nitrogen fixation in July, 1969, and very low rates on several other occasions, fixation in Newnan's Lake is associated with the winter and early spring growth of _Aphanizomenon_. Maximum fixation, rates during the _Aphanizomenon_ bloom were about 40 μmoles ethylene/$(m^3)(hr)$, roughly equivalent to 0.4 μg N/(liter)(hr). Fixation was highly correlated with primary production during the winter bloom period (Figure 5) in Newman's Lake, but no obvious correlation exists between fixation and concentrations of nitrogen forms.

Bivins Arm is an alkaline hardwater lake 58 ha in area and less than 2 m in mean depth. The algal flora in this

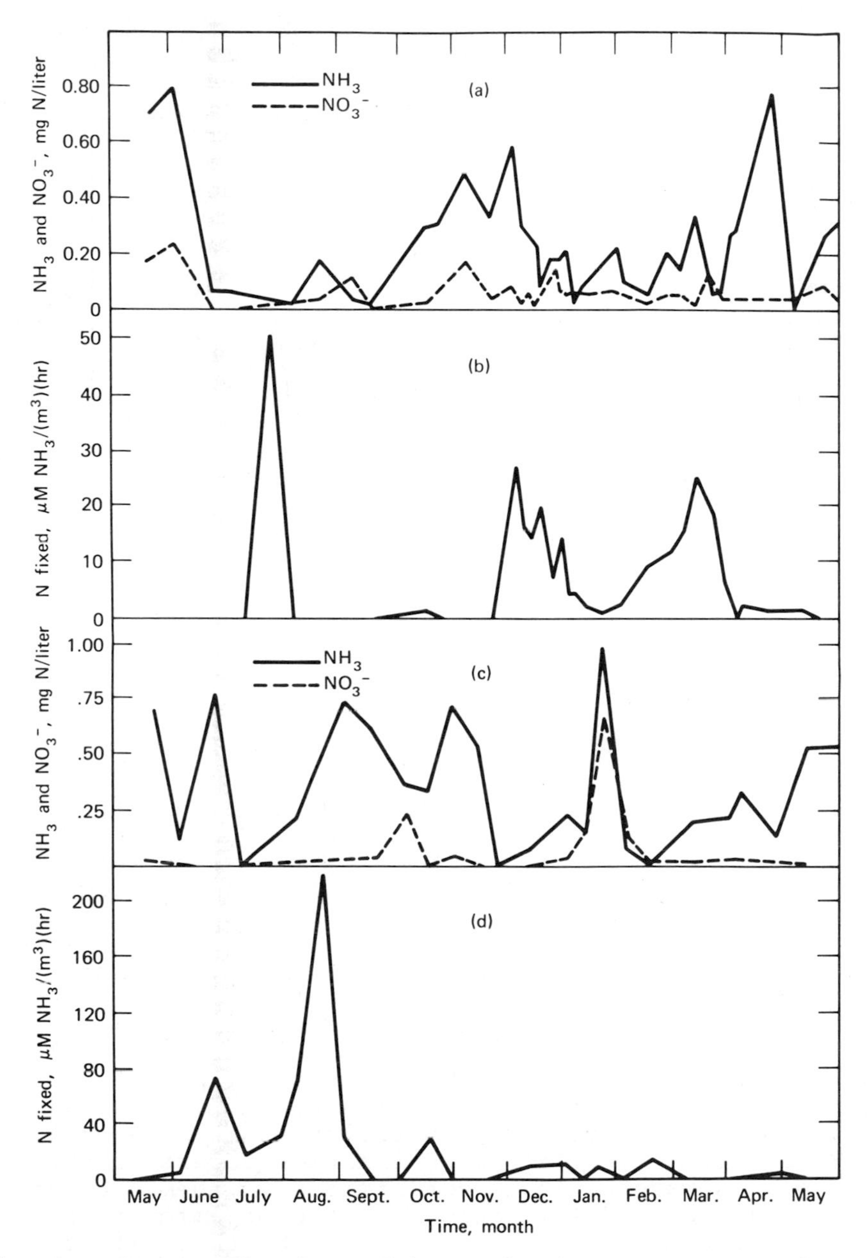

Fig. 4. Nitrogen fixation and inorganic nitrogen concentrations in surface waters of two eutrophic Florida lakes during 1969 and 1970: (a) and (b) Newnan's Lake; (c) and (d) Bivin's Arm.

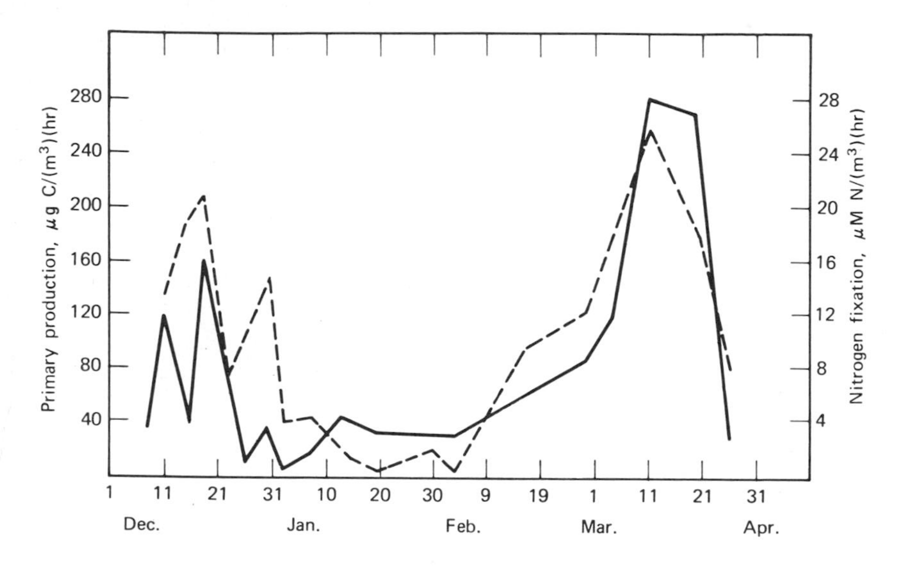

Fig. 5. Time course of primary production (solid curve) and nitrogen fixation (dashed curve) during winter Aphanizomenon bloom in Newnan's Lake, Florida.

lake is more diverse than in Newnan's Lake, and the apparent nitrogen fixing agent is Anabaena. Fixation occurred at moderate to high rates throughout the summer of 1969; a maximum of 330 μmoles ethylene/(m^3)(hr)--equivalent to about 3.1 μg N/(liter)(hr)--was found in late August. Lower rates were noted during winter, and fixation was detected on about 50% of the sampling dates. Fixation rates in this lake are not significantly correlated with either primary production or nitrogen concentrations.

It should be noted that algal fixation in these lakes occurred in the presence of moderate to high (for lake waters) concentrations of ammonia. In fact, the maximum rate in Bivins Arm occurred in the presence of about 0.5 mg NH_3-N/liter. Thus although fixation occurs in nutrient-depleted waters, it is not limited to them. Previous studies have shown simultaneous uptake of ammonia, nitrate, and molecular nitrogen (66,72); no doubt this situation occurs in the lakes under discussion. An obvious but unanswered question concerns the occurrence of fixation primarily as a winter phenomenon in one lake but as a summer phenomenon in another lake within 5 miles of the first.

Seasonal patterns of nitrogen fixation, nutrient concentrations, primary production, and other biogenic parameters are poorly defined in comparison with classic cycles reported for temperate lakes. Rather, production, nutrients, and algal populations are characterized by a bloom-crash cycle throughout the year. Pronounced differences were noted between sampling dates even when sampling was done as frequently as twice a week. The large changes in chemical and biological parameters from one sampling to the next were matched by obvious visual changes in the lakes, and many of the rapid fluctuations can be ascribed to rapidly changing weather conditions. Nitrogen fixation is apparently an important nitrogen source in these two lakes; but the conditions change so fast that it is impossible to make an accurate assessment of the total quantities of nitrogen fixed in the lakes over the period of a year.

Recent studies in our laboratory and elsewhere have shown that significant fixation in natural water systems is not limited to surface waters and cyanophycean agents. Using the acetylene reduction method, Brezonik and Harper (73) first reported rates of nitrogen fixation in anoxic, aphotic waters of two stratified dystrophic lakes: Mary, in northern Wisconsin, and Mize, near Gainesville. The maximum rate found in the former lake was 47 μg N/(m^3)(hr), whereas rates as high as 300 μg N/(m^3)(hr) were found in the hypolimnion of Lake Mize.

A detailed study of nitrogen fixation over a two-year period in Lake Mize (74) established a consistent seasonal and vertical pattern of fixation; positive rates were found only in anoxic hypolimnion. The depth profile of fixation is markedly dichotomic, and maximum rates occur in the 3 to 10 m zone of this 20-m-deep lake (Figure 6). Fixation occurs for only a short period during summer stratification (Figure 7), but the rate of fixation during this time is high--up to 3.3 μg N/(liter)(hr).

No evidence for nitrogen-fixing blue-green algae has been found at Lake Mize at any time. Enrichment methods indicate that at least two nitrogen-fixing groups of bacteria, one heterotrophic, the other autotrophic, exist in the depths of Lake Mize. A gram-positive, obligate anaerobic spore-forming rod (characterized as *Clostridium* sp.), capable of using N_2 as its sole nitrogen source, was isolated from all depths at which fixation occurred. Two photosynthetic bacteria (*Thiospirillum* and *Chromatium*) of the family Thiorhodaceae were isolated from samples taken at 7 m. Fixation in Lake Mize is thought to be primarily

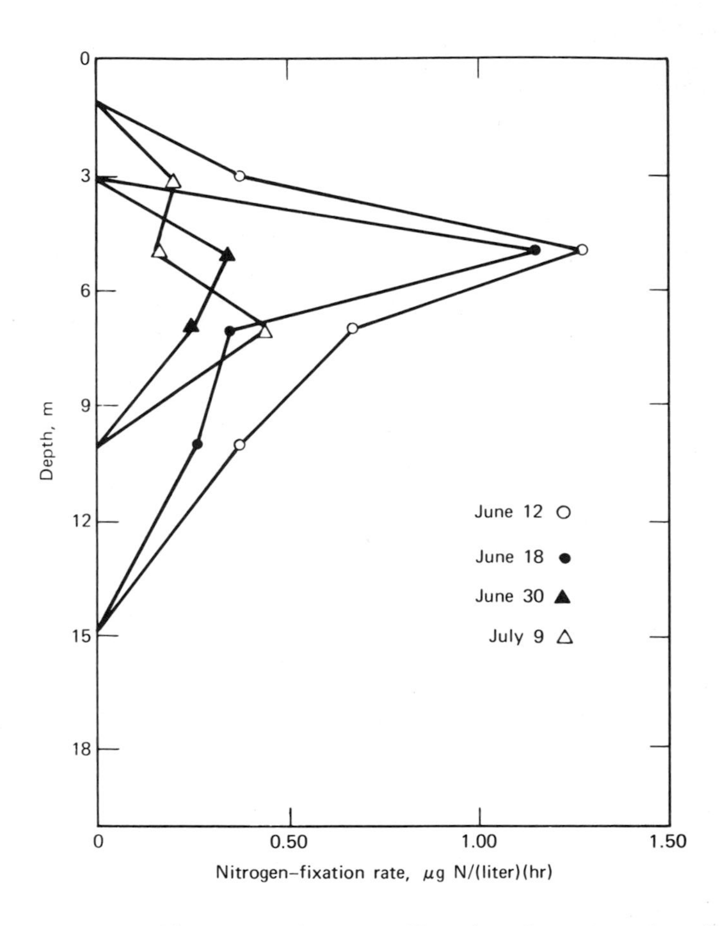

Fig. 6. Depth profiles of nitrogen fixation in Lake Mize during summer, 1970 (74).

heterotrophic based on the occurrence of high fixation rates in the dark and the comparative difficulty in obtaining isolates of photosynthetic bacteria. On the other hand, Stewart (18) has reported high rates of fixation which correlated with a bloom of the green photosynthetic bacterium Pelodictyon at the thermocline in a Norwegian fjord lake.

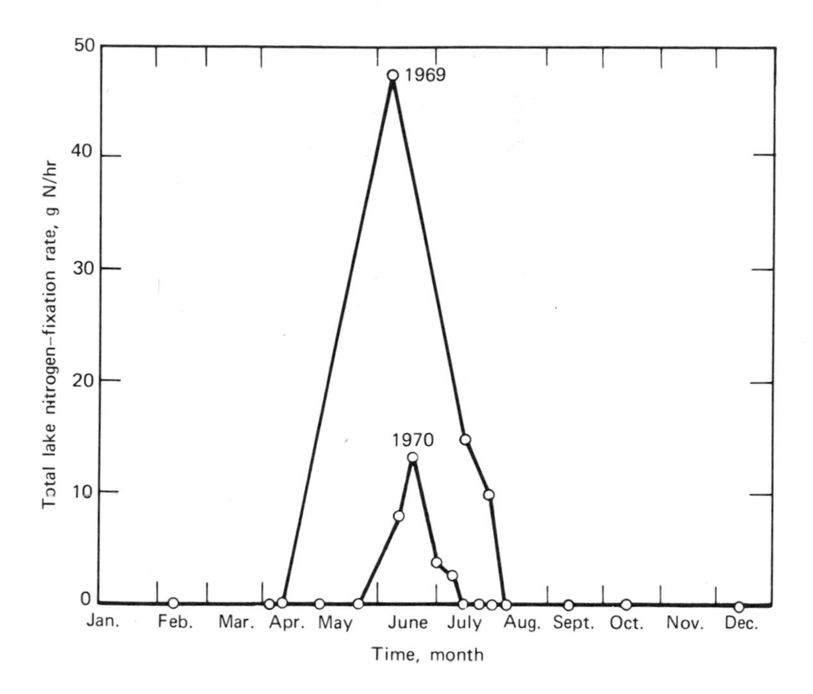

Fig. 7. Time course of total hourly nitrogen fixation at all depths in Lake Mize during 1969 and 1970 (74).

Nitrogen fixation also occurs at low but measurable rates in estuarine (75) and lacustrine (74,76) sediments. Brooks et al. (75) found acetylene reduction rates as high as 15 ng ethylene produced per gram dry weight of sediment per hour in the Waccasassa Estuary on the Florida Gulf coast, and reported that fixation is a consistent phenomenon both areally and seasonally in these sediments. A distinct layering occurred in sediment cores with no fixation in the flocculant surface sediment and maximum rates [1.6 to 11.4 ng of ethylene/(g of sediment)(hr)] in the next 2 to 5 cm of grey-black ooze. Much lower rates [0.03 to 0.04 ng of ethylene/(g of sediment)(hr)] were found in the coarse organic material of the bottom layer. A pure nitrogen-fixing Clostridium (or clostridialike) culture was isolated from Waccasassa Estuary sediments, but other nitrogen-fixing organisms could also be involved in the reaction in situ. Edmiston (77) has found nitrogen fixation in some sediments of Escambia Bay, near Pensacola,

Florida, but the frequency and distribution of estuarine sediment fixation on a broad geographical basis is wholly unknown.

In freshwater sediments, low rates of fixation were reported in Lake Erie sediments by Howard et al. (76). The distribution of nitrogen fixation in Florida lake sediments was studied by Keirn and Brezonik (74); acetylene reduction was detected in sediments from seven of 25 Florida lakes. Rates ranged from about 2 to 90 ng ethylene produced per gram of sediment per hour and decreased with depth in 30 to 50 cm cores.

High sucrose concentrations stimulated acetylene reduction in lake sediments, but a variety of other substrates did not. The significance of the low rates of fixation found in lake and estuarine sediments is not yet known. The matter is probably of minor importance to the nitrogen budgets of these water bodies, but presumably it is of some significance to the organisms involved. The biogeochemical implications may be considerable, however, if fixation is found to be as widespread in sediments as these early results imply.

Denitrification has been shown to be an important nitrogen sink in lakes (11,78). Its role in the nitrogen cycle of the oceans is less defined, but there is no doubt of its occurrence in some oceanic environments (9,10). Table 8 summarizes existing data on rates of denitrification in natural water systems. The measured rates are surprisingly comparable, considering the great diversity of aquatic environments represented. An important point is that not all the nitrate that is reduced is lost to the system as N_2, but a significant although variable amount of assimilatory reduction to ammonia and organic nitrogen occurs simultaneously. This was first noted by the author in denitrification experiments with sewage sludge digestors; only 50% of added nitrate was recovered as N_2 (79).

An early study of denitrification in lakes (78) suggested that assimilatory nitrate reduction was unimportant (based on low measured rates of ammonia production from labeled nitrate), but the authors did not measure incorporation of labeled nitrate into organic nitrogen. Brezonik and Lee (11) found that reduction to ammonia and organic nitrogen occurred at comparable but somewhat lower rates than denitrification in Lake Mendota, Wisconsin. Of the 4.43×10^7 g of nitrate-N present in the lake hypolimnion below 14 m in mid-summer of 1966, an estimated 2.81×10^7 g was lost by denitrification, and the balance (1.62×10^7 g) was reduced to ammonia and organic nitrogen. Goering (10) also reported ratios of N_2-N formed to

Table 8. Rates of Denitrification in Anoxic Natural Waters

Location	Depth, m	Nitrate,[a] μg N/liter	Rate of denitrification μg N/(liter)(day)	Rate of assimilatory nitrate reduction[b] μg N/(liter)(day)	Reference
Smith Lake, Alaska					
	1 (under ice cover)	140	15	2.8 (to NH_3)	78
	3 (bottom water plus sediment)	95	90	2.0 (to NH_3)	
	Average in lake		10.6 (first week of anoxic conditions)		66
			1.8 (average for entire winter)		
Lake Mendota, Wisconsin					
	14	600	8	9.1-11 (to TKN)	11

Table 8. continued

Location	Depth, m	Nitrate,[a] μg N/liter	Rate of denitrification μg N/(liter)(day)	Rate of assimilatory nitrate reduction[b] μg N/(liter)(day)	Reference
Lake Mendota, Wisconsin (continued)					
	23 (bottom)	170	26	3.1-9.1 (to TKN)	19
	Sediment interstitial water	2100 (added)[c]	330 (per liter of sediment)[c]	195 (to TKN)[c]	
Isla Genovesa Bay, Pacific Ocean					
	125	39	12		9
	175	0	18		
	225	6	13		
Tropical eastern Pacific near Mexico (two stations)					
	149-347	300-410	0.3-10.6	Ratio $-\Delta N_2$-N/ΔNO_3^--N	10
	94-417	392-450	1.8- 6.2	Varied from 1:10 to 8:10	

[a]Initial nitrate concentration in water.
[b]Rate of nitrate reduction to N form indicated (TKN = total Kjeldahl nitrogen).
[c]2100 μg NO_3^--N added to 1 liter of fresh sediment. Rates expressed per liter of sediment.

NO_3^--N lost ranging from 1:10 to 8:10 in the eastern Pacific, which indicates that assimilatory reduction is of great but variable importance.

Denitrification accounted for 11 to 15% of the total annual nitrogen input to Lake Mendota estimated by Lee et al. (22). If subsequent refinement of this budget leads to a smaller figure, as seems likely, denitrification will then assume greater relative importance as a nitrogen sink. For example, from the report of Keeney et al. (80) on denitrification in Lake Mendota sediments, we learn that the effect of this process is to decrease the estimated input of nitrogen to the lake by way of nitrate in groundwater seepage. Rapid disappearance of nitrate was found in sediment-lake water mixtures, but not all was denitrified. Nearly 40% of the labeled nitrate was later found in the ammonia and organic nitrogen fraction of the sediment.

Nitrogen reactions at the sediment-water interface have received little attention until recently. Several classic studies have discussed the role of oxygen and redox potential in controlling diffusion of ammonia from sediments. Mortimer (81) reported that little ammonia is released from sediments as long as an oxidized microzone exists, but that large quantities of this (and other nutrients) are liberated when the surface layer is reduced. He felt that this general increase could be explained by sorption of ammonia onto a complex floc containing iron (III). There are also possible partial explanations for the apparently low rate of nitrogen release from oxidized sediments: namely, that nitrification at the sediment surface causes nitrate rather than ammonia release, and that ammonia is unlikely to accumulate in oxygenated water, since it will either be oxidized or assimilated. The role of the oxidized microzone in influencing sediment-water nutrient cycling would be a fruitful area for further research.

For purposes of nutrient exchange, lake sediments can be divided into at least two main types based on the presence or absence of a defined water-sediment interface. In the former case, nutrient exchange may be limited by rates of diffusion, the presence of an oxidized surface layer, and the activities of burrowing animals (82). Many shallow, eutrophic Florida lakes have no defined sediment-water interface. A flocculant suspension covers the bottom in these lakes with gradual compaction from thin "soup" to consolidated sediment occurring over a depth of perhaps several feet. In this case, considerable exchange may be effected by wind-generated currents and turbulence, causing the sediment suspension to mix with overlying water. Rates of anaerobic decomposition of organic sediment are probably a primary factor limiting nutrient exchange in such sediments.

Sediment-water nitrogen interactions involve both biological and chemical processes, and rates of exchange may be further controlled by physical (e.g., diffusion) phenomena. Qualitative aspects of biological decomposition in sediments are fairly well understood (83,84) and need no further elaboration here. The importance of burrowing animals (e.g., worms, larvae, crustaceans) in aerating sediments and in physically transporting nutrients from the sediment and excreting them into the overlying water was suggested by Brooks (85). When surface sediment from the Waccasassa Estuary (Florida) was covered with estuary water in laboratory aquaria, a defined sediment-water interface and oxidized surface layer quickly arose, and within several days extensive burrowing and tube building activities were evident through the aquarium glass wall. However, quantitative data on regeneration rates induced by this mechanism are lacking.

Ammonia may be sorbed onto clays and organic colloids in sediments; if this occurs, an equilibrium between aqueous ammonia and ammonia in sediments would be set up. Principles of soil chemistry and cation-adsorption processes in soils should be applicable to the study of this phenomenon in sediments (86). A problem in defining the role of sorption in sediment-nutrient interactions is to separate the effects of biological activities. Use of poisoned or irradiated sediments is possible, but there is always a danger in changing the chemical nature of sediment by such treatment. The importance of chemical processes can be inferred if rapid uptake or release occurs during short (several hour) incubations, since biological decomposition and assimilation are inherently slow processes. That simple sorption is a significant reaction in situ is suggested by several simple experiments on estuarine and lacustrine sediments.

Table 9 shows the rapid uptake of ammonia by Waccasassa Esturay sediment over a 3-hr laboratory experiment (85). Initial sorption was very rapid, and the flask with added ammonia seemed to be approaching an equilibrium ammonia level comparable to that in the sediment-water flask with no addition.

In another experiment, fresh estuary sediment was sterilized by ^{60}Co irradiation and the uptake of ammonia was compared with unsterilized sediment in two aquaria. Similar rates of ammonia uptake occurred in both cases (Figure 8), implying a chemical rather than a biological mechanism. In some preliminary experiments conducted in our laboratory, rapid losses of ammonia from solution have

Table 9. Uptake of Ammonia by Waccasassa Estuary Sediment in Short-Term Shaker-Flask Experiment[a]

	Estuary water[b]	Estuary water with added NH_4Cl[b]
Control (no sediment)	0.11	6.80
5 min	0.11	2.70
1 hr	0.11	1.06
3 hr	0.10	0.35

[a]From Brooks (85). Four flasks had 300 ml of estuary water added. Ammonium chloride solution was added to two flasks in identical amounts and 22 g of fresh sediment was added to one of these. Of the two remaining flasks, one was a control and the other had 22 g of sediment added to it.
[b]Ammonia concentrations, mg N/liter; controls (no sediment) remained constant throughout the experiment.

been found with lake sediments.

In summary, the processes of nitrogen cycling and interchange in sediments are complicated and poorly understood. The mechanisms whereby nitrogen is exchanged between water and sediments are probably known, and in some cases qualitative rankings can be given to their importance. However, almost no information is available to establish the in situ rates and controlling factors for these processes.

INTERNAL NITROGEN CYCLING

Aside from nitrogen fixation and denitrification, which are really in situ source and sink reactions, the internal cycle of nitrogen in aquatic systems consists essentially of four types of reactions: (1) assimilation of inorganic nitrogen (primarily nitrate and ammonia), (2) regeneration of organic nitrogen to ammonia (ammonification), (3) oxidation of ammonia to nitrate (nitrification), and (4) heterotrophic conversion of organic nitrogen from one form or organism to another. The last reaction is of immense importance as the mechanism whereby zooplankton and higher trophic levels obtain all their nitrogen requirements. Algae are thought to have a limited capability to assimilate organic nitrogen (primarily urea and some amino acids). However, the complicated heterotrophic reactions are beyond the scope of the present

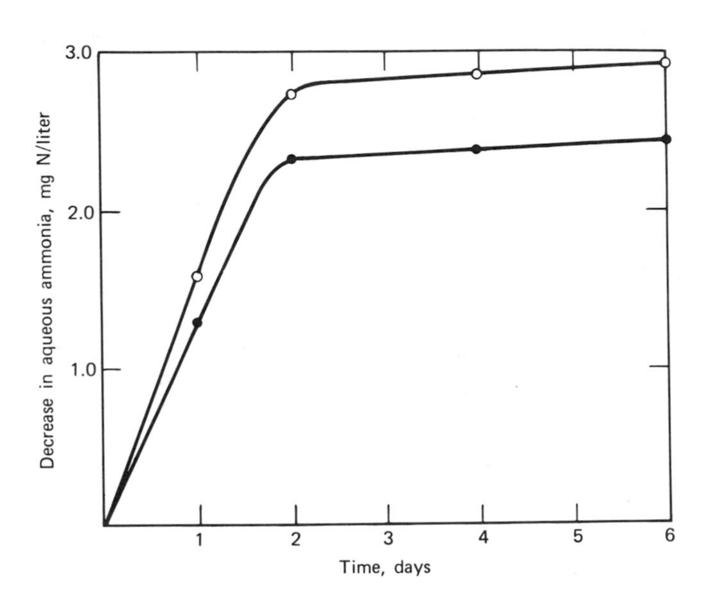

Fig. 8. Ammonia uptake by fresh and sterilized (irradiated) sediment from the Waccasassa Estuary, Florida: open circles, fresh sediment; solid circles, sterile sediment (85).

discussion, and in the remainder of this section we shall concentrate on the first three reactions.

Rates of ammonia and nitrate assimilation in aquatic environments have been determined by several workers in recent years. Such studies are relevant to an increased understanding of the metabolism of the aquatic community because the availability and utilization of nitrogen profoundly influence and in some cases control primary production.

Of interest in this regard is such information as the rates of inorganic nitrogen turnover, the relative nutritional importance of the various nitrogen forms, the effect of concentration on rates of utilization, and the relation between nitrogen assimilation and primary production. Although the ^{15}N tracer approach to studying aquatic nitrogen cycle dynamics is only about 10 years old, considerable gains have already been made through its use. The nitrogen cycle can no longer be viewed as a slow, primarily seasonal cycle. Turnover times for inorganic nitrogen are on the order of days or even hours in many waters, and this means that rapid organic nitrogen decomposition or a constant influx of new nitrogen into the system is essential.

Dugdale and Dugdale (72) were the first to report in situ ammonia and nitrate assimilation rates in surface waters. Using ^{15}N techniques, they found two main pulses of assimilation in Sanctuary Lake, Pennsylvania. During the first pulse, in June, ammonia was assimilated most rapidly with a maximum rate of about 110 μg N/(liter)(day). Maximum nitrate assimilation rate during the same period was about 40 μg N/(liter)(day). A second pulse of assimilation in early September utilized primarily N_2.

Some of the high reported assimilation rates may be an artifact of methodology. For example, the nitrate concentration at the time of the June assimilation maximum was only about 10 μg N/liter. The amounts of $^{15}NO_3^-$ and $^{15}NH_3$ added to samples were not specified, but a large amount of tracer relative to the initial unlabeled nitrogen would stimulate assimilation beyond the normal rate.

Figure 9 shows rates of ammonia and nitrate assimilation measured at various surface stations in Lake Mendota during spring and summer of 1966 (87). Ammonia assimilation was greater than nitrate assimilation in all cases, and highest rates for both occurred in late spring and in late summer. The highest measured ammonia assimilation rates are probably overestimates of the in situ rates for the same reason discussed above. Short incubations (4 hr) were used when possible, in order to minimize incubation effects. No correlation was found between ammonia and nitrate concentrations and assimilation rates. Considerable variations in assimilation rates occurred over short periods of time, but this feature is accentuated because the data represent three widely different stations in the lake.

Depth profiles of ammonia assimilation were also determined on several occasions (Figure 10). During holomixis and early phases of stratification, assimilation was fairly uniform with no trends in the depth profile. However, by early June a pronounced stratification of assimilation was found, displaying high rates in the epilimnion and much lower rates in the hypolimnion. As an approach to considering the turnover of newly assimilated nitrogen in natural waters, the appearance of ^{15}N from labeled ammonia in the particulate and soluble organic nitrogen fractions was determined on four occasions (Table 10). A surprising amount of ^{15}N appeared in the soluble fraction in 4-hr incubations. We do not know whether the mechanism involved was algal excretion (or "leakage") of low-weight compounds, zooplankton excretion, or cell decomposition,

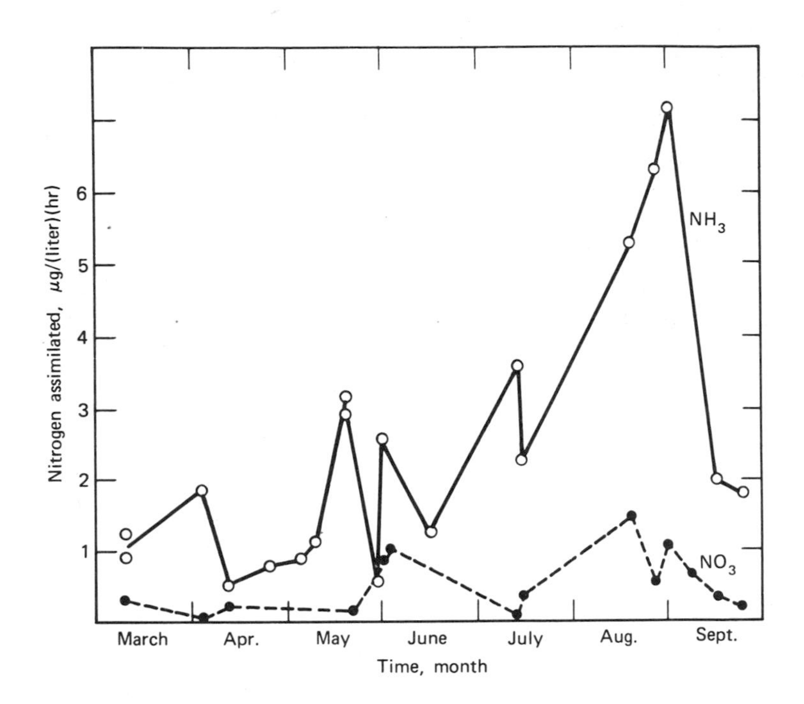

Fig. 9. Ammonia and nitrate assimilation rates in Lake Mendota surface waters during 1966 (87).

but the short incubation times should rule out the latter. A certain amount of cell lysis may have occurred during filtration, but even so the results suggest a rapid turnover of organic nitrogen. Dugdale (88) also found significant but lower ^{15}N enrichment in the soluble organic nitrogen fraction in Smith Lake, Alaska.

In nearly every assimilation study published thus far, ammonia was much more important than nitrate as a nitrogen source. In view of the ease with which ammonia can be assimilated this is not surprising; prior to the tracer studies cited, however, most marine biologists considered nitrate to be the only significant nitrogen source in the sea. It now seems certain that both marine and freshwater algae derive most of their nitrogen from ammonia, often in spite of higher nitrate concentrations.

There may be a certain amount of "wheel-spinning"

involved in ammonia assimilation; rather than representing new production, at least part of the uptake seems to derive from the necessity to recapture nitrogen from compounds that apparently continually leak through cell walls (89-91). Dugdale and Goering (92) derived a simple model for the nitrogen cycle in marine surface waters in which they considered primary production associated with ammonia assimilation to be "regenerated" production; primary production associated with nitrate assimilation was regarded as "new" production. Only the latter would be available for export to higher trophic levels, since nitrate input from the deep water is the principal source of nitrogen to the surface. Ammonia in seawater is the product of short-term regeneration, and primary production

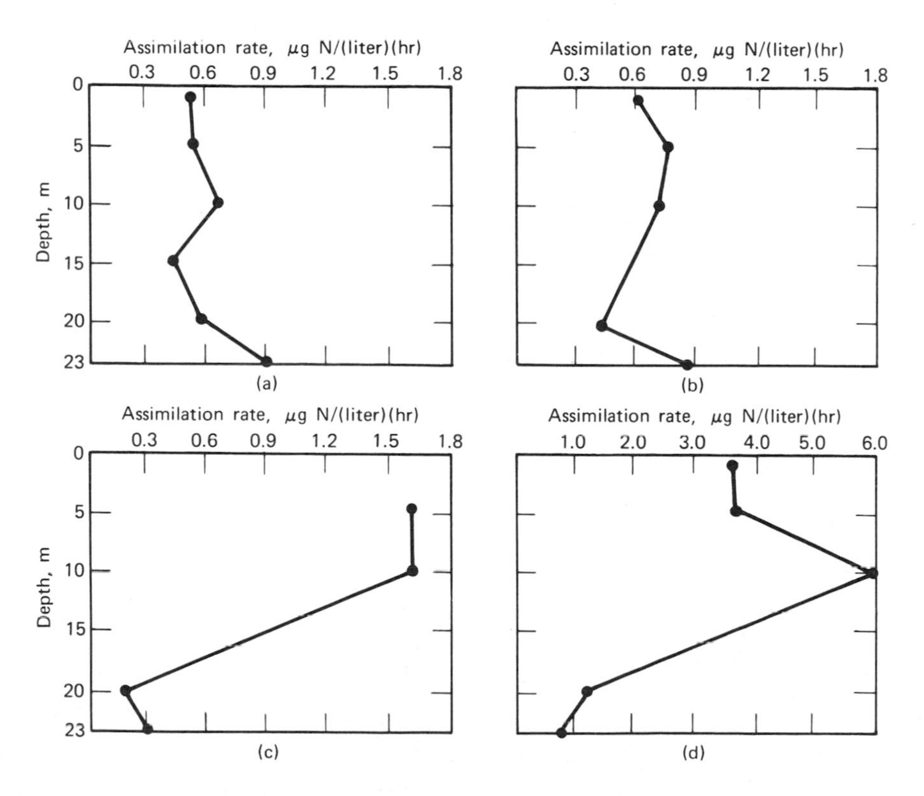

Fig. 10. Depth profiles of ammonia assimilation at deepest station in Lake Mendota spring and summer of 1966 (87): (a) April 12, (b) May 14, (c) June 3, (d) July 13.

Table 10. Rates of Ammonia Transfer to Particulate and Soluble Organic Nitrogen in Lake Mendota Surface Waters (87)

Date	PON,[a] mg N/liter	SON,[b] mg N/liter	NH_3 Assimilation PON, μg N/(liter)(hr)	NH_3 Assimilation SON, μg N/(liter)(hr)
Aug. 19	0.44	0.46	5.4	7.9
Aug. 26	0.42	0.62	6.3	9.0
Sept. 1	0.42	0.61	7.2	9.6
Sept. 8	0.46	0.56	4.8	6.4

[a]Particulate organic nitrogen.
[b]Soluble organic nitrogen.

associated with its assimilation is essentially that needed to maintain the standing crop. Extension of this approach to the more complicated nitrogen cycles in shallow lakes has not been attempted.

An exception to the general rule of more rapid ammonia than nitrate assimilation was recently reported by Goering et al. (93) in the eastern subtropical Pacific. Phytoplankton in a nutrient-rich discontinuity layer at the bottom of the euphotic zone were found to use nitrate primarily, whereas algae in the impoverished surface water used ammonia.

Dugdale (94) has developed a model for nutrient limitation in the sea in which assimilation of ammonia and nitrate were assumed to follow Michaelis-Menten kinetics:

$$v = \frac{VS}{K_t + S}$$

where v is the rate of nutrient assimilation; V is a maximum rate, constant under a given set of conditions; S is nutrient concentration; and K_t is a transport or half-saturation constant with units of concentration. The numerical value of K_t is equal to the nutrient concentration at which v is one-half the maximum for the system (1/2 V).

Subsequent ^{15}N measurements by MacIsaac and Dugdale (95) (95) and direct uptake measurements by Eppley et al. (96) have verified this model and estimated K_t values for various marine surface waters. Calculated K_t values ranged from less than 0.2 to 4.2 μg-at./liter for nitrate and from 0.1 to 1.3 μg-at./liter for ammonia (95). Lower transport constants correlated with oligotrophic (nutrient-depleted) areas, and higher K_t values were found in nutrient-rich waters. It was thus suggested that phytoplankton in oligotrophic waters are adapted to low ambient concentrations and can assimilate nutrients more rapidly under these conditions than can phytoplankton from nutrient-enriched regions.

Much less is known about the dynamics of other nitrogen cycle reactions. Ammonification is important in renewing a limited inorganic nitrogen supply, especially in impoverished surface waters. Rates of ammonification can be determined by an isotope-dilution method (87,88,97). Assuming steady-state conditions for assimilation and ammonification, ^{15}N label added to the ammonia pool will gradually become diluted as both labeled and unlabeled ammonia N is assimilated but comparatively unlabeled N is returned to the ammonia pool by ammonification of organic N. The decrease in ^{15}N enrichment in the ammonia can be quantified and related to the rate of ammonification (87, 97). Using such a technique, Dugdale (88) found rapid ammonification [25 μg N/(liter)(hr)] in a single determination in Smith Lake, Alaska, during August, 1964. In a more complete study Alexander (98) reported that higher rates of ammonification were associated with the spring bloom of *Anabaena* and the period of maximum ammonia assimilation rates in Smith Lake.

Ammonification rates were measured (87) at various depths in Lake Mendota during the summer of 1966 (Table 11). High rates were found, with a general trend toward higher rates in the bottom waters. Some of the high surface rates were probably overestimated because of the low ammonia concentrations in the water at the time, and this also seems to be likely for some of the rates reported for Smith Lake (88,98). The isotope dilution method is simple in theory, but it is more difficult in practice, especially when low ammonia levels are encountered. The assumptions of steady-state and constant ammonia levels during incubation often do not apply, and spurious data can result. Even considering the possibilities for error, the data in Table 11 imply that ammonia has a small turnover time in Lake Mendota. From steady-state kinetics, ammonia turnover times were calculated to range from 7 to 62 hr in the lake surface waters during the summer. Because steady-state conditions do not exactly apply, these calculations should be considered only approximate; nonetheless, they suggest a much more rapid cycling of nitrogen than measurement of concentration changes alone would suggest.

Table 11. Ammonification Rates in Lake Mendota in 1966 (87)

Date	Depth, m	TON,[a] μg N/liter	NH_3, μg N/liter	Ammonification Rate, μg N/(liter)(hr)
March 9	1	310[b]	180	3.7
	7	250[b]	210	1.1
	15	260[b]	230	6.9
July 13	20	640[b]	100	4.4
	23	710[b]	100	5.3
Aug. 19	1	900	40	16
Aug. 26	1	1040	40	2.1
	16	820	350	7.7
	22	1100	800	22.4
Sept. 24	1	900	20	13
	14	790	200	0.6
	15	1000	540	8.4

[a]Total organic nitrogen.
[b]Particulate organic nitrogen.

The rate at which newly fixed nitrogen is transferred from agents of fixation to other aquatic organisms is of great interest. Jones and Stewart (99) have reported that extracellular nitrogen liberated by the marine nitrogen fixer *Calothrix scopulorum* is assimilated by marine algae, fungi, and bacteria and could serve as the sole nitrogen source for a species of *Chlorella*. Based on limiting nutrient bioassays, Fitzgerald (100) has concluded that transfer of nitrogen newly fixed by *Aphanizomenon* is of little significance to colonies of *Microcystis* growing in the same water, but this is perhaps an overextension of the data. It seems clear from Fitzgerald's results that *Aphanizomenon* is unable to supply *Microcystis* with sufficient nitrogen to satisfy the organisms, but the actual magnitude of supply is still unclear.

Rates of nitrification in natural waters can also be measured by ^{15}N techniques. For this purpose $^{15}NH_3$ is added to a water sample and after incubation, the amount of label in the nitrate and nitrite fractions is determined by reduction of oxidized nitrogen to ammonia with Devarda's alloy (87,101) and subsequent conversion of ammonia to N_2 by hypobromite oxidation for isotope ratio analysis (102,103). Dugdale and Goering (92) found that nitrification was undetectable in surface waters of the Sargasso Sea when they used incubation periods as long as a week.

A few nitrification rates were measured in Lake Mendota during 1966 (Table 12). Rates increased with depth in spring but were very low in the surface water during summer. Nitrification was evidently unable to keep pace with algal assimilation of nitrate, as surface water concentrations declined to trace levels by mid-June. It should be noted that substantial nitrification evidently occurred at middepths (7 to 17 m) in Lake Mendota during late spring and early summer (11,87). Largest increases occurred in the 12 to 15 m zone, where nitrate increased from less than 0.2 mg N/liter in June to 0.5 to 0.7 mg N/liter in late July. Nitrification in the hypolimnion is especially significant because the water eventually becomes anoxic during late summer, and denitrification takes place (11). Thus nitrification increases the importance of denitrification as a nitrogen sink in Lake Mendota. Vollenweider (104) has also reported nitrification at middepths during summer stratification in Lake Orta, Italy.

Table 12. Nitrification Rates in Lake Mendota During Spring and Early Summer, 1966 (87)

Date	Depth, m	Nitrate, mg N/liter	Nitrification, μg N/(liter)(hr)
March 9	1	0.35	0.21
	7	0.40	0.60
	15	0.40	0.62
April 3	1	0.36	0.07
	7	0.36	0.03
	15	0.35	0.36
May 5	1	0.29	0.07
June 3	20	0.28	0.08

MISCELLANEOUS NITROGEN REACTIONS

The major nitrogen cycle reactions as presently known have now been discussed. Whether they in fact represent all significant pathways for nitrogen in aquatic systems is still questionable, for a variety of other reactions are thermodynamically feasible and some of these could have a substantial impact on the overall cycle. The mechanisms of N_2 formation from fixed nitrogen have been the subject of a long controversy, and many pathways have been proposed and rejected. Formation of nitrogen directly from ammonia was hypothesized to occur in sludge digestion (105) and in anoxic lake waters (106), but others have found no evidence for such a reaction (79,107). Wijler and Delwiche (107) used ^{15}N methods to show that all N_2 in denitrification derives from oxidized nitrogen (NO_3^-, NO_2^-) sources. Some workers have proposed that the so-called Van Slyke reaction of nitrous acid with amino groups under acid conditions

$$\underset{\displaystyle NH_2}{\underset{|}{RCHCOOH}} + HNO_2 \longrightarrow RCHOHCOOH + N_2 + H_2O$$

is significant in acid lakes (28) and soils (108), but others have rejected this idea (109, 110). Recent work has provided evidence for significant chemical decomposition of nitrite to a variety of gaseous products in soils. Bremner and Nelson (109) have demonstrated that nitrite decomposition in sterilized soils is due to reaction of nitrite with soil organic matter and to self-decomposition of nitrous acid ($2HNO_2 = NO + NO_2 + H_2O$) under acid conditions. The nitrite-organic matter reaction was found to involve phenols and polyphenols such as lignins and tannins. The mechanism is thought to involve reaction of nitrous acid under slightly acid conditions (pH 5) with phenolic compounds to form nitrosophenols which tautomerize to quinone oximes. The latter are decomposed by nitrous acid to form N_2 and N_2O (111). This reaction could be important in the nitrogen cycle of natural waters. Colored lakes are high in polyphenolic substances (tannins, lignins, "humic acid") and generally have acid pH values. Although nitrite levels in natural waters are normally low, such a reaction could act to decompose nitrite as rapidly as it is formed.

Photochemical reactions are highly important in the atmospheric chemistry of nitrogen, but photochemistry is presently thought to play an insignificant role in aquatic nitrogen transformations. Since most photochemical reactions are induced by UV light, such processes should be limited to a narrow surface layer. Photochemical nitrification in the sea was proposed by various workers in the 1930s (112-114) partly as a result of difficulties in culturing marine nitrifiers, but Hamilton (115) concluded that such a reaction is of no consequence in the marine nitrogen cycle. Hamilton found slight photoreduction of nitrate to nitrite but concluded that this too was insignificant.

Several thermodynamically possible nitrogen reactions have never been shown to occur as biological phenomena. For example, N_2 could be used as a terminal electron acceptor in organic carbon oxidation with the release of sufficient energy for organism growth:

$$\text{glucose} + 4N_2 + 8H^+ + 6H_2O = 6CO_2 + 8NH_4^+,$$

$$\Delta G_w^o = -84.5 \text{ kcal/mole}$$

The foregoing reaction of course occurs when heterotrophic organisms fix nitrogen, but it is not thought that any of this energy is trapped and retained by nitrogen-fixing cells for other uses. However, it is conceivable that this reaction could be a source of energy for some anaerobic bacteria. Oxidation of molecular nitrogen to nitrate by molecular oxygen is also exergonic at pH 7:

$$N_2 + \frac{5}{2}O_2 + H_2O = 2NO_3^- + 2H^+(w),$$

$$\Delta G_w^o = -15.2 \text{ kcal/mole}$$

With the large reservoir of available reactants for this process, it is surprising that no organism have evolved to take advantage of the situation. However, this is probably fortunate considering the nature of the product (nitric acid).

SUMMARY

The nitrogen cycle in natural waters is a highly complicated phenomenon. Qualitatively the reactions comprising the cycle and the sources of nitrogen to natural waters are well known; however, quantitative knowledge is lacking in many areas. The significance of various cultural and natural sources of nitrogen are described in the text and summarized in Table 2. Many contributions, especially from diffuse sources, are inadequately known, and in some cases they are essentially unknown. However, given such information we can estimate the nutrient budget of a lake from knowledge of the land use and population characteristics of the watershed. As more information on the nutrient contributions from various sources is obtained, the accuracy of such simulated nutrient budgets should approach that of budgets that are determined by actual measurement. Calculated nitrogen and phosphorus budgets for 55 Florida lakes exhibit significant correlations (Table 5) with the trophic states of the lakes as measured by a quantitative index. Critical loading rates, above which eutrophication problems are likely to occur, can be estimated by such regression relationships.

Nitrogen fixation, an in situ source of nitrogen, occurs fairly commonly in eutrophic surface waters, and several bloom-forming blue-green algae are the primary agents. Although fixation is of undoubted importance to these organisms, measured rates (Table 7, Figure 4) are too low and sporadic for this reaction to be of general significance for the overall nitrogen budgets of most lakes. Nitrogen fixation by heterotrophic and photosynthetic bacteria also occurs in the aquatic environment, and in at least one case (Figures 6 and 7), Lake Mize, Florida, heterotrophic fixation in the anoxic hypolimnion represents a significant source of nitrogen to the lake. Low rates of bacterial fixation have also been reported in lacustrine and estuarine sediments, but the significance of these rates is as yet unknown.

Denitrification may be an important nitrogen sink in stratified eutrophic lakes where anoxic conditions in the hypolimnion enable this reaction to occur. Hypolimnetic nitrification of ammonia regenerated from sinking plankton before the onset of anoxic conditions apparently increases the significance of denitrification by increasing the supply of nitrate. However, not all the nitrate reduced under anoxic conditions is lost to the system as N_2; Table 8 indicates that a variable but considerable fraction of the nitrate is reduced to organic nitrogen and ammonia.

Recent ^{15}N tracer studies of nitrogen cycling within lakes and marine waters have revealed the cycle as a highly dynamic phenomenon. In almost all cases ammonia is assimilated more rapidly than nitrate (e.g., Figures 9 and 10) and appears to be the principal nitrogen source for algae. Ammonia turnover times of a few hours have been measured, which implies a rapid recycling of nitrogen between organic compounds and the ammonia pool. Ammonification rates measured in Lake Mendota (Table 11) tend to bear this out.

The nitrogen cycle as presently understood is primarily a biological phenomenon. However, the possible importance of chemical reactions and of heretofore undetected (but potential) biological transformations should not be overlooked. The recently reported decomposition of nitrite in soils by reaction with phenolic compounds under slightly acid conditions (109) should serve to indicate that the nitrogen cycle is far from being a well-understood, predictable phenomenon.

ACKNOWLEDGMENTS

The work reported in this chapter has been supported in part by Federal Water Quality Administration (FWQA) research grant 16010 DCK, Office of Water Resources Research Grant DI 14-31-0001-3068 and the Game and Freshwater Fish Division of the State of Florida. Parts of the work reported in this paper were conducted in the laboratory of Limnology and Water Chemistry Laboratory of the University of Wisconsin. Support of these phases was given by FWQA training grant 5TI-WP-22-04 and FWQA fellowship 2F-/WP-26. Use of the University of Wisconsin Biochemistry Department mass spectrometer is gratefully acknowledged.

REFERENCES

1. C. I. Kuznetsov, "Die Rolle von Mikroorganisnien im Stoffkreislauf der Sein," Berlin, 1959.
2. J. D. H. Strickland, in "Chemical Oceanography," Vol. I, J. P. Riley and G. Skirrow, Eds. Academic Press, New York, 1965, p. 477.
3. R. W. Eppley, J. L. Coatsworth, and L. Solorzano, "Studies of Nitrate Reductase in Marine Phytoplankton," Limnol. Oceanogr., 14, 194-205 (1969).
4. T. von Brand, N. W. Rakestraw, and C. E. Renn, "The Experimental Decomposition and Regeneration of Nitrogenous Organic Matter in Sea Water," Biol. Bull. Wood's Hole, 72, 165-175 (1937).
5. R. E. Johannes, in "Advances in Microbiology of the Sea," M. R. Droop and E. J. Ferguson Wood, Eds., Academic Press, New York, p. 203.
6. H. R. Krause, "Zur Chemie und Biochemie der Zersetzung von Susswasser organismen unter besonderer Beruck sichtigung des Abbaues der organischen Phosphorkomponenten," Verh. Int. Verein. Limnol., 15, 549-561 (1964).
7. C. C. Delwiche, in "Microbiology and Soil Fertility," C. M. Gilmour and O. N. Allen, Eds., Oregon State University Press, Corvallis, 1965, p. 29.
8. J. M. Symons, S. R. Weibel, and G. G. Robeck, "Impoundment Influences on Water Quality," J. Am.

Water Works Assoc., 57, 51-75 (1965).
9. J. J. Goering and R. C. Dugdale, "Denitrification Rates in an Island Bay in the Equatorial Pacific Ocean," Science, 154, 505-506 (1966).
10. J. J. Goering, "Denitrification in the Oxygen Minimum Layer of the Eastern Tropical Pacific Ocean," Deep-Sea Res., 15, 157-164 (1968).
11. P. L. Brezonik and G. F. Lee, "Denitrification as a Nitrogen Sink in Lake Mendota, Wis.," Envir. Sci. Technol., 2, 120-131 (1968).
12. K. Wuhrman, "Nitrogen Removal in Sewage Treatment Processes," Verh. Int. Verein. Limnol., 15, 580-596 (1964).
13. W. K. Johnson and G. J. Schroepfer, "Nitrogen Removal by Nitrification and Denitrification," J. Water Pollut. Control Fed., 36, 1015-1036 (1964).
14. G. E. Hutchinson, "Nitrogen in the Biogeochemistry of the Atmosphere," Am. Sci., 32, 178-195 (1944).
15. P. W. Wilson, "First Steps in Biological Nitrogen Fixation," Proc. Roy. Soc., B 172, 319-325 (1969).
16. W. D. P. Stewart, "Nitrogen Fixation in Plants," Athlone Press, London, 1966.
17. W. D. P. Stewart, "Algal Fixation of Atmospheric Nitrogen," Plant and Soil, 32, 555-588 (1970).
18. W. D. P. Stewart, in "Algae, Man and the Environment," D. Jackson, Ed., Syracuse University Press, Syracuse, N. Y., 1968, p. 53.
19. P. L. Brezonik, W. H. Morgan, E. E. Shannon, and H. D. Putnam, Bull. 134, Engineering and Industrial Experiment Station, University of Florida, Gainesville, 1969.
20. G. A. Rohlich and W. L. Lea, Report of University of Wisconsin Lake Investigations Committee, Madison, (mimeo.), 1949; see A. D. Hasler, in "Limnology in North America," D. G. Frey, Ed., University of Wisconsin Press, Madison, 1963, p. 55.
21. R. A. Vollenweider, Rept. DAS/CSL/68.27 to Organization for Economic Cooperation and Development, Paris, 1968.
22. G. F. Lee (Chairman), Report of Nutrient Sources Subcommittee of Lake Mendota Problems Committee, Madison, Wis. (mimeo.), 1966.
23. E. E. Shannon, Ph.D. thesis, University of Florida, Gainesville, 1970.
24. C. F. Cooper, in "Eutrophication: Causes, Consequences, Correctives," National Academy of Science, Washington, D. C., 1969, p. 446.
25. R. O. Sylvester, in "Algae and Metropolitan Wastes," U. S. Public Health Serv. Rep. W61-3, 1961, p. 80.

26. D. W. Cole and S. P. Gessel, in "Forest-Soil Relationships in North America," C. T. Youngberg, Ed., Oregon State University Press, Corvallis, 1965, p. 95.
27. S. R. Weibel, R. B. Weidner, A. G. Christianson, and R. G. Anderson, Proc. 3rd Int. Conf., Water Pollut. Res., Munich, Pergamon Press, 1966.
28. G. E. Hutchinson, "Treatise on Limnology," Vol. I. Wiley, New York, 1957.
29. J. H. Feth, "Nitrogen Compounds in Natural Water--A Review," Water Resour. Res., 2, 41-58 (1966).
30. W. W. Sanderson, "Studies of the Character and Treatment of Wastes from Duck Farms," Proc. 8th Ind. Waste Conf., Purdue Univ. Ext. Series, 83 (3B), 170-176 (1953).
31. C. D. Gates, "Treatment of Long Island Duck Farm Wastes," J. Water Pollut. Control Fed., 35, 1569-1579 (1963).
32. A. A. Paloumpis and W. C. Starret, "An Ecological Study of Benthic Organisms in Three Illinois River Flood Plain Lakes," Am. Midland Natur., 64, 406-435 (1960).
33. C. R. Goldman, "The Contribution of Alder Trees (Alnus Tenuifolic) to the Productivity of Castle Lake, California," Ecology, 42, 282-288 (1961).
34. H. B. N. Hynes and N. K. Kaushik, "The Relationship Between Dissolved Nutrient Salts and Protein Production in Submerged Autumnal Leaves," Verh. Int. Verein. Limnol., 17, 95-103 (1969).
35. P. J. Richerson, G. A. Moshiri, and G. L. Godshalk, "Certain Ecological Aspects of Pollen Dispersion in Lake Tahoe (California-Nevada)," Limnol. Oceanogr., 15, 149-153 (1970).
36. K. V. Slack and H. R. Feltz, "Tree Leaf Control on Low Flow Water Quality in a Small Virginia Stream," Envir. Sci. Technol., 2, 126-131 (1968).
37. C. N. Sawyer, "Fertilization of Lakes by Agricultural and Urban Drainage," J. New Eng. Water Works Assoc., 61, 109-127 (1947).
38. K. M. Mackenthum, W. M. Ingram, and R. Porges, "Limnological Aspects of Recreational Lakes," Public Health Serv. Publ. 1167 (1964).
39. R. S. Englebrecht and J. J. Morgan, "Studies on the Occurrence and Degradation of Condensed Phosphate in Surface Waters," Sewage Ind. Wastes, 31, 458-478 (1959).
40. J. Montelaro, personal communication to E. E. Shannon, 1970.
41. C. E. Miller, "Soil Fertility," Wiley, New York, 1955.

42. W. R. Johnson, F. Ittihadieh, R. M. Daum, and A. F. Pillsbury, "Nitrogen and Phosphorus in Tile Drainage Effluent," Proc. Soil Sci. Soc. Am., 29, 287-289 (1965).
43. P. G. Moe, J. V. Mannering, and C. B. Johnson, "Loss of Fertilizer Nitrogen in Surface Runoff Water," Soil Sci., 104, 389-394 (1967).
44. J. W. Biggar and R. B. Corey, in "Eutrophication: Causes, Consequences, Correctives," National Academy of Science, Washington, D.C., 1969, p. 404.
45. R. Loehr, "Animal Wastes--A National Problem," J. Sanit. Eng. Div., Am. Soc. Civil Eng., 95, 189-221 (1969).
46. B. A. Stewart, "Volatilization and Nitrification of Nitrogen from Urine under Simulated Cattle Feed Lot Conditions," Envir. Sci. Technol., 4, 579-582 (1970).
47. G. L. Hutchinson and F. G. Viets, Jr., "Nitrogen Enrichment of Surface Water by Absorption of Ammonia Volatilized from Cattle Feed Lots," Science, 166, 514-515 (1969).
48. J. R. Miner, R. I. Lipper, L. R. Fina, and J. W. Funk, "Cattle Feed Lot Runoff--Its Nature and Variation," J. Water Pollut. Control Fed., 38, 1582-1591 (1966).
49. S. R. Weibel, in "Eutrophication: Causes, Consequences, Correctives," National Academy of Science, Washington, D.C., 1969, p. 383.
50. C. L. Palmer, "The Pollutional Effects of Storm-Water Overflows from Combined Sewers," Sewage Ind. Wastes, 22, 154-165 (1950).
51. C. N. Sawyer, J. B. Lackey, and A. T. Lenz, Report to Governor's Committee, State of Wisconsin, Madison, 1945.
52. R. C. Polta, in "Water Pollution by Nutrients--Sources, Effects and Control," Water Resources Research Center, University of Minnesota, Minneapolis, Bull. 13, 1969, p. 53.
53. S. R. Quasim, Ph.D. thesis, University of West Virginia, Morgantown, 1965.
54. F. H. Bormann, G. E. Likens, D. W. Fisher, and R. S. Pierce, "Nutrient Loss Accelerated by Clear-Cutting of a Forest Ecosystem," Science, 159, 882-884 (1968).
55. D. Benson, in New York State Conservationist Bull. L-132, N.Y. State Conservation Dept., 1965.
56. G. F. Lee, University of Wisconsin, Water Resources Center, Occasional Paper 2, 1970.
57. P. L. Brezonik, Report of University of Florida, School of Forestry, Austin Carey Forest Committee, 1971 (mimeo.).

58. E. E. Shannon and P. L. Brezonik, "Eutrophication Analysis: A Multivariate Approach," J. Sanit. Eng. Div., Am. Soc. Civil Eng., 98, 37-57 (1972).
59. E. E. Shannon and P. L. Brezonik (in preparation).
60. C. C. Chen, "Concepts and Utilities of Ecologic Model," J. Sanit. Eng. Div., Am. Soc. Civil Eng., 96, 1085-1097 (1970).
61. W. Stumm and J. J. Morgan, "Aquatic Chemistry." Wiley, New York, 1970.
62. G. E. Hutchinson, "Limnological Studies in Connecticut. IV. The Mechanisms of Intermediary Metabolism in Stratified Lakes," Ecol. Monogr., 11, 21-60 (1941).
63. R. C. Dugdale, V. A. Dugdale, J. C. Neess, and J. J. Goering, "Nitrogen Fixation in Lakes," Science, 130, 859-860 (1959).
64. V. A. Dugdale and R. C. Dugdale, "Nitrogen Metabolism in Lakes. II. Role of Nitrogen Fixation in Sanctuary Lake, Pennsylvania," Limnol. Oceanogr., 7, 170-177 (1962).
65. J. J. Goering and J. C. Neess, "Nitrogen Fixation in Two Wisconsin Lakes," Limnol. Oceanogr., 9, 530-539 (1964).
66. V. A. Billaud, "Nitrogen Fixation and the Utilization of Other Inorganic Nitrogen Sources in a Subarctic Lake," J. Fish. Res. Bd., Canada, 25, 2101-2110 (1968).
67. R. C. Dugdale, J. J. Goering, and J. H. Ryther, "High Nitrogen-Fixation Rates in the Sargasso Sea and the Arabian Sea," Limnol. Oceanogr., 9, 507-510 (1964).
68. W. D. P. Stewart, "Nitrogen Turnover in Marine and Brackish Habitats. I. Nitrogen Fixation," Ann. Bot., 29, 229-239 (1965).
69. W. D. P. Stewart, G. P. Fitzgerald, and R. H. Burris, "In Situ Studies on N_2 Fixation Using the Acetylene Reduction Technique," Proc. Nat. Acad. Sci. (U.S.), 58, 2071-2078 (1967).
70. E. E. Shannon and P. L. Brezonik, Limnol. Oceanogr. (in press).
71. P. L. Brezonik and M. A. Keirn (in preparation).
72. V. A. Dugdale and R. C. Dugdale, "Tracer Studies of the Assimilation of Inorganic Nitrogen Sources," Limnol. Oceanogr., 10, 53-57 (1965).
73. P. L. Brezonik and C. L. Harper, "Nitrogen Fixation in Some Anoxic Lacustrine Environments," Science, 164, 1277-1279 (1969).
74. M. A. Keirn and P. L. Brezonik, "Nitrogen Fixation by Bacteria in Lake Mize, Florida, and in Some Lacus-

trine Sediments," Limnol. Oceanogr., 16, 720-731 (1971).
75. R. H. Brooks, Jr., P. L. Brezonik, H. D. Putnam, and M. A. Keirn, "Nitrogen Fixation in an Estuarine Environment: The Waccasassa on the Florida Gulf Coast," Limnol. Oceanogr., 16, 701-710 (1971).
76. D. L. Howard, J. I. Frea, R. M. Beeister, and P. R. Dugan, "Biological Nitrogen Fixation in Lake Erie," Science, 169, 61-62 (1970).
77. J. Edmiston, personal communication, 1971.
78. J. J. Goering and V. A. Dugdale, "Estimates of the Rates of Denitrification in a Subarctic Lake," Limnol. Oceanogr., 11, 113-117 (1966).
79. P. L. Brezonik and G. F. Lee, "Sources of Elemental Nitrogen in Fermentation Gases," Int. J. Air Water Pollut., 10, 145-160 (1966).
80. D. R. Keeney, R. L. Chen, and D. A. Graetz, "Importance of Denitrification and Nitrate Reduction in Sediments to the Nitrogen Budgets of Lakes," Nature, 233, 66-67 (1971).
81. C. H. Mortimer, "The Exchange of Dissolved Substances Between Mud and Water in Lakes," J. Ecol., 29, 280-330 (1941); 30, 147-201 (1942).
82. A. R. Gahler, in Proc. Eutrophication-Biostimulation Workshop, University of California, Berkeley, 1969.
83. E. G. Foree, W. J. Jewell, and P. L. McCarty, presented at 5th Int. Water Pollut. Res. Conf., San Francisco, July 1970.
84. S. C. Rittenberg, K. O. Emery, and W. L. Orr, "Regeneration of Nutrients in Sediments of Marine Basins," Deep-Sea Res., 3, 23-45 (1955).
85. R. H. Brooks, Jr., Ph.D. thesis, University of Florida, Gainesville, 1969.
86. D. W. Toetz, "Experiments on the Adsorption of Ammonium Ions by Clay Particles in Natural Waters," Water Resour. Res., 6, 979-980 (1970).
87. P. L. Brezonik, Ph.D. thesis, University of Wisconsin, Madison, 1968.
88. V. A. Dugdale, Ph.D. thesis, University of Alaska, College, 1965.
89. R. H. Whittaker and P. P. Feeny, "Allelochemics: Chemical Interactions Between Species," Science, 171, 757-770 (1971).
90. J. A. Hellebust, "Excretion of Some Organic Compounds by Marine Photoplankton," Limnol. Oceanogr., 10, 192-206 (1965).
91. W. D. P. Stewart, "Liberation of Extra-Cellular Nitrogen by Two Nitrogen-Fixing Blue-Green Algae," Nature, 200, 1020-1021 (1963).

92. R. C. Dugdale and J. J. Goering, "Uptake of New and Regenerated Forms of Nitrogen in Primary Productivity," Limnol. Oceanogr., 12, 196-206 (1967).
93. J. J. Goering, D. D. Wallen, and R. M. Nauman, "Nitrogen Uptake by Phytoplankton in the Discontinuity Layer of the Eastern Subtropical Pacific Ocean," Limnol. Oceanogr., 15, 789-796 (1970).
94. R. C. Dugdale, "Nutrient Limit in the Sea: Dynamics, Identification, and Significance," Limnol. Oceanogr., 12, 685-695 (1967).
95. J. J. MacIsaac and R. C. Dugdale, "The Kinetics of Nitrate and Ammonia Uptake by Natural Populations of Marine Phytoplankton," Deep-Sea Res., 16, 45-57 (1969).
96. R. W. Eppley, J. N. Rogers, and J. J. McCarthy, "Half-Saturation Constants for Uptake of Nitrate and Ammonium by Marine Phytoplankton," Limnol. Oceanogr., 14, 912-920 (1969).
97. D. B. Zilversmit, C. Entenman, and M. C. Fishler, "On the Calculation of Turnover Time and Turnover Rate from Experiments Involving the Use of Labelling Agents," J. Gen. Physiol., 26, 325-331 (1943).
98. V. Alexander, Proc. 25th Ind. Waste Conf., Purdue University (1970) (in press).
99. K. Jones and W. D. P. Stewart, "Nitrogen Turnover in Marine and Brackish Habitats. IV. Uptake of the Extracellular Products of the Nitrogen-Fixing Alga Calothrix Scopulorum," J. Mar. Biol. Assoc. (U.K.), 49, 701-716 (1969).
100. G. P. Fitzgerald, "Field and Laboratory Evaluations of Bioassays for Nitrogen and Phosphorus with Algae and Aquatic Weeds," Limnol. Oceanogr., 14, 206-212 (1969).
101. J. M. Bremner, in "Methods of Soil Analysis. Part 2. Chemical and Microbiological Properties," C. A. Black, Ed., American Society of Argonomists, Madison, Wis., p. 1256.
102. D. Rittenberger, A. S. Keston, F. Roseburg, and R. Schoenheimer, "The Determination of Nitrogen Isotopes in Organic Compounds," J. Biol. Chem., 127, 291-299 (1939).
103. J. C. Neess, R. C. Dugdale, J. J. Goering, and V. A. Dugdale, "Nitrogen Metabolism in Lakes. I. Measurement of Nitrogen Fixation with ^{15}N," Limnol. Oceanogr., 7, 163-169 (1962).
104. R. A. Vollenweider, "Studi sulla Situazione Attuale del Regime Chimico e Biologico del Lago d'Orta," Mem. 1st Ital. Idrobiol., 16, 21-125 (1963).
105. J. S. Crane, M. S. thesis, University of Wisconsin,

Madison, 1962.
106. T. Koyama, in Proc. 2nd Int. Water Pollut. Res. Conf., Tokyo, 1964, Paper III-10, Discussion.
107. J. Wijler and C. C. Delwiche, "Investigations on the Denitrifying Process in Soil," Plant and Soil, 5, 155-169 (1954).
108. J. O. Reuss and R. L. Smith, "Chemical Reactions of Nitrites in Acid Soils," Proc. Soil Sci. Soc. Am., 29, 267-270 (1965).
109. J. M. Bremner and D. W. Nelson, Trans. 9th Int. Congr. Soil Sci., 2, 495 (1968).
110. D. W. Nelson and J. M. Bremner, "Factors Affecting Chemical Transformations of Nitrite in Soils," Soil Biol. Biochem., 1, 229-239 (1969).
111. A. T. Austin, "Nitrosation in Organic Chemistry," Sci. Progr., 49, 619-640 (1961).
112. C. E. ZoBell, "The Assimilation of Ammonium Nitrogen by Nitzschia Closterium and Other Marine Phytoplankton," Proc. Nat. Acad. Sci., (U.S.), 21, 517-522 (1935).
113. N. W. Rakestraw and A. Hollaender, "Photochemical Oxidation of Ammonia in Sea Water," Science, 84, 442-443 (1936).
114. L. H. N. Cooper, "The Nitrogen Cycle in the Sea," J. Mar. Biol. Assoc. (U.K.), 22, 183-204 (1937).
115. R. D. Hamilton, "Photochemical Processes in the Inorganic Nitrogen Cycle of the Sea," Limnol. Oceanogr., 9, 107-111 (1964).

Chapter II

PHOSPHORUS: ANALYSIS OF WATER, BIOMASS, AND SEDIMENT

James R. Kramer

Department of Geology, McMaster University,
Hamilton, Ontario, Canada

Stephen E. Herbes and Herbert E. Allen

Department of Environmental and Industrial Health,
University of Michigan, Ann Arbor, Michigan

Phosphorus is one of the major elements required for zooplankton and phytoplankton growth. Phosphorus is also an important element in soil, sediment, and rocks. It is a dynamic variable in all natural systems and, depending on the specific situation, it may be concentrated temporarily or permanently in one or more of the following regions: the biosphere, the hydrosphere, or the lithosphere. Rigler (1) and Hayes (2) who have measured turnover times in natural systems, found that residence times in aqueous, biological, and sediment sub-systems vary from a few minutes to tens of hours.

Phosphorus is distributed through all portions of the earth in measurable amounts. Table 1 is a summary of the relative atomic proportions of phosphorus with respect to other elements (C, N, Si) which are important in biomass uptake. The earth as a whole is concentrated in phosphorus compared with carbon and nitrogen with reference to average sediment, water, and biomass data. Nitrogen and carbon are more concentrated on the surface of the earth than phosphorus, and are concentrated in detritus and sediment relative to biomass. The stability sequence appears to favor carbon over nitrogen over phosphorus. The ocean N:P concentration ratio appears to be directly related to uptake of the elements by organisms. Carbon in the hydrosphere is not closely associated with organism content but is probably more nearly a function of the $CaCO_3$-H_2O equilibrium. In summary, nitrogen and phosphorus are taken up by organisms, but large portions of phosphorus first, followed by nitrogen, are released upon decay, burial, and diagenesis.

Table 1. Relative Atom Proportions of C, N, P, and Si in the Cosmos, Earth, Hydrosphere, and Organisms

Environment	P	C	N	Si	Reference
Cosmos	1	615	1200	77	3
Earth	1	0.7	0.002	100	3
Average sediment	1	96	2	525	3
Ocean (maxima)	1	780	16	47	4
Average river (soluble)	1	610	11	140	5
Average river (particulate)	1	300	26	-	4
Average organism	1	106	16	1-50	6

The importance of phosphorus in the process of eutrophication of lakes has been emphasized repeatedly (7,8). Phosphorus is a critical nutrient for algae, whose growth into massive, malodorous blooms is a major manifestation of the eutrophication process. A relationship between phosphorus levels and algal productivity has been demonstrated for many natural waters; significant correlations have been found between the amount of phosphorus available and the density of algal growth in 46 Swiss lakes (9). More important, man has exerted greater influence on the relative level of phosphorus in the environment than on any other nutrient element, and this effect is reflected in the decrease of the nitrogen to phosphorus ratio in Lake Erie from 35.0 in 1942 to 9.2 in 1966 (10).

In the United States (which consumes about 25% of the world's phosphorus), 76% of this element is used in agricultural fertilizers, 7% in detergents, 3% in plating and polishing, 3% for animal feed, and 11% in other applications (11). In 1968 total consumption in the United States was 3.5×10^6 short tons of phosphorus. Ample phosphorus sources are forecast for the next century, with increasing demands predicted especially for agricultural fertilizers.

The major supply of phosphorus in natural waters is from detergents and agricultural sources. An estimated 70% of the phosphorus entering Lake Erie is a result of municipal and industrial effluent (12). Gilchrist and Gillingham (13) have shown that intense rainfall (2 in./hr for 1 hr) could remove 25% of the fertilizer applied in test strips. On the other hand, Carter et al. (14) studied a 200-acre calcareous soil field in Idaho; they demonstrated that nitrogens increased in the irrigation effluent, but both natural and fertilizer phosphorus decreased in the effluent. Bernhardt (15) has pointed out that cultivated fields and manure piles, rather than fertilized fields, may be more significant factors in the release of phosphorus to water courses.

There has been a great deal of discussion with respect to nuisance phytoplankton growth and accelerated eutrophication in context with the "limiting" role of phosphorus and especially in context with detergents. Ryther and Dunstan (16) suggest that coastal marine eutrophication is limited by nitrogen and is not due to phosphorus or detergents. Vallentyne et al. (17), however, have shown that when sewage treated for phosphorus removal was added to Lake Erie water and incubated, it was greatly reduced in growth effect compared with sewage not treated for phosphorus removal. Sawyer (18) has perhaps used the proper perspective by considering urban and rural sources of phosphorus as quite separate. Vollenweider (19) has calculated loadings of phosphorus based on population, industry, and land use factors in a drainage area; he has also defined "permissible" loadings for lakes as a function of mean depth. The charge has been made (20) that phosphorus has been simplistically isolated as a single causal factor in a complicated system, whereas proponents of phosphorus removal have stated that emphasis has been placed on phosphorus because (1) it often is the "key" element for growth and (2) from a technical point of view phosphorus can be removed (in comparison to C, N, H, etc.).

Atmospheric inputs of phosphorus may be more significant than detergent-industrial-agricultural runoff, especially in low-population areas with waters of low conductivities. For example, aeolian inputs would be the most significant sources to the unpopulated areas of the Canadian Precambrian Shield. Vollenweider (19) carefully summarized the atmospheric fall data for nitrogen and phosphorus and found contributions of 0.3 to 3 mg-at. P/(m^2)(yr); Johnson et al. (21) discovered that quantities of soluble and insoluble dustfall in the greater Seattle area generally increased with population density and industrial activity. They found, for a 3-month study, ranges of 18 to 900 mg-at. N/(m^2)(yr) and 0 to 2.3 mg-at. P/(m^2)(yr) for water-soluble fractions. Phosphorus concentrations were generally independent of population density and industrial activity. If soil and unconsolidated sediment capable of nutrient adsorption are minimal in the drainage area ("bedrock lakes"), atmospheric input of nutrients would be maximized where lake drainage area/volume (mean depth) is a maximum.

SCOPE

Our purpose here is not to review all aspects of phosphorus as a nutrient but to concentrate on (a) the composition of phytoplankton, zooplankton, and detritus in context to the "limiting" concept; (b) biological utilization of phosphorus; (c) analysis of the forms of the element in solution; and (d) the mineral uptake and release of phosphorus. From the outset, it is obvious that we must consider phosphorus in context with other environmental factors, especially other nutrient concentrations. Whenever data permit, phosphorus is considered in context with nitrogen and carbon.

OTHER REVIEWS

The following list mentions some of the recent reviews of phosphorus in natural waters:

General	Van Wazer (22)
Annotated bibliography (1965)	Mackenthun (23)
General: N and P in water	McCarty (24)
Algal chemistry	Heilborn (25)
Life processes	Katchman (26)
Phytoplankton growth and nutrition	McCombie (27)
Phytoplankton nutrition	Ketchum (28)
Metabolism and P-algae	Kuhl (29)
Algal nutrition and ecology	Provasoli (30)
Phytoplankton periodicities	Tucker (31)
Physiology, fresh-water algae	Krause (32)
Eutrophication and water management	Vollenweider (19)
Pesticides and organo-phosphorus compounds	Quentin (33)
Organism composition in seawater	Redfield et al. (6)

The two-volume work of Van Wazer is probably the best single source covering all aspects of phosphorus. The critical analysis and review by Vollenweider is a good reference to nitrogen and phosphorus phytoplankton growth and to water management.

COMPOSITION OF BIOMASS

In most hydrologic conditions, phosphorus and other nutrients are constantly being incorporated in organisms and being depleted in water. Simultaneously (through excretion and decay) nutrients are being released to the system. To account for phosphorus in natural waters, we must be able to explain the quantitative mechanisms for the uptake and release of nutrients. If we assume that the major elements in phytoplankton and zooplankton growth are C, N, P, H, O we may write the following symbolic equation for the uptake:

$$aHPO_4^{2-} + bHCO_3^- + cNO_3^- + dH_2O + eH^+ = C_bN_cP_aH_qO_r + sO_2 \qquad (1)$$

where a, b, c, ... are constants or variables. Redfield et al. (6) have analyzed the rate of change of nutrients with respect to one another and, assuming that the change is due entirely to organism uptake and that there is a fixed organism composition, thus have defined an organism stoichiometry of $C_{106}N_{16}P_1H_{263}O_{110}$ with 138 at. O_2 for the oceans. Actual analyses of organisms do not always agree with this composition (see Table 2), although many of the analyses compiled were for culture studies that involved nutrient concentrations many orders of magnitude greater than those found in natural waters. On the other hand, there are analyses of net hauls which deviate significantly from the foregoing average formulation. The deviation may be due to "net trauma" (44), method of processing (dried or wet samples), and/or analytical accuracy in addition to real variations.

Extrapolations of the apparently constant ocean formulation to freshwater environments is perhaps unwise. The oceans have the capacity to average out variations in loadings because of their large size and relatively rapid mixing compared with rates of inputs. Hence an average value with certain deviations might be expected; the oceans then become a gross averager of the net inputs and removals of nutrients. This is probably not the case for

Table 2. Composition of Phytoplankton, Zooplankton, Bacteria, Detritus and Excreta in Natural Waters and Cultures

Organism	Setting	Organism composition[a]				Reference
		C	H	Si	N	
Phytoplankton-zooplankton-aquatic plants						
All	Ocean water	106	263	0-50	16	6
All	Sewage lagoon	112	--	--	18.8	34
All	Atlantic-Pacific Isthmus of Panama; #2, #20 nets	constant composition[b]				35
All	E. South Atlantic	--	--	--	18	36
	E. tropical Pacific	--	--	--	13.6	
All	N.W. North Sea	91	--	6	12.2	37
Nitzschia closterium	Culture	137	--	--	14.2	38
Ceratium tripos	Culture	150	--	--	28	
Scenedesmus	Culture	44	--	--	28	
Chlorella (normal)	Culture	57	--	--	6.7	
Chlorella (average)	Culture	53	--	--	6.2	
Chlorella (N-P deficient)	Culture	86	--	--	5.9	
Chlorella (N-P deficient)	Culture	119	--	--	6	
Myxophyceae (typical)	Culture	90	--	--	12.5	
Mixed population	Long Island Sound	152	--	--	15.5	

Natural hauls (1951)	Lake Lauzon, Quebec 11 m, #20 net					39
July 11-18		--	--	--	30	
July 18-25		--	--	--	29	
July 25-Aug. 1		--	--	--	26	
Aug. 1-8		--	--	--	25	
Aug. 8-15		--	--	--	27	
Angiosperm aquatic plants	Fertile lake	--	--	--	8-25	40
	Infertile lake				19-45	
	Artificial seawater[c]					41
Chlorophyceae						
Tetraselmis maculata		29	--	0.4	5.5	
Dunaliella salina		31	--	0.7	6.1	
Chrysophyceae						
Monochrysis lutheri		46	--	0.6	5.8	
Syracosphaera carterae		85	--	0.2	4.4	
Bacillarophyceae						
Chaetoceras sp.		53	--	--	8.3	
Skeletonema costatum		39	--	9.3[d]	7.7	
Coccinodiscus sp.		104	--	61[d]	15	
Phaeodactylum tricornatum		48	--	0.2	5.8	
Dinophyceae						
Amphidinium caerti		95	--	0.05	9	
Exuviella sp.		88	--	0.2	8.4	
Myxophyceae						
Agmenellum quadruplicatum		77	--	0.1	9.1	

Table 2. continued

Organism	Setting	Organism composition[a]				Reference
		C	H	Si	N	
	Ocean net tows, Atlantic off New England					42
Aug. 13	Water composition:					
#20 net; 0.2 copepods; 0.8 _Ceratium_ and diatoms	N:0.74; P:0.34 μmolar	--	--	--	18.8-19.7	
#10 net; 0.95 copepods		--	--	--	17.3	
Aug. 14	Water composition:					
#10 net, 0.6 copepods; 0.4 dinoflaggelates	N:0.51; P:0.37 μmolar	--	--	--	18.8	
#2 net; 0.95 copepods		--	--	--	20	
Aug. 14	Water composition:					
#20 net; 0.3 copepods; 0.7 _Ceratium_	N:0.54; P:0.3 μmolar	--	--	--	15-16	
#10 net; 0.97 copepods		--	--	--	17-19.5	
#2 net; 0.97 copepods		--	--	--	16-17	
All	Millipore filter top 10 m					
August 1962	N:1.45; P:0.38 μmolar	--	--	--	11.2	
April 1962	N:2.8; P:0.22 μmolar	--	--	--	16.5	

<u>Bacteria</u>						
	Culture experiment: N:0.0187; P:0.0088 μmolar					43
<u>Pseudomona</u> <u>Nitroreducens</u> B96						
Pellicle		--	--	--	5.2	
Water-phase cells		--	--	--	5.7	
Polar liquid		--	--	--	5.5	
<u>Corynebacterium</u> <u>sp</u>. B162						
Pellicle		--	--	--	6.2	
Water-phase cells		--	--	--	2.5	
Polar liquid		--	--	--	31	
<u>Excreta and Detritus</u>						
Sediment trap 1951 av.	Lake Luzon, Quebec, 11 m depth with total depth 27 m	--	--	--	31	39
July 11-18		--	--	--	22	
July 18-25		--	--	--	2	
July 25-Aug. 1		--	--	--	1.7	
Aug. 1-8		--	--	--	1.5	
Aug. 8-15		--	--	--	4.1	
<u>Calanus</u> <u>finmarchias</u>	Clyde Sea area					44
Range excreta		--	--	--	10.4-11.8	
Spring, excreta		--	--	--	10.8	
Winter, excreta		--	--	--	14.6	
Fecal pellets, Spring diatom bloom		--	--	--	27.7	
Fecal pellets		--	--	--	17.8	

[a]Molar ratios of elements relative to phosphorus.
[b]Organisms were also analyzed for Ca, Fe, Mn, Se, Sn, and Zn.
[c]Artificial seawater composition: 500 μmolar N, 50 μmolar P, 200 μmolar Si, and 2.3 μmolar HCO_3^-. Organism composition determined on a dry weight basis.
[d]Organisms cultured in artificial seawater of the same composition as given in footnote c, with the exception of Si which was 1000 μmolar.

most freshwater bodies, especially since loadings of nutrients can change rapidly because of natural or cultural inputs, and production rates (and decay) are higher with great variations.

The "limiting" nutrient concept is based on a fixed or approximately fixed organism composition. Hence the limiting substance is that material which is used up first in the list of "reagents" (equation 1). If the composition of organisms can vary quite widely, the limiting nutrient concept (Liebig's law) is not valid. It is therefore important to define the range of compositions of organisms by examining organism compositions in natural waters of widely varying nutrient concentrations. The problem may be analyzed by considering the composition of organisms as fixed, as a "solid solution" within limits, or as widely varying.

Fixed Composition

If we assume a constant organism stoichiometry as defined by Redfield and if we assume that only carbon, nitrogen, phosphorus, and silicon are potentially limiting, we can assess various natural waters to determine the potential limiting effect with respect to each element by examining nutrient concentrations and ratios. The main source of carbon in limestone-drained lakes and the oceans is HCO_3^- from the $CaCO_3$-H_2O equilibrium; HCO_3^- varies little from 2×10^{-3} moles/liter; in drainage areas void of limestone, the HCO_3^- may be as low as 10^{-5} moles/liter and soluble organic carbon may be greater than HCO_3^-. Phosphorus occurs in all natural waters as $H_2PO_4^-$ - HPO_4^{2-} and as inorganic-organic complexes. In the deep oceans, phosphorus may reach 4×10^{-6} moles/liter and may be barely detectable at the surface at 10^{-8} moles/liter. Nitrogen and silicon follow similar trends and deep ocean to surface water concentrations range from 5×10^{-5} to 10^{-7} moles/liter for N and 1.4×10^{-4} to 10^{-8} moles/liter for Si.

Lakes show the same general range of values for nitrogen and phosphorus, but the maximum encountered for silicon may be slightly higher. Using the figures just cited and the constant organism composition concept, we can ascertain where limiting conditions may develop. Table 3 summarizes the maxima-minima ratios for nutrients with respect to phosphorus. By comparing these ratios to the ratios for organisms, the following conclusions can be stated:

1. No element is universally limiting, although silicon should not be limiting except where diatoms (high Si

Table 3. Maximal and Minimal Ratios for Nutrients with Respect to Phosphorus[a]

Ratio	Maximum	Organism	Minimum	Comments
C:P	10^5	106	10	Maximum approached in oceans; minimum possible in nonlimestone lakes
N:P	10^3	16	10^{-2}	Typically near 10; extremes not common unless nonnatural inputs
Si:P	10^4	1-50	1	Extremes not common unless nonnatural inputs; typical values are 100 to 1000

[a]A nutrient may be limiting if its concentration ratio in natural waters does not bracket the composition of the organism, and it is possible to have a limiting nutrient condition when the minimum is less than the organism ratio.

content) are present and nutrients other than silicon are in abundance.

2. Carbon would be limiting for nonlimestone lakes if other nutrients are in abundant supply. Carbon would not be limiting in the oceans or in limestone lakes.

3. Nitrogen can be limiting when phosphorus and carbon are in excess. This would occur in limestone lakes (or the oceans) with abundant phosphorus.

4. Similarly phosphorus can be limiting when nitrogen and carbon are in excess. This could occur in a limestone lake or oceans with abundant nitrogens or with organisms capable of fixing N_2 (19, Table 2).

In summary, if a constant stoichiometry is assumed, limiting conditions may appear, but no overall generalization can be made from the range of water chemistry data available for actual situations.

Varying Organism Composition

The rationale for arguing that organism composition is variable is associated with the basic assumption that organisms, up to a maximum nutrient concentration, assimilate nutrients at a rate proportional to the concentration of the nutrient in the water. If the nutrient concentration is large, "luxury consumption" of the nutrient may occur. This concept is incorporated in the Michaelis-Menten function:

$$\frac{d\ Nu'}{dt} = \frac{kNu}{s + Nu} \tag{2}$$

where Nu is the nutrient concentration in water; Nu' is the nutrient assimilated in time, dt; k is the maximum uptake rate for experimental conditions; and s is the nutrient concentration when the rate is k/2.

Both k and s should be expected to vary with concentration changes of other nutrients in the environment as well as with other environmental factors such as temperature and light. Therefore, equation 2 should be considered to be a partial derivative function. If we imagine two or more nutrients involved, a series of uptake equations can be written for the nutrients. If we then consider a finite interval, we see that the nutrient uptake ratios define the composition of the biomass (if mineral excretion occurs), and it is evident that the composition of the organism will change depending on all the nutrient concentrations. This implies that not only phosphorus composition will be altered by phosphorus concentration changes in natural waters but also by changes in C, N,... concentrations in water, light, temperature, and other variables that will alter the value of k and s. The limit at which change would occur in each case would be between zero and k, the maximum uptake rate. The modification of equation 2, considering N and P together over a finite time interval Δt, is an example of the foregoing:

$$(N/P)_o = \frac{k'(N/P)_w(s_p + P)}{(s_N + N)} \tag{3}$$

where the subscripts o, w, p, and N refer to organisms, water, phosphorus concentration, and nitrogen concentration, respectively; k' is the ratio of k's for nitrogen and phosphorus uptake curves; and N and P are nitrogen and phosphorus concentrations in water in 10^{-6} mole/liter.

It is evident from this line of argument that organism composition requires information on nutrients in the water as well as other environmental parameters. Of course changes of k and s in equation 2 with respect to other environmental parameters may be negligible, but even with this assumption, the nutrient concentrations and ratios will significantly alter the biomass ratio. This argument confirms the caution that Vollenweider (19) attaches to the extrapolation from culture data to natural environments when the nutrient concentrations in culture experiments are many orders of magnitude higher than natural conditions. This approach would also qualitatively explain the relatively low C:N and C:P ratios of organism in Table 2 when nitrogen and phosphorus concentrations were high compared with carbon.

The best documented study of nutrient uptake as a function of nutrient concentration using concentrations approximating natural values is that of Ketchum (45); Figure 1 is a summary of his data (Table 6 in Ref. 45). It is immediately apparent that a series of curves compatible to the form of equation 3 result for constant nutrient concentrations of phosphorus. Further inspection of the data suggests that the maximum uptake rate (k) for each curve is linearly related to one/P concentration in solution. Modifying equation 3 accordingly, we obtain an equation of the following form:

$$(N/P)_o = \frac{(a + b/P)(N/P)_w(s_p + P)}{(s_N + N)} \quad (4)$$

where a and b are constants and the other symbols have the same significance as before. A least-squares fit to all the data of Ketchum (range of organism N:P - 6 to 40; and range of water concentrations N:P - 0.2 to 140) gives the following coefficients for equation 4:

$a = 8.6672$

$b = 4.517$

$s_p = -0.00626$

$s_N = -0.04944$ (N and P concentrations in micromoles per liter).

The multiple correlation coefficient for 42 samples is .9991 with a standard error of estimate with respect to the mean of the dependent variable of about 6% (±0.9 for N:P organism ratio). The fit statistically and in terms of form appears to be more than coincidental and coincides with the Michaelis-Menten concept of uptake. There are, however, some defects in the statistical results in context with the real significance of coefficients in the function. Both s_p and s_N have a real positive value in concept; yet both turn out to be negative

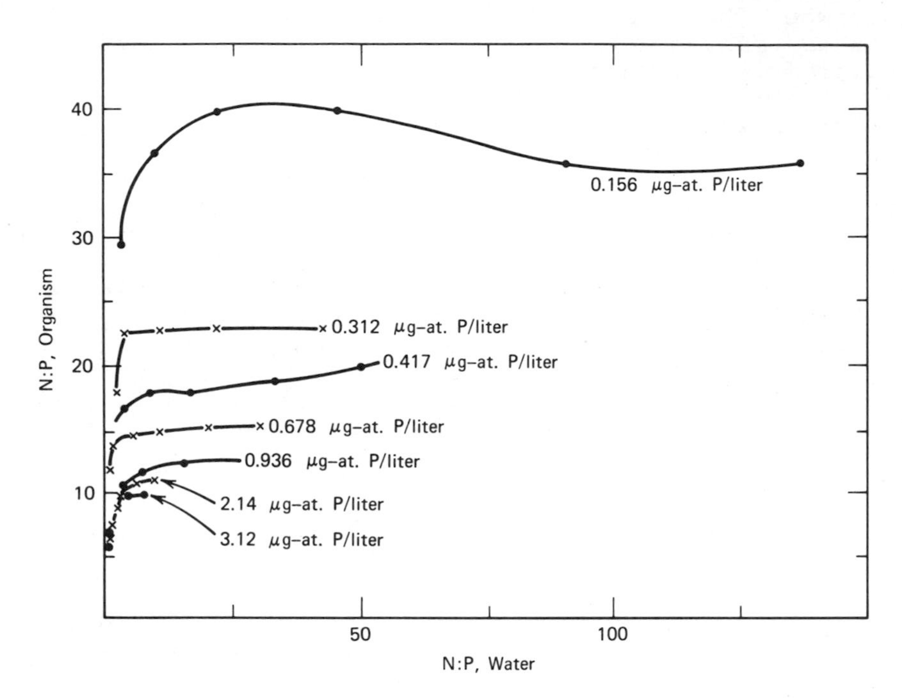

Fig. 1. Atomic ratio of uptake of N and P as a function of N and P nutrient concentrations for various concentrations of P.

in the least-squares fit.

An explanation may be that, since s_p and s_N are so near zero in value, deviations of the larger numbers will bias their determination. (In a stepwise regression, s_p and s_N terms are the least significant of all the terms.) The coefficient s_p is so nearly zero that it may be assumed to be zero, and the following function results:

$$(N/P)' = \frac{(a' + b'/P)(N/P)_w}{s_N'/P + (N/P)_w} \tag{5}$$

and from the least-squares fit, the coefficients are:

$a' = 9.16$

$b' = 4.12$

$s'_N = 0.160$ (concentration of P in micromoles per liter).

The fit statistically is slightly better than before with a multiple correlation coefficient of .9996 for 42 samples and a standard error of estimate about the mean of the dependent variable of 6% (±0.9 for N:P of organism). It is evident that more data below maximum uptake rates are required to better define the coefficients s_N and s_P.

A third way of fitting an equation to the data is to use the basic definitions of the Michaelis equation in defining k and s and fit the values to the variable one/P. Maximum values (k) for each P concentration can be rapidly ascertained, as well as an interpolation for s. Graphical plots of k against one/P are linear, and the resulting least-squares fit for the data is:

$$k = 9.475 + 4.289/P \quad \text{and} \quad s = 0.531 + 0.345/P$$

where P is in micromoles per liter. Substituting these relations into an equation of the Michaelis form gives:

$$(N/P)_o = \frac{(9.475 + 4.289/P)\ (N/P)_w}{(0.531 + 0.345/P) + (N/P)_w}$$

where $(N/P)_o$ is the atom ratio in the organism of nitrogen and phosphorus, and $(N/P)_w$ is the atom ratio in water of nitrogen and phosphorus.

The significance implied by the linear function for k is that a minimum organism ratio of 9.5 should be attained when nutrient concentrations are high and the ratio of nitrogen to phosphorus in water is large. As far as the linear function of s is concerned, we would expect the term to vanish when the phosphorus concentration is large; yet a rather large value (0.53) persists under these conditions.

The work of Ketchum was carried out on *Nitzschia closterium* in normal salinity seawater at room temperature with only additions of nitrogen and phosphorus to the culture solution. It is worthwhile to see to what extent the function derived from this culture experiment can be generalized to other environmental conditions. Table 4 compares the N:P ratio of the organism predicted by equation 5 with that found or calculated by other means. The equation statistically reproduces the data to ±0.9 of the N:P of the organism; with the exception of only a few cases, the analytical data are explained by this function with nearly the same tolerance. The analyses represent

Table 4. Comparison of Measured N:P Ratios for Organisms with Ratio Predicted by Equation 5

Setting	Measured	Predicted	Reference
Lake Erie, 1970; differences in weighted values	15.2	14.7	
Lake Ontario, 1969; using differences in values	18.3	17	
Lake Erie; differences in hourly measures			
August 0900	4.5	9.3	46
1000	9.7	9.9	
1100	13	14	
1300	11	11	
Gulf of Main			
Net hauls	17.3-19.7	17.5	42
	17.1-18.1	15.5	
	18.8-20.1	15.5	
	14.9-16.1	17.6	
	17.4-19.4	17.6	
	15.7-17.4	17.6	
particulate matter	11.2	16.6	
	16.5	21.8	
Various phytoplankton in 500 μmolar N, 50 μmolar P 25-35 °/₀₀ salinity, 200 or 1000 μmolar Si; measures are for dried cells	Mean of 11: 7.8 4.4-15	9.2	41
Bacteria in 0.0187 molar N, 0.0098 molar P	Mean of 6: 5.8 2.6-9.0	9.2	43

freshwater and oceanic environments as well as culture experiments.

Undefinable Organism Composition

Without exception, organisms show a maximum rate of assimilation for a particular nutrient. The maximum may be a function of a number of environmental factors, including the concentration of other nutrients; yet the condition is always obtained where the uptake rate of a nutrient reaches a maximum constant. In this context, the compositions of organisms are always defined within broad limits. It is important to define the conditions that define maxima, and this can only be done in context with other variables. With attention focused on composition, it is best to determine the maximum ratio of two nutrients as a function of their concentration ratios for a constant value of one of the nutrients. This could also be repeated for varying temperatures and light intensities. In the case of Ketchum's studies, the N:P-organism maxima are related to the phosphorus concentration, again in micromoles per liter, by

$$(N/P)_{o}^{max} = 9.48 + 4.29/P$$

This implies that waters of high N:P and high concentrations of phosphorus ("luxuriant" waters) should approach an organism composition of 9.5.

Minimum Cell Concentration

Another approach to organism composition in context with the "limiting" nutrient is to ascertain the minimum concentration of P/cell, P/cell volume, or P/cell weight (all are interrelated for constant cell size and density) for growth. A standard technique is to add varying amounts of phosphorus to a culture and determine the growth rate. Regression of the relation between growth rate and P/cell and extrapolation to zero growth rate gives a minimum value for the phosphorus for growth to commence. Goldberg et al. (47) found a minimum cell content of 3×10^{-14} moles P/cell for a seawater culture of *Asterionella Japonica*; Lund found a value of 7×10^{-15} moles P/cell for the freshwater *Asterionella Formosa*; for a culture medium of *Chlorella Pyrenoidosa*, Al Kholy (48) found a minimum cell phosphorus for growth of 3×10^{-16} moles P/cell; and for *Asterionella Formosa*, Mackereth (49) found 2×10^{-15} moles P/cell.

Vollenweider (19, p. 25) has defined a sequence of organisms based on their minimal phosphorus requirements for growth; this sequence is *Asterionella*, *Fragillaria*, *Tabellaria*, *Scenedesmus*, *Oscillatoria*, and *Microcystic* in order from lowest to highest phosphorus requirement/cell volume. The requirements are, respectively, < 0.65, 0.65 to 1.1, 1.4 to 1.9, > 1.6, > 1.6, and > 1.6×10^{-8}

moles P/mm^3 cell volume. Using the first set of data and assuming that a bloom is defined by approximately 10^6 cells/liter, up to 0.03 μmoles P/liter "in excess" would be required for the minimal values for phosphorus for growth. Using the second set of data and assuming that a bloom contains about 10 mm^3 cell vol./liter of water, up to 0.2 μmoles P/liter "in excess" would be required for a bloom. Both figures defining the minimum phosphorus water concentration for a bloom are somewhat below the commonly quoted value of 0.3 μmoles P/liter. This is probably because during rapid growth, organisms assimilate phosphorus at levels greater than minimum requirements. The first estimate is low by an order of magnitude because of the incorrect use of cell count rather than cell volume, as pointed out by Vollenweider (19, p. 35), in defining an aquatic bloom.

Detritus and Excrement

There are two general approaches to the discussion of release of nutrients from an organism into natural waters. The first is to assume "congruency," meaning that organisms excrete or decay in direct proportion to their composition. Equation 1 therefore can be considered to be reversible. The second approach is to assume that nutrients are differentially removed with respect to their composition; that is, they show incongruency in excretion and decay.

The second approach appears to be more nearly what actually happens. Butler et al. (44) have studied the excretion patterns of zooplankton in the Clyde Estuary. They found the excretion ratio did not vary much among males and females and was less than the composition proposed by Redfield for oceanic organisms. They did find small but statistically significant deviations from spring to winter (see Table 2). The soluble excreta of *Calanus* for N:P atom ratios ranged from 10.4 to 11.8 when tested among sexes; the average spring ratio for all stages was 10.8 compared with a winter value of 14.6. During a diatom bloom in the spring, however, the ratio reached 27.7.

Johannes (50) has examined the rate of excretion of marine zooplankton and has found that the rate of excretion of dissolved phosphorus per organism weight increased as the body weight decreased. This may be due to the increased surface area for smaller organisms for a constant total biomass. The function relating excretion rate to body weight is:

$$\log B = 3.2 + 0.67 \log W$$

where B is the time required (hr) for an organism to release an amount of dissolved phosphorus equivalent to its total phosphorus content, and W is the dry weight of the organism.

Pomeroy et al. (51) determined that zooplankton daily excrete between 0.2 and 0.46 μmoles P/g of dry zooplankton, or nearly the total body weight of phosphorus in one day. Kuenzler (52) examined the nature of the excretion products of marine plankton and found the amount of dissolved organic phosphorus excreted by different species to be proportional to light intensity and to phosphorus and nitrogen concentrations, and independent of salinity. *Chlorella* cultures required iron for dissolved organic phosphorus excretion. Alkaline phosphatase availability was important to the use by certain algae of the excreted dissolved organic phosphorus.

Vaccaro et al. (42) subjected net hauls of copepods and diatoms (found at various depths off New England and Nova Scotia) to analysis for nitrogen and phosphorus. There is a definite increase in the N:P:detritus ratio in deeper waters compared with hauls at the surface. If we define

$$\Delta = \frac{(N/P)_d - (N/P)_o}{(N/P)_o}$$

where $(N/P)_d$ is the net haul composition ratio at depth d meters, and $(N/P)_o$ is the net haul composition at the surface, there is a simple linear relationship between the difference ratio Δ and the logarithm of the depth of the form:

$$\Delta = -1.22 + 0.9 \log_{10} d$$

The data (5 points) have a standard estimate of error of 30% with respect to the mean of Δ. The difference ratio is zero when the depth is 22 m, suggesting this is the depth at which loss of phosphorus starts to become significant because of decay. At 100-m depth, the change in N:P ratio would be increased by 0.6 by differential release of phosphorus, and at 1000 m the ratio would be increased by 1.5.

BIOLOGICAL UTILIZATION OF ORGANIC PHOSPHORUS

Although phosphorus occurs in a variety of forms in natural waters, the role of inorganic phosphorus (orthophosphate) has generally been stressed; detergent polyphosphates are important due to their ability to produce orthophosphate by hydrolysis in natural waters (53). The role of organic phosphorus, however, appears to have been largely neglected.

Dissolved organic phosphorus comprises a significant fraction of the total phosphorus present in natural waters; 25 to 32% of the phosphorus present in eight Ontario lakes was organic (four to six times as much as was inorganic), and the remainder was sestonic (54). Dissolved organic phosphorus plus polyphosphate levels of 70 μg PO_4/liter were determined in the surface waters of a small eutrophic Michigan lake during the summer and fall, in contrast to the orthophosphate concentration of 5 to 10 μg PO_4/liter (55). The ratio of organic to inorganic phosphorus increases at times of maximal algal growth, and although organic phosphorus levels remain nearly constant (54) or increase slightly (56), inorganic phosphorus may decrease to undetectable levels (57). The utilization of dissolved organic phosphorus, therefore, is of great importance, particularly during periods when inorganic phosphorus levels may become low enough to limit algal growth.

The ability of algae to utilize phosphorus from organic sources for metabolic functions and growth has been demonstrated; glycerophosphate, phytin (58), hexose 1,6-diphosphate, adenosine phosphate (59), and adenylic, guanilic, and cytiditic acids (30) all satisfy algal requirements. In contrast, neither lecithin, a nucleic acid, nor pyrophosphate produced measurable growth in a bacteria-free algal culture (58). Controls demonstrated that none of the compounds producing growth released orthophosphate by spontaneous hydrolysis at a measurable rate. Marine diatoms were able to utilize 40% of the organic phosphorus excreted by a diatom culture, whereas marine bacteria were able to use 92% (60). Species of marine algae cultured in media containing glucose-6-phosphate as the sole phosphorus source experienced different rates of growth corresponding to the degree of utilization of the organic phosphorus (61).

Although organic phosphorus is known to be present in significant amounts in natural waters, the nature of the compounds has received little attention. Fractionation of particulate phosphorus in seawater demonstrated wide variations of acid-soluble, lipid, inorganic, polyanionic, and nucleic acid phosphorus content (62). More than 75% of the phosphorus present in dead algae was liberated axenically; 90% of the refractory material was associated with nucleic acids. Phospholipids were enzymatically hydrolyzed to orthophosphate; sugar-phosphate was not altered (63). Six fractions of organic phosphorus were detected on paper chromatograms for each of three different seawater samples inoculated with $^{32}PO_4$ (64). Preliminary extractions of eutrophic lake water have indicated that less than 5% of dissolved organic phosphorus present was nucleic acid and phospholipid (55).

The utilizability of organic phosphorus by algae is apparently dependent on the activity of phosphatase enzymes, whose synthesis is induced in some species by low environmental orthophosphate levels (65). An inverse relationship has been demonstrated between the total phosphorus present in a lake and the specific phosphatase activity of the biomass (66). The ability to use glucose-6-phosphate as a phosphorus source in 12 algae was directly related to the amount of phosphatase activity observed (61). Although acid phosphatase may be present within the cell, the activity of the induced enzyme is generally greatest under alkaline conditions (61), which would be expected to predominate in most natural waters.

Although Kuenzler's study revealed identical rates of phosphorus uptake by each algal species from three different organic phosphorus substrates, the phosphomonoesterase activity of Neurospora sp. has been shown to be dependent on the phosphorus compound present; relative activities ranged from 100 for α-glycerophosphate through 27 for fructose-1,6-diphosphate to zero for ATP, pyrophosphate, and phospholipids (67). Only the phosphate group enters the algal cell; the organic moiety remains in the medium after cleavage (65). The induced enzyme is localized at the cell surface (68). Little phosphatase appears to be released into the surrounding water: an average of 84% of the enzyme activity in Lake Kinneret was removed by filtration through a 0.45-μ membrane filter, suggesting that most of the phosphatase is firmly bound to the organism (66).

Several considerations arise from the demonstrated ability of some phosphorus-deficient algal species to utilize certain organic phosphorus compounds for growth. First, because not all algae are capable of the induced synthesis of phosphatases, if orthophosphate limits the growth of algae in a particular lake, those species possessing an inducible enzyme will enjoy a distinct competitive advantage. The second point may prove of paramount importance to the effort to control the influx of phosphorus to natural waters; virtually nothing is known of the nature and composition of the organic phosphorus that is released in wastewater treatment plant effluents. Although removals as high as 95% of the total phosphorus in wastewater are theoretically possible by precipitation with alum or lime (69,70), the removal efficiency is far greater for orthophosphate than for organic phosphorus species.

Sawyer suggests that even though 97% of orthophosphate can be removed by lime precipitation, only 47% of the organic phosphorus will be removed by the same process (71). Similar values of 93 and 38% removal efficiencies for dissolved orthophosphate and organic phosphorus, respectively, were reported for lime treatment of anaerobic digester effluent (72). Although 88% of the total phosphorus in the influent to a combination trickling-filter activated-sludge unit was precipitated, a mean of 55% of the organic phosphorus remained in solution. The effluent contained a mean organic phosphorus level of 50% of the total phosphorus (73). Thus the potential importance of the organic phosphorus fraction in wastewater treatment plant effluent, as a source of biologically utilizable phosphorus, increases as the total phosphorus removal efficiency rises. Neither the composition nor the biological utilizability of the organic phosphorus fraction of wastewater treatment effluent is known. Research is clearly necessary to assess accurately the impact of organic phosphorus in treated wastes on algal growth in natural waters.

ANALYSIS OF PHOSPHORUS IN NATURAL WATERS

Phosphorus may be present in dissolved, colloidal, or particulate form and may exist as inorganic orthophosphate or polyphosphate or in organic compounds. Since phosphorus in its different forms does not undergo chemical, biochemical, or geochemical transformations at the same rates or to the same extents, it is often essential to utilize techniques capable of the differentiation of phosphorus species and forms. This challenge alone is great; when it is coupled with the requisite level of sensitivity required for the analysis of many unpolluted bodies of water, it would appear that current analytical procedures cannot achieve the complete analysis of phosphorus in the aquatic environment.

Soluble Reactive Phosphate (Orthophosphate)

Little information was available about phosphate in natural waters prior to 1923, when Atkins (74) introduced Denige's (75) colorimetric phosphomolybdic acid method for the determination of phosphate. All methods of phosphate determination commonly used today are modifications of this procedure. Much of the literature on the analysis of phosphate has been previously reviewed by Olsen (76) and, therefore, this discussion is focused on papers that have appeared since Olsen's review, and those dealing with comparisons of different analytical procedures and with automated or novel analytical techniques.

Colorimetric techniques are based on the complexation of one molecule of phosphate with twelve molecules of molybdate in an acidified solution to form molybdophosphoric acid. Subsequent reduction of the complex to form the heteropoly blue compound can be achieved by using a variety of reducing agents, including stannous chloride, ascorbic acid, 1-amino-2-napthol-4-sulfonic acid, hydrazine sulfate, and p-methylamino-phenol sulfate. Most attention, however, has been devoted to the use of stannous chloride and ascorbic acid as reducing agents.

Stannous chloride reduction requires precise control of reaction conditions, since the intensity of the blue color formed is dependent on the reaction temperature and reaction time. The fading is due to aggregation of the blue micelles, which can be prevented by the addition of a protective colloid such as gelatin (77). AutoAnalyzer procedures for the automated analysis of phosphate in water using stannous chloride reduction have been reported by Coote et al. (78), EPA (79), Gaddy (80), Gales and Julian (81), and Lee (82); manual procedures are given in both Standard Methods (83) and in the ASTM manual (84).

Ascorbic acid reduction of phosphomolybdic acid was first introduced to oceanography by Greenfield and Kalber (85) in 1954 and was subsequently modified by Murphy and Riley (86) to incorporate the necessary reagents into a single solution. A later modified procedure (87) incorporated potassium antimonyl tartrate in the reagent solution and decreased the reaction time from 24 hr to 10 min. Strickland and Parsons (88), in their excellent manual of seawater analysis, have characterized this method as "so superior to other methods in terms of the rapidity and ease of the analysis that it probably represents the ultimate in sea-going techniques." The method has been modified by the addition of ethyl or isopropyl alcohol, which increases the sensitivity and extends the range of the procedure by preventing precipitation of the blue color attributable to high chloride concentration. This procedure is listed in Standard Methods (83).

Increasing the amount of antimony added has been reported (90) to extend the range of the Murphy and Riley (87) procedure. Procedures have been automated using ascorbic acid as one reagent and a mixture of sulfuric acid, ammonium molybdate, and potassium antimonyl tartrate as the second (91,92); other procedures use a single mixed reagent (79,93,94). An automated ascorbic acid reduction procedure that does not utilize antimony has been developed by Grasshoff (95). An automated ascorbic acid reduction procedure incorporating alcohol to increase sensitivity and extend the range of the method has been reported (96).

Increased sensitivity in the analysis of phosphate can be achieved by extraction of molybdophosphoric acid or the heteropoly blue into an organic solvent. Fifteen solvents were tested by Wadelin and Mellon (97) to determine their selectivities in the extraction of the heteropoly acids of phosphorus, arsenic, silicon, and germanium. Ross and Hahn (98) used radiochemical techniques to study the separation of arsenic and phosphorus heteropoly acids by a chloroform-butanol mixed solvent. A solvent containing 10% butanol was found to give the best separation, together with a high yield of phosphorus.

Solvent extraction was first applied to the analysis of phosphate in water by Proctor and Hood (99), who achieved one order of magnitude in sensitivity over stannous chloride reduction methods. Stephens (100) used isobutanol extraction of the heteropoly blue complex formed by the method of Murphy and Riley (87) to decrease the limit of detection to 0.006 μg-at. P/liter (0.2 μg P/liter) (Stephens increased the concentration of ascorbic acid). A 1:1 mixture of isobutanol and chloroform has been used both to increase the sensitivity of analysis and to eliminate the interference of suspended matter in the colorimetric estimation of phosphate concentration (101). An automated extraction procedure capable of detecting 1 μg P/liter using stannous chloride reduction and isobutanol extraction was reported by Henriksen (102).

Jones and Spencer (103) compared five methods of estimating dissolved inorganic phosphate in seawater: Burton and Riley (104), Harvey (105), Murphy and Riley (86), Proctor and Hood (99), and Wooster and Rakestraw (106). Jones (107) repeated the study using the modified Murphy and Riley method (87) rather than the earlier method of these authors (86). Compared with the method of Harvey (105) for 18 analyses, Jones (107) found that Wooster and Rakestraw's method (106) gave higher results, Proctor and Hood's method (99) gave lower results, and the results obtained from Burton and Riley's method (104) produced no statistically significant trend. A more extensive comparison was made between the methods of Harvey (105) and Murphy and Riley (87); here the former method gave higher results in most cases. This discrepancy did not appear to be due solely to the more significant arsenate interference in Harvey's method, nor was it due to greater hydrolysis of organic phosphorus by Harvey's method, since the differences between the results of the two analytical procedures was not significant when surface and deep-water samples were compared.

The methods of Murphy and Riley (87) and Edwards et al. (89) were compared to test their applicability to wastewater analysis, and no significant difference was found between their results (108). The methods of Murphy and Riley (87) and Edwards et al. (89) were compared in an eight-laboratory round-robin test based on the recovery of soluble orthophosphate added to distilled and natural water samples at the 0.2 mg P/liter level (109). The method of Murphy and Riley had better precision and accuracy and less bias. Murphy and Riley's method (87) was evaluated by a 26-laboratory round-robin study (110). At the 0.029 and 0.038 mg/liter levels, a negative bias of 5 to 13% was found; this value decreased to 1.2 to 2.7% at the 0.335 and 0.383 mg/liter levels.

A method capable of detecting approximately 0.03 μg P/liter has been reported (111). The phosphomolybdic acid is extracted into a butanol-chloroform mixture and decomposed by the addition of base, after the organic phase has been washed free of excess molybdate. The molybdate in the aqueous alkali solution is proportional to the original phosphate concentration and the liberated molybdenum (12 Mo/P) is determined spectrophotometrically as the thiocyanate complex.

Hurford and Boltz (112) have reported indirect UV spectrophotometric and atomic absorption methods based on the analysis of molybdate or molybdenum, respectively, after complexation and extraction of phosphomolybdic acid. After the molybdate has been stripped from the organic phase by a basic buffer, the molybdate is determined by measurement of the absorbance at 230 mμ using a UV spectrophotometer or measurement of the absorbance at the 313.3-mμ resonance line of molybdenum, with the aid of an atomic absorbance spectrometer. The optimum concentration ranges were found to be approximately 0.1 to 0.4 mg P/liter for the indirect UV spectrophotometric method and 0.1 to 1.2 mg P/liter for the indirect atomic absorption spectrometric method.

Activation analysis has been used for the simultaneous determination of sulfur and phosphorus with a detection limit of 20 μg P/liter (113). An indirect activation analysis procedure with a detection limit of 4 μg P/liter based on the analysis of tungsten in tungstomolybdophosphoric acid has been reported (114).

An enzyme inhibition assay for phosphate in waste

effluents and streams has been described by the Corning Glass Works (115). The procedure is not affected by carbonate, sulfate, nitrate, acetate, or chloride.

Through a radiobiological analysis, Rigler (116) concluded that chemical measurement may overestimate the concentration of orthophosphate in lake water by as much as one-hundredfold. Further research (117) provided direct chemical evidence that heteropoly blue methods measure a component of lake water other than orthophosphate. Hydrous zirconium oxide columns, which have a high affinity for phosphate, were used to remove orthophosphate from samples of lake water. Phosphate analyses were performed on both the influent to and the effluent from the columns. In eight of ten samples, the influent was determined to have a higher concentration of phosphorus than the effluent, and ratios of measured influent to effluent concentrations as great as 21-fold were observed. In another experiment, the orthophosphate was eluted from the column with sodium hydroxide. Nineteen samples were analyzed, and in all cases the direct phosphate analysis gave a higher result than that for the phosphate eluted from the column. The results (column × 100/direct) ranged from 0 to 89%.

Arsenic interference is reduced by solvent extraction of phosphomolybdate. Chamberlain and Shapiro (118) have proposed a procedure in which the sample was acidified for only 30 sec during the extraction. This was felt to minimize both the interference of arsenic and the hydrolysis of organic phosphorus compounds. Arsenic concentrations as great as 224 μg/liter have been reported in lake waters (119). In the sample containing 224 μg As/liter, a phosphate concentration of 104 μg/liter was determined by Harvey's method (120), but it was estimated to be less than 1 μg/liter by a bioassay method (118). A method for the simultaneous analysis of phosphorus and arsenic has been developed (121). In one aliquot, arsenate is reduced to unreactive arsenite by the addition of a metabisulfitethiosulfate reducing agent. The Murphy and Riley procedure is used for color development of both a reduced and a nonreduced aliquot, and the difference in absorbance at 865 mμ is due to arsenate.

Total Phosphorus

In addition to the analysis of orthophosphate, the analyst may need to acid hydrolyze or digest a sample aliquot of either a filtered or unfiltered sample to determine the phosphorus fraction of interest (Figure 2).

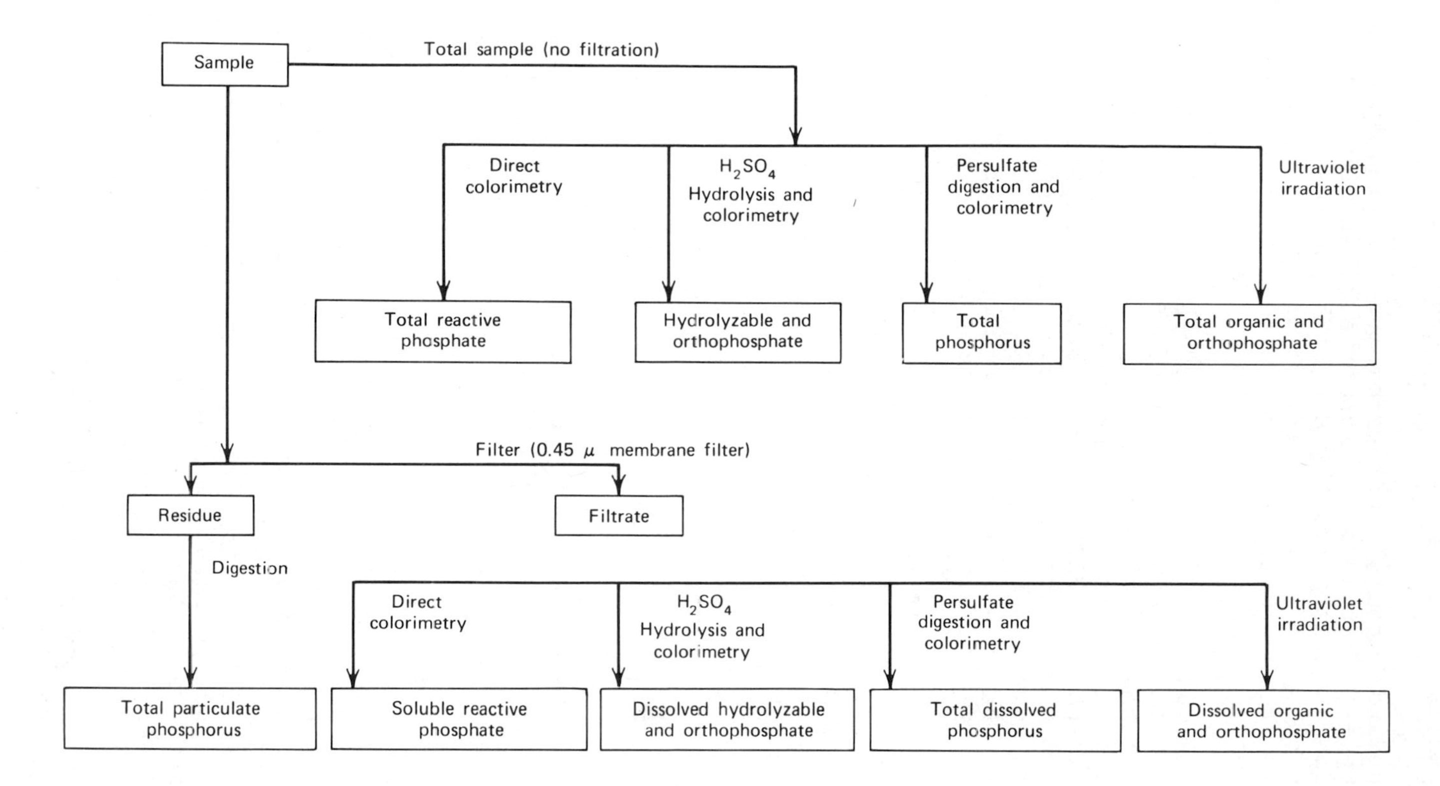

Fig. 2. Analytical scheme for differentiation of phosphorus forms.

"Standard Methods" (83) recommends boiling an acidified sample for at least 90 min or autoclaving at 15 to 20 psi for 30 min to hydrolyze polyphosphates.

Many phosphorus hydrolysis and digestion procedures are cited in the review by Olsen (76). Persulfate oxidation is gaining acceptance as the oxidant choice. The procedure of Menzel and Corwin (122) uses a final potassium persulfate concentration of 0.69% and digests the samples by either autoclaving for 0.5 hr at 15 lb/in.^{2} or by placing them in a water bath for 1 hr. An acidified persulfate digestion procedure requiring a 30-min boiling has been reported (123). The persulfate digestion procedure has been automated using 3.5 atm pressure and a digestion temperature of 125°C (124).

Armstrong et al. (125) have reported that UV radiation oxidizes organic matter and converts organic phosphorus to orthophosphate with the addition of no reagents except for one or two drops of 30% hydrogen peroxide. Ribonucleic acid (RNA) exposed for 20 min at a distance of 7 cm from a 1200-W quartz-jacketed mercury arc lamp was completely degraded to produce the theoretical yield of orthophosphate. The same exposure degraded β-glycerophosphate, ribose-5-phosphate, RNA, choline phosphate, and 2-aminoethanephosphonic acid. Polyphosphate and ATP did not produce orthophosphate. This digestion procedure can then be used in the estimation of phosphate present in organic esters as shown in Figure 2. Later research indicated (126) that triphenyl phosphine (containing carbon-phosphorus bonds) was rapidly decomposed while adenosine diphosphate (ADP) was capable of slowly producing orthophosphate. Rate constants for the decomposition of pure phosphorus compounds and of seawater were determined, and the orthophosphate, total organic phosphorus, and organic polyphosphate concentrations in the English Channel were ascertained by initial phosphate measurements, irradiation, and acid hydrolysis. Henriksen (127) obtained consistently lower values using UV digestion than he found with persulfate oxidation. However, it was not determined whether this result was due to the presence of significant levels of iron (50 to 1000 μg/liter), which could react with the liberated orthophosphate, or to high levels of polyphosphate.

Grasshoff (128) has developed an automated phosphate procedure incorporating UV digestion of samples. A 30-turn quartz coil placed 5 cm away from a 900-W high-pressure mercury lamp was used. This provided a 15-min digestion period. A stream of air was utilized to keep the sample cool.

An interlaboratory study was reported by the Analytical Reference Service in 1966 (129). For samples containing 0.5 mg/liter phosphate added as a polyphosphate, a mean recovery of 0.52 mg/liter with a standard deviation of .099 was reported for 32 laboratories using a stannous chloride method after acid hydrolysis. For 20 laboratories using an aminonaphtholsulfonic acid method, the mean recovery and standard deviations were .65 and .314, respectively.

Scanning (130) compared acid hydrolysis and persulfate digestion for the analysis of 34 sediment samples. Of these, 18 gave the same results for phosphorus by both methods, 10 gave results by the persulfate method, and 6 gave lower values. The study also included the analysis of 96 water samples ranging in concentration from less than 0.01 to 652 mg P/liter. Six samples gave the same result by both methods, 79 gave from 3 to 308% higher results by the persulfate method, and 16 were from 1 to 67% lower by the persulfate method.

In a study of phosphorus digestion methods for the analysis of total phosphorus in sewage, it was found that potassium persulfate was a superior digestant when compared with potassium permanganate, potassium perchlorate, or perchloric acid (108).

In a comparison of five methods for the determination of total organic phosphorus in lake sediments (131), it was concluded that the extraction procedure of Mehta et al. (132) was the best for routine determination of total organic phosphorus in lake sediments.

Kjeldahl treatment of particulate matter and plankton has been used for the liberation of orthophosphate from organic matter (133). Because persulfate acts poorly on particulate matter, Fuhs (134) has dissolved particulate biological material in hot 8N alkali prior to persulfate oxidation.

In addition to the automated persulfate technique for total phosphorus (124) and the automated UV digestion method for organic phosphorus (128), several automated procedures using the Technicon Continuous Digestor have appeared in the literature. A persulfate-sulfuric acid digestant has been employed prior to automated colorimetric analysis using a hydrazine reductant (135). A detection limit of 5 μg/liter of total phosphorus was achieved. A combined sulfuric acid-perchloric acid digestant with a hydrazine sulfate-stannous chloride reduction of the phosphomolybdic acid was found to give recoveries of 88 to 96% from samples of the refractory compound disodium phenyl phosphate at phosphorus levels of 0.004 to 2.0 mg/liter (136). Sulfuric acid-perchloric acid was also used by Tyler and Eiles (137). Reduction of the phosphomolybdate complex by ascorbic acid served for the analysis of the liberated orthophosphate. It is possible that the inclusion of antimony in the colorimetric analysis (87) could improve the authors' reported detection limit of 50 μg P/liter.

A method for the simultaneous automated digestion of sewage samples for total phosphorus and total Kjeldahl nitrogen using a perchloric acid-sulfuric acid mixture and amino napthyl sulfonic acid reduction of the phosphomolybdic acid had a detection limit of 0.3 mg/liter total phosphorus (138). Twenty-one replicate analyses of a primary effluent sample with 94 mg/liter suspended solids averaged 11.7 mg P/liter with a standard deviation of .6. The three automated digestion methods that utilize perchloric acid-sulfuric acid (136-138) also use vanadium pentoxide as a catalyst in replacement of the selenium dioxide catalyst employed for Kjeldahl nitrogen analysis (139). A potassium persulfate-sulfuric acid digestant was used for the digestion followed by ascorbic acid reduction of the phosphomolybdate with isopropyl alcohol being added to prevent precipitation (140). Modifications of the manifold are given for analyses in the 0 to 1, 0 to 6, and 0 to 20 mg P/liter ranges. The sensitivity is 0.01 mg P/liter using the most sensitive manifold and 0.1 mg P/liter using the least sensitive manifold.

Automated separation and quantitation of mixtures of orthophosphate, pyrophosphate, and tripolyphosphate has been accomplished by the use of gradient elution ion-exchange chromatography (141).

PHOSPHORUS IN SEDIMENT

The exchange of phosphorus between sediment and water is perhaps the most important aspect of phosphorus in natural waters. In most lakes, there appears to be a net movement of phosphorus into the sediment; in Lake Erie as much as 80% of the total phosphorus is removed from the waters by natural processes and is presumably in the sediment (12). Similarly, the sediments of the shelf areas of the oceans appear to concentrate phosphorus (142). There are, however, conditions when phosphorus is released from the sediment and is available for organism use. Abbott (143) has shown that the soils entering the reservoir, Lake Houston, were a major source of phosphorus; Zicker et al. (144) found that agitation of muds of bog lakes doubled the amount of soluble phosphorus. Gumerman (145) studied samples of the upper few centimeters of sediment of Lake Erie under anoxic conditions; he learned that although the muds adsorbed phosphorus rapidly, almost as much phosphorus could be desorbed as sorbed.

The important master chemical variables determining the adsorption-desorption of phosphorus from sediment appear to be the oxidation-reduction potential, the pH values, the calcium concentration, and the degree of agitation of sediment in water. The redox potential defines the iron (II)/iron (III) ratio and indicates the ammonia concentration, as well. A large proportion of phosphorus is adsorbed on ferric hydroxides and oxides; when the redox potential decreases, ferric hydroxides and oxides dissolve, and phosphorus is released. Phosphorus may be also bound as ferric phosphate minerals, and a decrease in redox potential would cause mineral solution. Ammonia may be important in the formation of the mineral struvite (146). Mortimer (147), who recently reevaluated the exchange of nutrients between sediment and water, concluded that as long as the oxygen concentration at the sediment surface is greater than 2 mg/liter, sediment release of nutrients is nil; below 2 mg/liter, major amounts of phosphorus, ammonium, and silicon are released. Factor analysis of sediment surface samples from Lake Mendota (148) show a good correlation among phosphorus, iron, and manganese, with an atom ratio of iron to phosphorus of about 10:1.

Stevenson (149) agitated sediment samples from the oceans and recorded variable results; sometimes phosphorus was adsorbed, and sometimes it was desorbed. He found that an increase in pH generally causes an increase in phosphorus in solution.

Macpherson et al. (150) agitated muds from a variety of lakes and found that phosphorus in solution was a minimum at a pH of 5 to 7; phosphorus increases rapidly above and below this range. Studies on phosphorus release of Lake Windermere muds as a function of pH showed a minimum between pH 6 and 7 (49).

Lake sediment was equilibrated with calcium chloride solutions containing 0.05 to 0.42 ppm phosphorus (151). All the experimental sediments removed phosphorus under calcium concentrations of 226 to 325 ppm. Calcium was also removed from solution.

Olsen (154) has determined that the net adsorption of phosphorus on sediment under aerobic and anaerobic conditions follows the function:

$$a = KC^{v} - K_{b}/C^{v_{b}}$$

where a = net phosphorus adsorbed (10^{-9} g P/g dry sediment), C = solution concentration (10^{-9} g P/ml solution), and K, K_b, v, and v_b are constants. This first term of the function is the gross P adsorbed and the second term is the exchange term. Constants for fine-grained sediment are as follows:

	Oxidized	Reduced
K	2.6	1.8
K_b	2.2	2.4
v	0.26	0.26
v_b	-0.05	-0.05

These results indicate that the major difference in uptake of phosphorus in oxidizing and reducing environments is due to the adsorption term (first term) and in particular differences in K.

Most studies show that agitated solutions of sediment release more phosphorus from the sediment than unstirred solutions. Laboratory studies using polyethylene sheets and layers of sand as diffusion barriers greatly reduced the flow of nutrients from sediment to water (152). Diffusion coefficients for many substances are at least a factor of 10 lower for mud than solutions (153).

MacKereth (155) has emphasized that the efficiency of removal of phosphorus by incorporation into planktonic algae is followed by incorporation into sediment by inorganic and/or biological mechanisms. Holden (156) has found that the daily rate of removal of phosphate from water is about 6.5 μg P/(cm^2) per mg P/liter, whereas uptake of phosphorus from mud cores under aerated unstirred conditions is about 0.1 this rate. Agitation in natural systems may be an important factor in maintaining similar rates between biological removal and sediment uptake in the chemical removal of phosphorus if environmental conditions encourage adsorption.

Many analytical schemes have been devised to define the various fractions of phosphorus in sediment (157-160). One of these schemes is discussed in more detail in Chapter IX. The major limiting factor to all these analytical techniques appears to be lack of precise information regarding the surface chemistry and mineralogy of the phases involved in phosphorus reactions. Grain size, nature of mineral surface, and degree of crystallinity, as well as the mineralogy, are factors that can profoundly effect the reaction of a solid containing phosphorus.

Mineralogy

There are three phosphorus mineral groups by stoichiometry: the calcium phosphates, the iron phosphates, and the aluminum phosphates. The mineralogy and the stoichiometry of all these groups are complicated.

The calcium phosphates are found predominantly as apatites and whitlockite (161-165). They occur as fine-grained subcrystalline solids and are common in ocean sediments and areas of high calcium concentration. Carbonate substitutes in the apatite structure and changes the free energy of the solid; in addition, sulfate, sodium, silicate, and rare earths are common substitutions. A general compositional formula for apatite is:

$$(R)_{10}(S)_6(T)_2$$

where R = Ca, Na, rare earths

S = PO_4, CO_3, SiO_4, AsO_4, SO_4

T = OH, F.

When CO_3 substitutes for PO_4, there appears to be a parallel increase in F, suggesting that CO_3 and F substitute for PO_4 (166). Kramer (167) has determined the ion activity product of a carbonate-fluor-apatite--$Na_{0.29}Ca_{9.56}(PO_4)_{5.37}(SO_4)_{0.3}(CO_3)_{0.33}F_{2.04}$--to be 105, whereas the ion activity product of $Ca_{10}(PO_4)_6(OH)_2$ is 114 and that of $Ca_{10}(PO_4)_6F_2$ is 119 at 25°C. Apatite crystallizes in the hexagonal system. Major reflections are similar for all apatites; cell parameters are (168):

	a_o	c_o
Fluor-apatite	9.364	6.879
Hydroxy-apatite	9.422	6.883
CO_3-fluor-apatite	9.344	6.881

Large variations should be expected in the unit cell values, however, because of the many substitutions possible.

The iron phosphate minerals constitute a large series of unique mineral phases. Moore (169-171), who summarized the structures, found that the basic iron phosphate minerals comprise a homologous series in which their structures are related. In turn, the pressure-temperature stability of these minerals can be classified by their structure. The minerals strengite, metastrengite ($FePO_4 \cdot 2H_2O$) and vivianite ($Fe_3(PO_4)_2 \cdot 8H_2O$) are the probable minerals to form under lacustrine conditions. Vivianite consists of insular, octahedral, edge-sharing doublets and insular singlets to form slabs perpendicular to the b-axis by phosphate tetrahedra (172). The slabs are connected by hydrogen bonds. The strengite and metastrengite structures are even more open than the vivianite structure. The structure is composed of a three-dimensional phosphate tetrahedral network linked by $FeO_4(H_2O)_2$ octahedra in which the water molecules are always fixed to the same apex of the octahedron. The difference in the strengite and metastrengite occurs in the tilt of the phosphate tetrahedra about a structural octahedron.

Strengite and metastrengite have been synthesized under conditions comparable to those found in lake systems (173). Vivianite has been reported in lacustrine sediments, and MacKereth (155) suggested a mechanism of formation involving local conditions of reduction for iron mobility and concentration. Nriagu (174) measured the ion activity product of strengite, $(Fe^{+3})(H_2PO_4)(OH)^2$, pK = 34.88, and from calorimetric data (175), pK = 36.4.

Variscite ($AlPO_4 \cdot 2H_2O$) is the major aluminum phosphate mineral found in conditions similar to lacustrine environments. Variscite is considered to be isostructural with strengite, and quite commonly some ferric ion substitutes for aluminum. A complete substitution series exists between iron and aluminum (168), but substitution in natural samples is normally confined to compositions near the end members.

Cole and Jackson (173) determined the solubility product of $(Al)(OH)_2(H_2PO_4)$ to have a pK of 28.55. Hsu (176) has recently ascertained that the same product for variscite has a pK of 29.55.

In 1971 Shapiro and Edmundson (177) defined the iron phosphate, phosphoferrite, from Lake Washington. Unfortunately, their results do not follow the predictions of Moore (170).

Equilibrium Relationships

It is possible to calculate the relative stability relationships of the apatites, variscite, and strengite using current data of pK = 29.55 for variscite, pK = 34.88 for strengite, pK = 105 for $Na_{0.29}Ca_{9.56}(PO_4)_{5.37}(SO_4)_{0.3}(CO_3)_{0.33}F_{2.04}$, and pK = 114 for $Ca_{10}(PO_4)_6OH_2$. Assuming concentrations appropriate to freshwater lakes and streams [pF = 5 (0.19 ppm), pNa = 3.5 (7.4 ppm), pSO_4 = 3.5 (30 ppm), pCa = 3 (40 ppm), $pHCO_3$ = 3 (61 ppm)], it is possible to calculate the mineral equilibria stability as a function of pH and soluble orthophosphate; in so doing, however, it is also necessary to assume that concentrations of iron and aluminum are controlled by oxide equilibration and that the system is oxidizing (ferric/ferrous ion ratio is large). Data for gibbsite ($Al_2O_3 \cdot 3H_2O$) solubility of pK = 65.4 and for precipitated ferric hydroxide ($Fe(OH)_3$) solubility of pK = 38.0, are used. All calculations are done for 25°C and 1 atm total pressure. Figure 3 shows the results of the calculations. The diagram is very similar to an earlier calculation of Lindsay and Moreno (178), although recent thermodynamic data have been used to construct the diagram (175).

It is interesting to note that all solid phases [$(Fe(OH)_3$, $Al_2O_3 \cdot 3H_2O$, $FePO_4 \cdot 2H_2O$, $AlPO_4 \cdot 2H_2O$, carbonate-fluor-apatite, and aqueous solution] are in equilibrium at a pH of about 5.5 and a soluble phosphate concentration of about 50 ppb PO_4-P. The pH is coincident with the minimum soluble phosphate determined by Macpherson et al. (150) for equilibrium studies with various lake sediments.

Adsorption

Iron and aluminum hydroxides and oxides, as well as silicates of these elements, adsorb phosphorus (179). Kaolinite is a common clay and adsorbs both polyphosphates (181) and orthophosphate from solution (180). Adsorption of aluminum oxides and hydroxides is rapid--half-life for equilibrium is about 10 min (182). The capacity is rather high (about 10^{-4} moles P/g of mineral); the degree of removal can be complete (less than 10^{-7} moles P/liter of solution), and the process can be irreversible (176,183). The adsorption data available for iron oxides suggest characteristics similar to those for aluminum oxides, except that the iron oxide does not settle readily.

The minerals just mentioned occur as very fine-grained

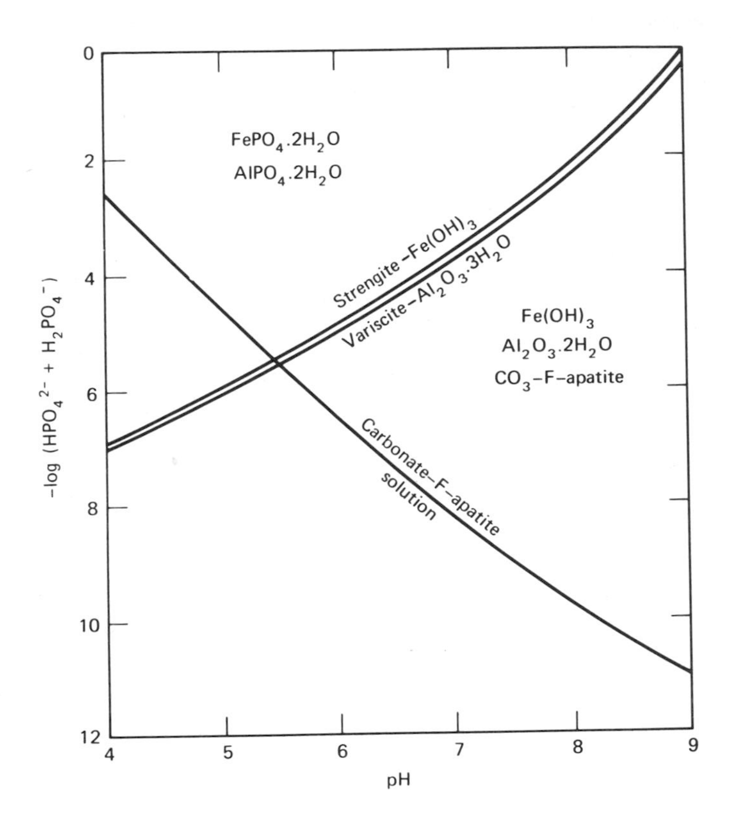

Fig. 3. Stability relationships of strengite, variscite, carbonate, fluor-apatite, precipitated ferric hydroxide, and gibbsite.

substances and hence have a high degree of surface reactivity. One of the major parameters determining the adsorption characteristics is the measure of the pH of neutral charge. Parks (184) has defined a mechanism for estimating the pH of the isoelectric points (zero point of charge) of common minerals. In addition, adsorption on iron oxides and hydroxides as well as aluminum oxides and hydroxides is a maximum when the concentration of the metal in solution is a minimum (179). A comparison of pH of zero point of charge, minimum pH of solubility, and pH of minimum phosphorus dissolved from sediment suggests that a pH of 5 to 7 represents the optimum lake condition for phosphorus adsorption onto iron and aluminum oxides and hydroxides and clay minerals:

Mineral	pH, zero pt of charge	pH, min solubility
Kaolinite	4.6	
Gibbsite	5.1	6
$Fe(OH)_3$	8.5	8.5

Minimum solution of phosphorus at pH 5 to 7 (150) corresponds closely to minima of solubility and the pH of zero point of charge for aluminum-bearing minerals. These may be more significant than iron oxides and hydroxides in the adsorption of phosphorus.

REFERENCES

1. F. H. Rigler, "A Trace Study of the Phosphorus Cycle in Lake Water," Ecology, 37, 550-562 (1956).
2. F. R. Hayes and N. R. Becket, "The Flows of Minerals through the Thermocline of a Lake," Arch. J. Hydrobiol., 51, 391 (1956).
3. Brian Mason, "Principles of Geochemistry," 2nd ed. Wiley, New York, 1958.
4. H. Y. Sverdrup, M. W. Johnson, and R. H. Fleming, "The Oceans, Their Physics, Chemistry, and General Biology." Prentice-Hall, Englewood Cliffs, N. J., 1942.
5. D. A. Livingstone, "Chemical Composition of Rivers and Lakes, Data of Geochemistry," U.S. Geological Survey Prof. Pap. 440-G, 1 (1963).
6. A. C. Redfield, B. H. Ketchum, and F. A. Richards, "The Influence of Organisms on the Composition of Sea Water," in M. N. Hill, ed., "The Sea, Vol. 2. Interscience, New York, 1963, pp. 26-77.
7. A. F. Bartsch, "Role of the Federal Government in Controlling Nutrients in Natural Waters," presented at the American Chemical Society Symposium on Nutrients in Natural Waters, Los Angeles, March 28-April 2, 1971.
8. J. R. Vallentyne, "Phosphorus and the Control of Eutrophication," Can. Res. Dev., 49, 36-43 (1970).

9. R. A. Vollenweider, unpublished note, Canada Centre for Inland Waters, Fisheries Research Board Detachment, 1970.
10. J. Verduin, in "Agricultural Practices and Water Quality," Water Pollution Control Research Series, DAST-26, 1969, pp. 63-71.
11. R. W. Lewis, "Phosphorus," in "Mineral Facts and Problems," U.S. Dept. Interior, Bur. Mines, Washington, D.C., 1970.
12. A. T. Prince and J. P. Bruce, "Development of Nutrient Control Policies in Canada," presented at the American Chemical Society Symposium on Nutrients in Natural Waters, Los Angeles, March 28-April 2, 1971.
13. A. N. Gilchrist and A. G. Gillingham, "Phosphate Movement in Surface Run-Off Water," N.Z. J. Agric. Res., 13, 225-231 (1970).
14. D. L. Carter, J. A. Bondwrant, and C. W. Robbins, "Water-Soluble NO_3-Nitrogen, PO_4-Phosphorus, and Total Salt Balances on a Large Irrigation Tract," Soil Sci. Soc. Am. Proc., 35, 331-335 (1971).
15. H. Bernhardt, W. Such, and A. Wilhelms, "Nutrient Content of Water from Watersheds with Mainly Agricultural Utilization and Rural Population," Sewage Wastes, 72, 293 (1970).
16. J. H. Ryther and W. M. Dunstan, "Nitrogen, Phosphorus and Eutrophication in the Coastal Marine Environment," Science, 171, 1008-1013 (1971).
17. J. R. Vallentyne, W. E. Johnson, and A. J. Harris, "A Visual Demonstration of the Beneficial Effects of Sewage Treatment for Phosphate Removal of Particulate Matter Production in Waters of Lakes Erie and Ontario," J. Fish. Res. Bd. (Canada), 27, 1493-1496 (1970).
18. C. N. Sawyer, "Problem of Phosphorus in Water Supplies," J. Am. Water Works Assoc., 57, 1431-1439 (1965).
19. R. A. Vollenweider, "Scientific Fundamentals of the Eutrophication of Lakes and Flowing Waters, with Particular Reference to Nitrogen and Phosphorus as Factors in Eutrophication," Organization for Economic Co-operation and Development, DAS/CSI/6827, 1968.
20. P. J. Weaver, "Phosphates in Surface Waters and Detergents," J. Water Pollut. Control Fed., 41, 1647-1653 (1969).
21. R. E. Johnson, A. T. Rossano, Jr., and R. O. Sylvester, "Dustfall as a Source of Water Quality Impairment," J. Sanit. Eng. Div., 92, 245-256 (1966).

22. J. R. Van Wazer, "Phosphorus and its Compounds," Vols. I and II. Interscience, New York, 1958 and 1961.
23. K. M. Mackenthun, "Nitrogen and Phosphorus in Water." U.S. Dept. Health, Education and Welfare, Washington, D.C., 1965.
24. P. L. McCarty, "Chemistry of Nitrogen and Phosphorus in Water," J. Am. Water Works Assoc., 62, 127-140 (1970).
25. I. M. Heilbron, "Some Aspects of Algal Chemistry," J. Chem. Soc., 79-89 1942.
26. B. J. Katchman, "Phosphates in Life Processes," in J. R. Van Wazer, Ed., "Phosphorus and its Compounds," Vol. II. Interscience, New York, 1961.
27. A. M. McCombie, "Factors Influencing the Growth of Phytoplankton," J. Fish. Res. Bd. (Canada), 10, 253-282 (1953).
28. B. H. Ketchum, "Mineral Nutrition of Phytoplankton," Ann. Rev. Plant Physiol., 5, 55-74 (1945).
29. A. Kuhl, "Inorganic Phosphorus Uptake and Metabolism," in R. A. Lewin, Ed., "Physiology and Biochemistry of Algae." Academic Press, New York and London, 211-224 1962.
30. L. Provasoli, "Nutrition and Ecology of Protozoa and Algae," Ann. Rev. Microbiol., 12, 279-308 (1958).
31. A. Tucker, "The Relation of Phytoplankton Periodicity to the Nature of the Physico-Chemical Environment with Special Reference to Phosphorus," Am. Mid. Natur., 57, 334-370 (1957).
32. R. W. Krause, "Physiology of the Fresh Water Algae," Ann. Rev. Plant Physiol., 9, 207-244 (1958).
33. K. E. Quentin, G. Philipp, and L. Weil, "Behavior of Organophosphorus Pesticides in Water," Wasser, 36, 54-63 (1970).
34. R. Nilsson, "Phosphate Separation in Sewage Treatment," Process Biochem., 4, 49-52 (1969).
35. J. H. Martin, "Distribution of C, H, P, Fe, Mn, Zn, Ca, Sr, and Sc in Plankton Samples Collected off Panama and Colombia," Biol. Sci., 19, 898-901 (1969).
36. D. W. Menzel, "The Role of 'in situ' Decomposition of Organic Matter on the Concentration of Nonconservative Properties in the Sea," Deep-Sea Res., 17, 751-764 (1970).
37. F. Schott and M. Ehrhardt, "On Fluctuations and Mean Relation of Chemical Parameters in the North-western North Sea, Kiel," Meeres., 25, 272-278 (1969).
38. J. D. H. Strickland, "Measuring the Production of Marine Phytoplankton," Fish. Res. Bd.,(Canada), Bull. 122, (1966).

39. H. Kleerekoper, "The Mineralization of Plankton," J. Fish. Res. Bd. (Canada), 10, 283-291 (1953).
40. G. C. Gerloff, and P. H. Krombpolz, "Tissue Analysis as a Measure of Nutrient Availability for the Growth of Angiosperm Aquatic Plants," Limnol. Oceanog., 11, 529-537 (1966).
41. T. R. Parsons, K. Stephens, and J. D. H. Strickland, "On the Chemical Composition of Eleven Species of Marine Phytoplankton," J. Fish. Res. Bd. (Canada), 18, 1001-1011 (1961).
42. R. F. Vaccaro, "Available Nitrogen and Phosphorus and the Biochemical Cycle in the Atlantic off New England," J. Mar. Res., 21, 284-301 (1963).
43. W. B. Engel and R. M. Owen, "The Chemical Composition of Bacterial Kerosene-Water Interfacial Pellicles," Dev. Ind. Microbiol., 10, 234-246 (1969).
44. E. I. Butler, E. D. S. Corner, and S. M. Marshall, "On the Nutrition and Metabolish of Zooplankton, VII. Seasonal Survey of Nitrogen and Phosphorus Excretion by Calanus in the Clyde Sea Area," J. Mar. Biol. Assoc. (U.K.), 50, 525-560 (1970).
45. B. H. Ketchum, "The Development and Restoration of Deficiencies in the Phosphorus and Nitrogen Composition of Unicellular Plants," J. Cell. Comp. Physiol., 13, 373-381 (1939).
46. J. R. Kramer, H. E. Allen, G. W. Baulne, and N. M. Burns, "Lake Erie Time Study (LETS)," Can. Cent. Inland Waters Pap., 4, 1970.
47. E. D. Goldberg, T. J. Walker, and Alice Whisenand, "Phosphate Utilization by Diatoms," Biol. Bull., 101, 274-284 (1951).
48. A. A. AlKholy, "On the Assimilation of Phosphorus in Chlorella Pyrenoidosa," Physiol. Plants, 9, 137-143 (1956).
49. F. J. Mackereth, "Phosphorus Utilization by Asterionella Formosa Hass," J. Exp. Bot., 4, 296-313 (1953).
50. R. E. Johannes, "Phosphorus Excretion and Body Size in Marine Animals: Microzooplankton and Nutrient Regeneration," Science, 146, 923-924 (1964).
51. L. R. Pomeroy, H. M. Mathews, and H. S. Min, "Excretion of Phosphate and Soluble Organic Phosphorus Compounds by Zooplankton," Limnol. Oceanogr., 8, 50-55 (1963).
52. E. J. Kuenzler, "Dissolved Organic Phosphorus Excretion by Marine Phytoplankton," J. Phycol., 6, 7-13 (1970).
53. N. L. Clesceri and G. F. Lee, "Hydrolysis of Condensed Phosphates. I: Nonsterile Environment,"

Int. J. Air Water Pollut., 9, 723-742 (1965).
54. F. H. Rigler, "The Phosphorus Fractions and the Turnover Time of Inorganic Phosphorus in Different Types of Lakes," Limnol. Oceanogr., 9, 511-518 (1954).
55. S. Herbes, unpublished report, Department of Environmental and Industrial Health, University of Michigan, 1972.
56. B. H. Ketchum and N. Corwin, "The Cycle of Phosphorus in a Plankton Bloom in the Gulf of Maine," Limnol. Oceanogr., 10, Suppl. R148-R161 (1965).
57. E. J. Kuenzler, R. R. L. Guillard, and N. Corwin, "Phosphate-Free Sea Water for Reagent Blanks in Chemical Analyses," Deep-Sea Res., 10, 749-755 (1963).
58. S. P. Chu, "Utilization of Organic Phosphorus by Phytoplankton," J. Mar. Biol. Assoc. (U.K.), 26, 285-295 (1946).
59. G. E. Fogg and J. D. A. Miller, "The Effect of Organic Substances on the Growth of the Freshwater Alga Monodus subteraneus," Verh. Int. Verein. Limnol., 13, 892-895 (1958).
60. R. E. Johannes, "Uptake and Release of Dissolved Organic Phosphorus by Representatives of Coastal Marine Ecosystem," Limnol. Oceanogr., 9, 224-234 (1964).
61. E. J. Kuenzler, "Glucose-6-phosphate Utilization by Marine Algae," J. Phycol., 1, 156-164 (1965).
62. D. L. Correll, "Pelagic Phosphorus Metabolism in Antarctic Waters," Limnol. Oceanogr., 10, 364-370 (1965).
63. H. L. Golterman, "Mineralization of Algae under Sterile Conditions by Bacterial Breakdown," Verh. Int. Verein. Limnol., 15, 544-548 (1964).
64. W. D. Watt and F. R. Hayes, "Tracer Study of the Phosphorus Cycle in Sea Water," Limnol. Oceanogr., 8, 276-285 (1963).
65. E. J. Kuenzler and J. P. Perras, "Phosphatases of Marine Algae," Biol. Bull., 128, 271-285 (1965).
66. T. Berman, "Alkaline Phosphatases and Phosphorus Availability in Lake Kinneret," Limnol. Oceanogr., 15, 663-674 (1970).
67. Mau-Huai Kwo and H. J. Blumenthal, "Purification and Properties of an Acid Phosphomonoesterase from Neurospora crassa," Biochim. Biophys. Acta, 52, 1529-1537 (1961).
68. D. Brandes and R. N. Elston, "An Electron Microscopical Study of the Histochemical Localization of Alkaline Phosphatase in the Cell Wall of Chlorella vulgaris," Nature, 177, 274-275 (1956).
69. J. C. Buzzell, Jr. and C. N. Sawyer, "Removal of Algal Nutrients from Raw Wastewater with Lime," J. Wat. Pollut. Control Fed., 39, R16-R24 (1967).

70. "Phosphate Removal Processes Prove Practical," Environ. Sci. Technol., 2, 182 (1968).
71. C. N. Sawyer, "Problem of Phosphorus in Water Supplies," presented at the Annual Conference, Portland, Ore., June 29, 1965.
72. G. E. Bennett, in "Development of a Pilot Plant to Demonstrate Removal of Carbonaceous, Nitrogenous, and Phosphorous Materials from Anaerobic Digester Supernatant and Related Process Streams," Water Pollution Control Research Series, FKA-17010, p. 16, 1970.
73. "Phosphate Study at the Baltimore Back River Waste-Water Treatment Plant," Water Pollution Control Research Series, DFV-17010, 1970.
74. W. R. G. Atkins, "The Phosphate Content of Fresh and Salt Water in its Relationship to the Growth of the Algal Plankton," J. Mar. Biol. Assoc., (U.K.), 13, 119-150 (1923).
75. G. Deniges, "Reaction de Coloration Extremement Sensible des Phosphates et des Arseniates; ses applications," C. R. Acad. Sci. (Paris), 171, 802-804 (1920).
76. S. Olsen, "Recent Trends in the Determination of Orthophosphate in Water," in "Chemical Environment in the Aquatic Habitat," H. L. Golterman and R. S. Clymo, Eds., N. V. Noord-Hollandsche Uitgevers Maatschappij, Amsterdam, 1967, pp. 63-105.
77. H. W. Harvey, "The Estimation of Phosphate and of Total Phosphorus in Sea Waters," J. Mar. Biol. Assoc. (U.K.), 27, 337-359 (1948).
78. A. R. Coote, I. W. Duedall, and R. S. Hiltz, "Automatic Analysis at Sea," Advance in Automated Analysis, Technicon Int. Congr. 1970, Vol. II. Thurman Assoc., Miami, 1971, pp. 347-351.
79. "Methods for Chemical Analysis of Water and Wastes, 1971," Environmental Protection Agency, Analytical Quality Control Laboratory, Cincinnati, Ohio.
80. R. H. Gaddy, Jr., "Automated Analyses for the Characterization of Heavy Water Moderator," Automation in Analytical Chemistry, Technicon Symposium 1965. Mediad Inc., White Plains, N.Y., 1966, pp. 210-214.
81. M. E. Gales, Jr. and E. C. Julian, "Determination of Inorganic Phosphate or Total Phosphate in Water by Automatic Analysis," Automation in Analytical Chemistry, Technicon Symposium 1966, Vol. I, Mediad Inc., White Plains, N.Y., 1967, pp. 557-559.
82. G. F. Lee, "Automatic Methods for the Analysis of Natural Waters," in "Chemical Environment in the Aquatic Habitat," H. L. Golterman and R. S. Clymo, Eds., N.V. Noord-Hollandsche Uitgevers Maatschappij, Amsterdam, 1967, pp. 169-179.

83. American Public Health Association (APHA), American Water Works Association, and Water Pollution Control Federation, "Standard Methods for the Examination of Water and Wastewater," 13th ed. APHA, New York, 1971.
84. American Society for Testing and Materials, (ASTM), "1971 Annual Book of ASTM Standards," Pt. 23, "Water; Atmospheric Analysis." ASTM, Philadelphia, 1971.
85. L. J. Greenfield and F. A. Kalber, "Inorganic Phosphate Measurement in Seawater," Bull. Mar. Sci. Gulf Carib., 4, 323-335 (1954).
86. J. Murphy and J. P. Riley, "A Single-Solution Method for the Determination of Soluble Phosphate in Sea Water," J. Mar. Biol. Assoc. (U.K.), 37, 9-14 (1958).
87. J. Murphy and J. P. Riley, "A Modified Single Solution Method for the Determination of Phosphate in Natural Waters," Anal. Chim. Acta, 27, 31-36 (1962).
88. J. D. H. Strickland and T. R. Parsons, "A Practical Manual of Seawater Analysis," Fish. Res. Bd. (Canada), Bull. 167, Ottawa, 1968.
89. G. P. Edwards, A. H. Molof, and R. W. Schneeman, "Determination of Orthophosphate in Fresh and Saline Waters," J. Am. Water Works Assoc., 57, 917-925 (1965).
90. J. E. Harwood, R. A. van Steenderen, and A. L. Kuhn, "A Rapid Method for Orthophosphate Analysis at High Concentrations in Water," Water Res., 3, 417-423 (1969).
91. M. Bernhard and G. Macchi, "Application and Possibilities of Automatic Chemical Analysis in Oceanography," Automation in Analytical Chemistry, Technicon Symposium 1965. Mediad Inc., White Plains, N.Y., 1966, pp. 255-259.
92. K. Grasshoff, "Automatic Determination of Fluoride, Phosphate, and Silicate in Sea Water," Automation in Analytical Chemistry, Technicon Symposium 1965. Mediad Inc., White Plains, N.Y., 1966, pp. 304-307.
93. P. G. Brewer, K. M. Chan, and J. P. Riley, "Automatic Determination of Certain Micronutrients in Sea Water," Automation in Analytical Chemistry, Technicon Symposium 1965. Mediad Inc., White Plains, N.Y., 1966, pp. 308-314.
94. K. M. Chan and J. P. Riley, "The Automatic Determination of Phosphate in Sea Water," Deep-Sea Res., 13, 467-471 (1966).
95. K. Grasshoff, "A Simultaneous Multiple-Channel System for Nutrient Analysis in Seawater with Analog and Digital Data Record," Advances in Automated Analysis, Technicon International Congress 1969, Vol. II, Mediad Inc., White Plains, N.Y., 1970, pp. 133-145.

96. A. H. Molof, C. P. Edwards, and R. W. Schneeman, "An Automated Analysis for Orthophosphate in Fresh and Saline Waters," Automation in Analytical Chemistry, Technicon Symposium 1965. Mediad Inc., White Plains, N.Y., 1966, pp. 245-249.
97. C. Wadelin and M. G. Mellon, "Extraction of Heteropoly Acids," Anal. Chem., 25, 1668-1673 (1953).
98. H. H. Ross and R. B. Hahn, "A Study of the Separation of Phosphate Ion from Arsenate Ion by Solvent Extraction," Talanta, 7, 276-280 (1961).
99. C. M. Proctor and D. W. Hood, "Determination of Inorganic Phosphate in Sea-Water by an Iso-Butanol Extraction Procedure," J. Mar. Res., 13, 122-132 (1954).
100. K. Stephens, "Determination of Low Phosphate Concentrations in Lake and Marine Waters," Limnol. Oceanogr., 8, 361-362 (1963).
101. A. B. Isaeva, "Determination of Small Amounts of Phosphates in Sea Water after their Preliminary Extraction as Phosphomolybdic Acid, and Determination of Phosphates in Turbid Water," Zh. Anal. Khim., Khim., 24, 1854-1858 (1969).
102. A. Henrikson, "An Automatic Method for Determining Low-Level Concentrations of Phosphates in Fresh and Saline Waters," Analyst, 90, 29-34 (1965).
103. P. G. W. Jones and C. P. Spencer, "Comparison of Several Methods of Determining Inorganic Phosphate in Sea-Water," J. Mar. Biol. Assoc. (U.K.), 43, 251-273 (1963).
104. J. D. Burton and J. P. Riley, "Determination of Soluble Phosphate and Total Phosphorus in Sea Water and of Total Phosphorus in Marine Muds," Mikrochim. Acta, 9, 1350-1365 (1956).
105. H. W. Harvey, "The Chemistry and Fertility of Sea Waters," Cambridge University Press, Cambridge, England, 1955, 224 pages.
106. W. S. Wooster and N. W. Rakestraw, "The Estimation of Dissolved Phosphate in Sea Water," J. Mar. Res., 10, 91-100 (1951).
107. P. G. W. Jones, "Comparison of Several Methods of Determining Inorganic Phosphate in Oceanic Sea Water," J. Mar. Biol. Assoc. (U.K.), 46, 19-32 (1966).
108. S. G. Jankovic, D. T. Mitchell, and J. G. Buzzell, Jr., "Measurement of Phosphorus in Wastewater," Water and Sewage Works, 114, 471-474 (1967).
109. J. A. Winter and R. L. Booth, "Method Selection Study 1: A Comparison of Three Modifications of the Single Reagent Method for Soluble Orthophos-

phate," Federal Water Pollution Control Administration, Cincinnati, Ohio, 1969.
110. J. A. Winter and M. R. Midgett, "Method Study 2, Nutrient Analyses, Manual Methods," Environmental Protection Agency, Cincinnati, Ohio, 1970.
111. K. Sugawara and S. Kanamori, "Spectrophotometric Determination of Submicrogram Quantities of Orthophosphate in Natural Waters," Chem. Soc. Jap. Bull., 34, 258-261 (1961).
112. T. R. Hurford and D. F. Boltz, "Indirect Ultraviolet Spectrophotometric and Atomic Absorption Spectrometric Methods for Determination of Phosphorus and Silicon by Heteropoly Chemistry of Molybdate," Anal. Chem., 40, 379-382 (1968).
113. C. H. Wayman, "Simultaneous Determination of Sulfur and Phosphorus in Water by Neutron Activation Analysis," Anal. Chem., 36, 665-666 (1964).
114. H. E. Allen and R. B. Hahn, "Determination of Phosphate in Natural Waters by Activation Analysis of Tungstophosphoric Acid," Environ. Sci. Technol., 3, 844-848 (1969).
115. "Enzyme Inhibition Used as Assay for Phosphates," Environ. Sci. Technol., 4, 883 (1970).
116. F. H. Rigler, "Radiobiological Analysis of Inorganic Phosphorus in Lakewater," Verh. Int. Verein. Limnol., 16, 465-470 (1966).
117. F. H. Rigler, "Further Observations Inconsistent with the Hypothesis that the Molybdenum Blue Method Measures Orthophosphate in Lake Water," Limnol. Oceanogr., 13, 7-13 (1968).
118. W. Chamberlain and J. Shapiro, "On the Biological Significance of Phosphate Analysis; Comparison of Standard and New Methods with a Bioassay," Limnol. Oceanogr., 14, 921-927 (1969).
119. J. Shapiro, "Arsenic and Phosphate: Measured by Various Techniques," Science, 171, 234 (1971).
120. H. W. Harvey, "The Estimation of Phosphate and of Total Phosphorus in Sea Waters," J. Mar. Biol. Assoc. (U.K.), 27, 337-359 (1948).
121. D. L. Johnson, "Simultaneous Determination of Arsenate and Phosphate in Natural Waters," Environ. Sci. Technol., 5, 411-414 (1971).
122. D. W. Menzel and N. Corwin, "The Measurement of Total Phosphorus in Sea Water Based on the Liberation of Organically Bound Fractions by Persulfate Oxidation," Limnol. Oceanogr., 10, 280-282 (1965).
123. M. E. Gales, Jr., E. C. Julian, and R. C. Kroner, "Method for Quantitative Determination of Total Phosphorus in Water," J. Am. Water Works Assoc.,

58, 1363-1368 (1966).
124. M. Bernhard, E. Torti, and G. Rossi, "Automatic Determination of Total Hydrolyzable P-PO_4 in Sea-water and Algae Cultures," Automation in Analytical Chemistry, Technicon Symposium 1967, Vol. II. Mediad Inc., White Plains, N.Y., 1968, pp. 395-400.
125. F. A. J. Armstrong, P. M. Williams, and J. D. H. Strickland, "Photooxidation of Organic Matter in Sea Water by Ultraviolet Radiation, Analytical and Other Applications," Nature, 211, 481-483 (1966).
126. F. A. J. Armstrong and S. Tibbitts, "Photochemical Combustion of Organic Matter in Sea Water, for Nitrogen, Phosphorus and Carbon Determination," J. Mar. Biol. Assoc. (U.K.), 48, 143-152 (1968).
127. A. Henriksen, "Determination of Total Nitrogen, Phosphorus and Iron in Fresh Water by Photooxidation with Ultraviolet Radiation," Analyst, 95, 601-608 (1970).
128. K. Grasshoff, "Results and Possibilities of Automated Analysis of Nutrients in Seawater," Automation in Analytical Chemistry, Technicon Symposium 1966, Vol. I., Mediad Inc., White Plains, N.Y., 1967, pp. 573-579.
129. R. J. Lishka, L. A. Lederer, and E. F. McFarren, "Water Nutrients No. 1," U.S. Dept. Health, Education, and Welfare, Cincinnati, Ohio, 1966, 76 pages.
130. D. E. Sanning, "Phosphorus Determination--A Method Evaluation," Water and Sewage Works, 114, 131-133 (1967).
131. L. E. Sommers, R. F. Harris, J. D. H. Williams, D. E. Armstrong, and J. K. Sayers, "Determination of Total Organic Phosphorus in Lake Sediments," Limnol. Oceanogr., 15, 301-304 (1970).
132. N. C. Mehta, J. O. Legg, C. A. I. Goring, and C. A. Black, "Determination of Organic Phosphorus in Soils. I. Extraction Method," Soil Sci. Soc. Am. Proc., 18, 443-449 (1954).
133. V. U. Fossato, "Nitrogen and Phosphorus Determination in Plankton and Particulate Matter," Arch. Oceanogr. Limnol., 16, 189-193 (1969).
134. G. W. Fuhs, "Determination of Particulate Phosphorus by Alkaline Persulfate Digestion," Int. J. Environ. Anal. Chem., 1, 123-129 (1971).
135. H. Bernhardt and A. Wilhelms, "The Continuous Determination of Low Level Iron, Soluble Phosphate and Total Phosphate with the Auto-Analyzer," Automation in Analytical Chemistry, Technicon Symposium 1967, Vol. I. Mediad Inc., White Plains, N.Y., 1968, pp. 385-389.

136. J. J. McNulty and L. Johnson, "Automated Determination of Total Phosphorus in Estuarine Waters," Advances in Automated Analysis, Technicon International Congress 1970, Vol. II. Thurman Assoc., Miami, 1971, pp. 353-355.
137. L. P. Tyler and A. F. Biles, "Complete Automation of Low-Level Total Phosphorus Analysis in Water," Advances in Automated Analysis, Technicon International Congress 1970, Vol. II. Thurman Assoc., Miami, 1971, pp. 313-316.
138. W. F. Milbury, V. T. Stack, Jr., and F. L. Doll, "Simultaneous Determination of Total Kjeldahl Nitrogen in Activated Sludge with the Technicon Continuous Digestor System," Advances in Automated Analysis, Technicon International Congress 1970, Vol. II. Thurman Assoc., Miami, 1971, pp. 299-304.
139. R. W. Bide, "An Automated Method for the Estimation of Total Phosphate in Biological Materials," Anal. Biochem., 29, 393-403 (1969).
140. G. H. Stanley and G. R. Richardson, "The Automation of the Single-Reagent Method for Total Phosphorus," Advances in Automated Analysis, Technicon International Congress 1970, Vol. II. Thurman Assoc., Miami, 1971, pp. 305-311.
141. G. W. Heinke and H. Behmann, "Ion-Exchange Chromatographic Analysis of Condensed Phosphates in Lake Water and Waste Water," Advances in Automated Analysis, Technicon International Congress 1969, Vol. II. Mediad Inc., White Plains, N.Y., 1970, pp. 115-120.
142. Y. M. Senin, "Phosphorus in Bottom Sediments of the South West African Shelf," Litol. Polez. IsKop., No. 1, 11-26 (1970).
143. W. Abbott, "Unusual Phosphorus Source for Plankton Algae," Ecology, 38, 152 (1957).
144. E. L. Zicker, K. C. Berger, and A. D. Hasler, "Phosphorus Release from Bog Lake Muds," Limnol. Oceanogr., 1, 296-303 (1956).
145. R. C. Gumerman, "Aqueous Phosphate and Lake Sediment Interaction," Proc. 13th Conf. Great Lakes Res., 673-682, 1970.
146. P. G. Malone, and K. M. Towe, "Microbial Carbonate and Phosphate Precipitates from Sea Water Cultures," Mar. Geol., 9, 301-309 (1970).
147. C. H. Mortimer, "Chemical Exchanges Between Sediments and Water in the Great Lakes--Speculations on Probable Regulatory Mechanisms," Limnol. Oceanogr., 16, 387-404 (1971).
148. J. J. Delfino, G. C. Bortleson, and G. F. Lee, "Distribution of Mn, Fe, P, Mg, K, Na, and Ca in the

Surface Sediments of Lake Mendota, Wisconsin," Environ. Sci. Technol., 3, 1189-1192 (1969).
149. W. Stephenson, "Certain Effects of Agitation upon the Release of Phosphate from Mud," J. Mar. Biol. Assoc. (U.K.), 28, 371-380 (1949).
150. L. B. Macpherson, N. R. Sinclair, and F. R. Hayes, "Lake Water and Sediment. III. The Effect of pH on the Partition of Inorganic Phosphate between Water and Oxidized Mud or its Ash," Limnol. Oceanogr., 3, 318-326 (1958).
151. J. J. Latterell, R. F. Holt, and D. R. Timmons, "Phosphate Availability in Lake Sediments," J. Soil Water Conserv., 26, 21-24 (1971).
152. H. B. N. Hynes and B. J. Greib, "Movement of Phosphate and Other Ions from and through Lake Muds," J. Fish. Res. Bd. (Canada), 27, 653-668 (1970).
153. E. K. Duursma, "The Mobility of Compounds in Sediments in Relation to Exchange between Bottom and Supernatant Water," in "Chemical Environment in the Aquatic Habitat," H. L. Golterman and R. S. Clymo, Eds., N. V. Noord-Hollandsche Uitgevers Maatschappij, Amsterdam, 1967, p. 288-296.
154. S. Olsen, "Phosphate Equilibrium between Reduced Sediments and Water. Laboratory Experiments with Radioactive Phosphorus," Verh. Int. Verein. Limnol., 15, 333-341 (1964).
155. F. J. H. Mackereth, "Some Chemical Observations on Post-Glacial Lake Sediments," Phil. Trans. Roy. Soc. London, Ser. B, 250, 165-213 (1966).
156. A. V. Holden, "The Removal of Dissolved Phosphate from Lake Waters by Bottom Deposits," Verh. Int. Verein. Limnol., 14, 247-251 (1961).
157. J. D. H. Williams, J. K. Syers, and R. F. Harris, "Adsorption and Desorption of Inorganic Phosphorus by Lake Sediments in a 0.1 M NaCl System," Environ. Sci. Technol., 4, 517-519 (1970).
158. D. A. Wentz and G. F. Lee, "Sedimentary Phosphorus in Lake Cores-Analytical Procedure," Environ. Sci. Technol., 3, 750-754 (1969).
159. J. D. H. Williams, J. K. Syers, R. F. Harris, and D. E. Armstrong, "Fractionation of Inorganic Phosphate in Calcareous Sediments," Soil Sci. Soc. Am. Proc., 35, 250-255 (1971).
160. S. S. Shukla, J. K. Syers, J. D. H. Williams, D. E. Armstrong, and R. F. Harris, "Sorption of Inorganic Phosphate by Lake Sediments," Soil Sci. Am. Proc., 35, 244-249 (1971).
161. R. A. Gulbrandsen and M. Cremer, "Coprecipitation of Carbonate and Phosphate from Sea Water," U.S.

Geol. Survey Prof. Pap. 700-c, 125-126 (1970).
162. O. Flint, "Scales from Phosphate-Treated Sea Water," Desalination, 6, 319-334 (1969).
163. D. B. Porcella, E. J. Middlebrooks, and J. S. Kumagi, "Biological Effects of Sediment--Water, Nutrient Interchange," presented at American Chemical Society Meeting, New York, September 7-12, 1969.
164. M. G. de Belinko, "Conditions Oceanographiques de la genese des Phosphates Sedimentaires," C.R. Acad. Sci. (Paris), Ser. D, 269, 875-877 (1969).
165. M. G. de Belinko, "Conditions Oceanographiques de la genese des Phosphates Sedimentaires Marins," C. R. Acad. Sci. (Paris), Ser. D, 269, 935-938 (1969).
166. R. A. Gulbrandsen, J. R. Kramer, L. B. Beatty, and R. E. Mays, "Carbonate-Bearing Apatite from Faraday Township, Ontario, Canada," Am. Mineral., 51, 819-824 (1966).
167. J. R. Kramer, "Sea Water: Saturation with Apatites and Carbonates," Science, 146, 637-638 (1964).
168. C. Palache, H. Berman, and C. Frondel, "The System of Mineralogy," Vol. II. Wiley, New York, 1124 pages.
169. P. B. Moore, "Structural Hierarchies among Minerals Containing Octahedrally Coordinating Oxygen," Neues Jahrb. Mineral. Monatsch., 163-173, 1970.
170. P. B. Moore, "Crystal Chemistry of the Basic Iron Phosphates," Am. Mineral., 55, 135-169 (1970).
171. P. B. Moore, "A Crystal-Chemical Basis for Short Transition Series Orthophosphate and Orthoarsenate Parageneses," Neues Jahrb. Mineral. Monatsch., 39-44, 1970.
172. P. B. Moore, "The Fe_3^{+2} $(H_2O)_n(PO_4)_2$ Homologous Series; Crystal-Chemical Relationships and Oxidized Equivalent," Am. Mineral., 56, 1-7, 1971.
173. C. V. Cole and M. L. Jackson, "Solubility Equilibrium Constant of Dihydroxy Aluminum Dihydrogen Phosphate Relating to a Mechanism of Phosphate Fixation in Soils," Soil. Sci. Am. Proc., 15, 84-89 (1951).
174. J. O. Nriagu, "Stability of Strengite," Am. J. Sci. (in press), 1972.
175. D. D. Wagman, W. H. Evans, V. B. Parker, I. Halow, S. M. Bailey, and R. H. Schumm, "Selected Values of Chemical Thermodynamic Properties," U.S. Nat. Bur. Std. Tech. Notes 270-1 (1965), 270-3 (1968), 270-4 (1969).
176. P. H. Hsu and D. A. Rennie, "Reactions of Phosphate in Aluminum Systems. I and II," Can. J. Soil Sci., 42, 197-221 (1962).
177. J. Shapiro, W. T. Edmundson, and D. E. Allison, "Changes in the Chemical Composition of Sediments

of Lake Washington," Limnol. Oceanog., 16, 437-452 (1971).
178. W. L. Lindsay, and E. C. Moreno, "Phosphate Phase Equilibria in Soils," Soil Sci. Soc. Am. Proc., 24, 177-182 (1960).
179. W. Stumm, and J. J. Morgan, "Aquatic Chemistry," Wiley, New York, 1970, 583 pages.
180. P. A. Burns, and M. Salomon, "Phosphate Adsorption by Kaolin in Saline Environments," Proc. Nat. Shellfish. Assoc., 59, 121-125 (1969).
181. J. I. Bedwell, W. B. Kepson, and G. L. Toms, "The Interaction of Kaolinite with Polyphosphate and Polyacrylate in Aqueous Solutions--Some Preliminary Results," Clay Miner., 8, 445-460 (1970).
182. K. R. Kar, "Radioactive Tracer Study of the Adsorption of Phosphate Ions by Aluminum Oxide," J. Sci. Ind. Res., 17B, 175-178 (1958).
183. W. C. Yee, "Selected Removal of Mixed Phosphates by Activated Alumina," J. Am. Water Works Assoc., 58, 239-247 (1966).
184. G. A. Parks, "Aqueous Surface Chemistry of Oxides and Complex Oxide Minerals," in "Equilibrium Concepts in Natural Water Systems," W. Stumm, ed., American Chemical Society Spec. Publ. 67, pp. 121-160.

Chapter III

THE CARBON CYCLE IN AQUATIC ECOSYSTEMS

Pat C. Kerr, D. L. Brockway, Doris F. Paris, and
W. M. Sanders III

National Pollutants Fate Research Program,
Southeast Water Laboratory, Environmental
Protection Agency, College Station Road, Athens,
Georgia

The cycling process is similar for all nutrients in that each cycle involves the incorporation of inorganic materials into organic material and the subsequent mineralization of this organic material to the original inorganic species. The same molecular species can be involved in both chemical and biochemical reactions, so biological availability of any nutrient is determined by the relative rates of these competitive reactions.

Autotrophic organisms utilize CO_2 and/or HCO_3^- for a carbon source, so the availability of these forms of carbon is essential to the reactions resulting in the conversion of light to usable chemical energy. Since this energy transfer is essential to the functioning of the entire biota, any reactions affecting the amount of available carbon dioxide can affect the biology of all ecosystems.

Figure 1 is a schematic representation of the carbon cycle showing the various components of the system. Carbon dioxide is readily soluble in water, but the amount of the gas absorbed from any source is dependent on the partial pressure of carbon dioxide and the pH of the water.

In general, there are four sources of inorganic carbon in aquatic ecosystems: atmosphere, carbonates, alloch-

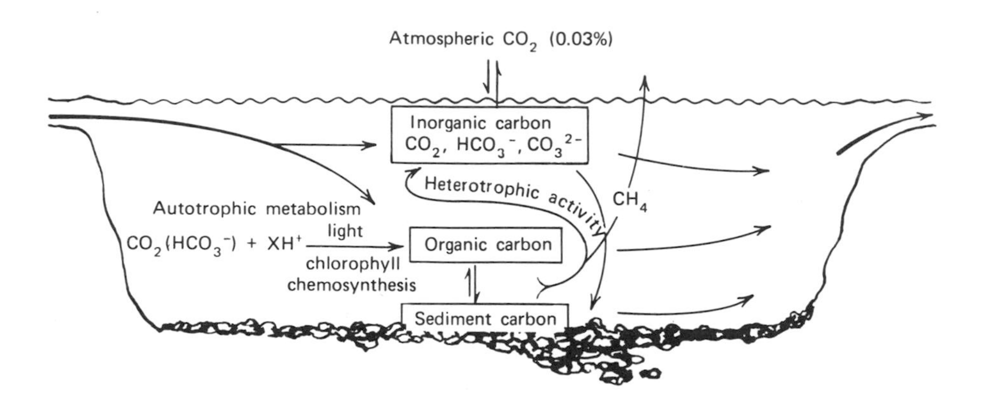

Fig. 1. Schematic representation of the carbon cycle.

thanous inorganic carbon, and biological cycling of autochthonous and allochthonous materials. The relative importance of these sources will vary with place and with time. Carbon dioxide, HCO_3^-, or both, are incorporated into organic carbon by autotrophic organisms. This autochthonous organic carbon, plus any allochthonous organic material, is available as a pool of suspended organic carbon which tends to sediment. Aerobic and anaerobic heterotrophic decomposition of both suspended and sedimented organic material results in the production of carbon dioxide, which is then available as a carbon source for autotrophs. Loss of inorganic or organic carbon from the system through flow or loss to the atmosphere results in a diminution of the amount of material that can be cycled in the system. Since, in any given system, the atmospheric and carbonate sources are considered to be constant, the rate of biological cycling of inorganic carbon represents the fertility of that system. Nitrogen, phosphorus, and other nutrients cycle concomitantly with the carbon; thus the decomposition of organic material results in increased availability of all nutrients. The relative rates at which the various nutrients cycle are the determinants of nutrient regulation of biological activity.

Birge and Juday (1) published a classic description of the general dynamics of the carbon cycle in soft- and hard-water lakes in Wisconsin. These authors discussed the various sources of specific molecular forms of carbon and the roles of algae and bacteria in removing and regenerating inorganic carbon.

> The free carbon dioxid [sic] of a water, however, will be affected by the presence of chlorophyl-bearing organisms when there is enough light for photosynthesis, and also, by the presence of decaying organic matter.*

Ohle (2) who has also discussed the carbon cycle in lakes, emphasized the role of anaerobic bacteria. Ohle (2) and Conger (3) stressed the importance of bubbles produced by anaerobes both as sources of carbon and as vehicles for the transport of carbon and other nutrients through the water column. Perhaps Ohle's (2,4) greatest contribution to aquatic ecology is his emphasis on the study of lake metabolism and "internal fertilization."

THE CARBON CYCLE

Fixation of Inorganic Carbon to Organic Carbon

Biological nutrient cycling results from the opposite but closely linked activities of autotrophic and heterotrophic organisms. Green plant photosynthesis is generally the major source of organic production in aquatic ecosystems (Table 1), but there are a few data that indicate rather large amounts of bacterial fixation. Culver and Brunskill (7) estimated that 83% of annual primary production in Greens Lake (Fayetteville, N. Y.) was due to photosynthetic sulfide-oxidizing bacteria in the chemocline at the 18- to 20-m depth. Takahashi and Ichimura (8) reported that in some Japanese lakes rich in hydrogen sulfide, photosynthetic sulfur bacteria produced from 9 to 25% of the total organic material. In lakes poor in hydrogen sulfide, 3 to 5% of the total annual organic production was by photosynthetic bacteria. Kuznetsov (9) and Ohle (2) both considered fixation of carbon dioxide by chemosynthetic bacteria to be insignificant, but no data were given by either author.

Heterotrophs can also fix carbon dioxide into organic carbon. Kuznetsov (9) reported that 6% of the total carbon of some heterotrophic bacteria comes from carbon dioxide fixation and that, in deep oligotrophic reservoirs,

---

*From E. A. Birge and C. Juday, "The Inland Lakes of Wisconsin. The Dissolved Gases of the Water and Their Biological Significance." Wisconsin Geological and Natural History Survey, Bull. XXII, pp. 64-92, 1911. Reprinted with permission of the Wisconsin Geological and Natural History Survey.

Table 1. A Comparison of Primary Production by Different Communities in Various Aquatic Ecosystems[a]

Type of community	Production g $C/(m^2)(yr)$	Method of measurement	Reference[b]
Marion Lake			
Epibenthic	40	Oxygen	Present study
Phytoplankton	8	^{14}C	Efford (1967)
Macrophytes	18	^{14}C	Davies (MS, 1968)
Borax Lake			
Periphyton	267	^{14}C	
Phytoplankton	91	^{14}C	Wetzel (1964)
Macrophytes	28	^{14}C	
Ocean phytoplankton	80		Westlake (1963) review of published data on plant productivity
Lake phytoplankton	80		
Coastal phytoplankton	120		
Submerged macrophytes	240		
Epibenthic algae, Georgia salt marsh	200	Oxygen	Pomeroy (1959)
Epibenthic Intertidal sandflat	143-226	Oxygen	Pamatmat (1968)
Intertidal marine sediments	115-178[c]	^{14}C	Grontved (1962)

[a]From "Epibenthic Algal Production and Community Respiration in the Sediments of Marion Lake," by Barry T. Hargrave, in Journal of the Fisheries Research Board of Canada, Vol. 26, p. 2023, 1969. Reprinted with permission of the Fisheries Research Board of Canada.

[b]Reference dates refer to publications cited in the paper by Hargrave (5).

[c]Re-calculated by Pamatmat (6).

heterotrophic carbon dioxide assimilation accounted for 24% of the total production of organic matter. Even though the organisms cannot utilize the gas as a sole carbon source, it has been demonstrated to be essential for the growth of many heterotrophic bacteria (10,11).

The amount of available carbon dioxide (and HCO_3^-) in an environment can act as a determinant for both species selection and production of biomass. Carbon dioxide is probably the most generally utilized carbon source for plants, but some species are reported to utilize HCO_3^-. Krauss (12) discussed some of the literature relative to this subject, including Osterlind's observation that apparently the relative uptake of CO_2 or HCO_3^- is dependent on the age of cells. Bristow (13) reported that the higher aquatic plants *Ranunculus flabellaris* and *Myriophyllum brasilense* can utilize only carbon dioxide. Conversely, Martin, Bradford, and Kennedy (14) reported that the HCO_3^- content of Pickwick Reservoir was important in regulating the growth of *Najas sp*. The availability of CO_2 and/or HCO_3^- is believed to regulate species distribution (15,16) and morphogenic responses (13,17), as well as biomass. Hatcher and Schmidt (18) reported that HCO_3^- stimulates nitrification by *Aspergillus flavus*.

Gladstone et al. (10) reviewed the literature and described some experiments that reveal the inhibitory effect of low carbon dioxide on growth of heterotrophic bacteria. Neither $NaHCO_3$ or $MgCO_3$ was utilized by the bacteria, nor was the dissociation of these salts, in systems sparged with CO_2-free air, enough to maintain adequate supplies of carbon dioxide to sustain bacterial division. The authors postulated that inhibition of growth by low carbon dioxide is responsible for the lag phase of bacterial growth. Lag phase of growth begins when the respiratory activity of the organisms has produced enough carbon dioxide to overcome the growth inhibition.

Hes (19) reported that carbon dioxide is essential for the normal functioning of the oxidation-reduction catalysts in heterotrophic cells. He found that complete removal of carbon dioxide leads to an inhibition of methylene blue reduction. When organisms like yeast and some fungi, which contain reserve carbohydrate, are placed in CO_2-free environments, there is no ultimate decrease of methylene blue reduction, but the reaction takes longer. *Blastocladia pringsheimii* and *Aqualinderella fermentans* require high concentrations of 5 to 20% of carbon dioxide for maximal growth and can tolerate 80 to 99% CO_2 (20). Oxygen availability has no real effect on the growth of these organisms (20).

Durbin (21) has shown the effect on growth of *Escherichia coli* of concentrations of carbon dioxide ranging from 0 to 90% (O_2 content was 20% except at 85 and 90% CO_2, where it was 15 and 10%, respectively). Inhibition ranged from approximately 10% at 20% CO_2 to 70% at 90% CO_2. This author also reported data showing growth of six fungi in concentrations of carbon dioxide ranging from 0 to 80%. The data indicate that organisms normally growing in lower soil depths are more tolerant to carbon dioxide than organisms usually confined to upper soil levels or aerial environments. Since the carbon dioxide concentration of subterranean environments normally ranges from 3 to 10%, the ability to tolerate higher concentrations allows these organisms to escape competition with many organisms that exist at or near the surface.

Krauss (12) listed references indicating that high carbon dioxide levels serve as growth inhibitors of some algae through toxic effects on photosynthesis.

Stout (22) regards carbon dioxide as the primary adverse factor limiting distribution of ciliates; he sees the ability to tolerate carbon dioxide as the distinctive feature of species associated with pollution. Tolerance to high concentrations of carbon dioxide may be just as important in regulating the growth of specific fungal, bacterial, protozoan, and algal populations in enriched situations.

The physiological mechanism(s) underlying the response to carbon dioxide are not known in detail, but evidence that is available indicates that regulation is at the enzymatic level. Nelson et al. (23) report that carbonic anhydrase activity in *Chlamydomonas reinhardtii* was regulated by the amount of carbon dioxide in the environment. Levels of the enzymes were twentyfold higher on both a protein and a chlorophyll basis when algae were grown on 0.03% CO_2 as compared with algae grown on 1% CO_2. If this enzyme is functional in carbon dioxide uptake, increased activity at lower carbon dioxide concentrations could allow efficient uptake of carbon from a dilute source. If complete repression were possible at high concentrations, the inhibitory effect of carbon dioxide could also be explained.

Nelson and Tolbert (24) have also demonstrated that the amount of carbon dioxide in the environment regulates the secretion of glycolate by *C. reinhardtii*. Excess carbon dioxide represses the synthesis of glycolate: 2,6-dichlorophenolindophenol oxidoreductase and the alga secretes glycolate. Removal of excess carbon dioxide by transferring the cells to air derepressed the level of this enzyme and no glycolate was released from the cells even though it was being formed by the alga. Thus *C. reinhardtii* appears to be capable of conserving carbon--that is, increasing the possible enzymatic sites for uptake of inorganic carbon and decreasing secretion of organic carbon--when placed under conditions where little carbon dioxide is available.

Conversion of Organic Carbon to Inorganic Carbon

Organic material secreted by autotrophic organisms, organics secreted and excreted by heterotrophic organisms, and cellular constituents of all organisms at death constitute the carbon and energy source for the heterotrophic populations. Secretion of organic materials by algae during photosynthesis has been demonstrated many times under both laboratory and field conditions, since the secretion of glycolate was first described by Tolbert and Zill (25). The amount of organic material secreted by algae differs from alga to alga and from environment to environment, but some indication of the range is given by Hellebust (26). He reported that most of the 22 algae studied excrete 3 to 6% of their photoassimilated carbon during logarithmic growth; a few species excreted as much as 10 to 25% under the same conditions. Apparently-healthy populations of natural phytoplankton excreted 4 to 16% of their photoassimilated carbon; 17 to 28% was excreted by a sample taken at the end of a diatom bloom.

Sieburth and Jensen (27) reported that total exudation by a littoral marine alga, _Fucus vesiculosus_, accounts for some 40% of the net carbon fixed. Extracellular organic matter production of _Laminaria_ appears to be some 30% of total carbon fixed. Since the standing crop of these algae is 1000 to 2000 g C/m^2 and they can fix up to 20 g $C/(m^2)$ (day), the extracellular organic production must approach 7 to 14 g $C/(m^2)$(day). Higher aquatic plants and benthic organisms are responsible for much of the primary production of certain aquatic ecosystems and presumably excrete organic compounds, although few quantitative data are available.

Allen (28) has demonstrated release of ^{14}C-labeled compounds by both planktonic algae and macrophytes in a New Hampshire pond. Hargrave (5) compiled data on primary production by different communities in various aquatic ecosystems, and his figures are reproduced in Table 1. Kajak et al. (29) reported comparable data from Polish lakes.

Secretion and excretion of organic phosphorus and nitrogen compounds by various protozoans, rotifers, and crustaceans have been documented (30-35). Although the composition and amount of organic compounds excreted were variable, these animals also contributed organic material to aquatic ecosystems.

The organic constituents of all organisms become available for biological utilization when the organisms die. Rapid release of water-soluble compounds follows rupture of cells [Vinogradov (36) states that approximately 50% of nutrient-containing compounds in freshwater algae are water soluble]. This autolysis and solution accomplishes 25 to 75% of the nutrient regeneration that occurs in the presence of microorganisms (35). Data reported by Golterman (37) indicate that 90 to 95% of the phosphorus incorporated into diatoms is released within 2 to 3 days. Only the nitrogen and phosphorus incorporated into nucleic acid and protein are not readily released.

Antia et al. (38), Grill and Richards (39), and Waksman et al. (40) also report data indicating rapid release of organic phosphorus from algae and its subsequent regeneration into inorganic phosphate. Measurements by Renn (41) indicate rapid return (2 to 3 days) of phosphorus to the water by lysis of bacteria. In all these cases, only regeneration of nitrogen and phosphorus was determined; there were no direct measurements of the amount of carbon returned to the system. Phosphorus cycled much faster than nitrogen in every case.

Bacterial uptake of glycolic acid, an important excretory product of algae, has been demonstrated by Wright (42). In a Massachusetts lake, glycolate was rapidly and totally removed from the water in the epilimnion. Wright's data on this lake agree with those of Wright and Hobbie (43) and Wetzel (44): namely, that at depths below 10 m, glucose and acetate uptake increased and uptake of glycolate could not be detected. Handa (45) reported that carbohydrate decomposes more rapidly than proteins and amino acids in the marine euphotic zone.

Allen's data (28,46) indicate rapid removal of dissolved organic material by bacteria. Maximum rates of uptake of acetate were 10 to 13 times greater than those of glucose. Changes in maximum velocities of substrate uptake are directly proportional to fluctuations in total bacterial numbers.

Kuznetsov (9) and Ohle (47) postulated that 90 to 98% of the organic constituents of algal cells are decomposed in the epilimnion. Kajak et al. (29) found that 63% of the total primary production in several Polish lakes was decomposed in the epilimnion. Only 30% of the total primary production reached the hypolimnion as seston.

Reports indicate that bacteria (48, Abernathy, personal communication) and fungi (49-51) play an important role in the decomposition of organic seston. Bacteria invade algal fragments and other detritus as they sink through the epilimnion and hypolimnion of Lake Hartwell, South Carolina (Abernathy, personal communication). Measurements indicate bacterial numbers in the range of 10^{10} cells per gram dry weight of detritus; 80% of the oxygen demand of the detritus of this lake is due to biological oxygen demand. Wojtalik (48) has demonstrated bacterial invasion of detritus in Cherokee Reservoir and reports that his data are in basic agreement with Hutchinson's computation showing a direct relation between organic seston and oxygen deficit. Triska (51) demonstrated that aquatic hyphomycetes play a significant role in the disappearance of leaf litter from a woodland stream.

Suspended bacteria, attached bacteria, and fungi are important in aquatic ecosystems as food sources as well as in the regeneration of inorganic nutrients. Butterfield (52) postulated that bacteria act as accumulators or concentrators of nutrients from dilute food sources, since protozoa could grow in dilute medium only when bacteria were present. Fredeen (53) proved that bacterial biomass plays a roll in the growth of black fly larvae, and McConnell (54) produced similar results for tadpoles. Fungi can increase the protein content of detritus eaten by omnivorous and herbivorous stream animals.

Biological nutrient cycling in sediments is difficult to study, and to date few quantitative data are available. Hargrave (5) and Pamatmat (6) extensively studied the benthic metabolism of a lake and a tidal salt flat, respectively, and pointed out many of the difficulties encountered in such studies. Indirect evidence indicates that large numbers of bacteria and/or an apparent accumulation of inorganic nutrients occur in or immediately over sediments of lakes and ponds (1,2,55-57). The data of Kajak et al. (29) indicate that in Polish lakes decomposition in the hypolimnion equals that in bottom deposits; 19 and 18% of the total primary production was decomposed in the hypolimnion and sediments, respectively. Data collected by Abernathy (personal communication) indicate that 80% of the sediment oxygen demand of Lake Hartwell, South Carolina, is biochemical oxygen demand (BOD). Comparable data from other lakes and streams will be necessary in order to ascertain the relative contribution of chemical and biological nutrient cycling in sediments.

The decomposition of both autochthonous and allochthonous organic material occurs by the same processes in the same places. Allochthonous organic material, particularly deciduous leaves, constitutes a significant proportion of the total energy flow of many aquatic ecosystems (2,58-60). Ohle (4) pointed out both direct and indirect effects on lake metabolism following addition of sewage.

In summary:

1. The molecular forms of carbon are constantly changing in aquatic ecosystems.
2. The rate(s) of these changes determine the instan-

taneous availability of specific molecular forms of carbon and thus regulate species selection and growth of those organisms dependent on any given molecular form of carbon.

3. Autochthonous organic carbon synthesized by autotrophs and chemosynthetic bacteria, carbon dioxide fixed by heterotrophs, and allochthonous organic carbon serve as a carbon and energy source for heterotrophs.

4. Suspended heterotrophs can utilize secreted and/or excreted organics as well as organic material resulting from the autolysis of cells.

5. Heterotrophs can invade organic particles as they sediment and oxidize the components of the particles to carbon dioxide.

6. Data indicate that from 63 to 98% of the total primary production of some lakes is oxidized in the epilimnion.

7. Approximately the same amount of heterotrophic oxidization (17 to 19%) occurred in the hypolimnion as in the sediments of some Polish lakes.

8. Carbon dioxide resulting from the biogenic oxidation of organic material plus allochthonous inputs of inorganic carbon from the watershed and atmosphere serve as the carbon source for autotrophs and chemosynthetic bacteria.

FACTORS AFFECTING NUTRIENT CYCLING

Nutritional

Biological cycling of any nutrient is dependent on adequate supplies of each essential nutrient. Since many classes of organisms with very different metabolic rates are involved in nutrient cycling, an infinite number of possible specific nutrient regulations can occur.

Our working hypothesis has been that because of their rapid metabolic rates, the heterotrophic bacteria and protozoa would require large amounts of nutrients and would be the populations most directly affected by increased nutrient availability (56,57). As a result of this increased oxidative metabolism, more carbon dioxide and HCO_3^- would be available to support algal growth.

Observations and data presented by fish culturists indicate the importance of this concept of biological interactions, or nutrient cycling, in practical fish management. Neess (61) emphasized the importance of the bottom soil, organic material (plant debris and/or added organics), and an environment suitable for heterotrophic activity; all these factors would allow rapid cycling of nutrients in fishponds and thereby increase productivity. Swingle (62) postulated that algal growth was regulated by the availability of carbon dioxide to algae. Satomi (63) advanced the same concept for Japanese fishponds.

This interaction between heterotrophic and autotrophic populations has been demonstrated in sewage oxidation ponds (64), in the laboratory (56,65), and in farm fishponds and lakes (56,57); it has been discussed by Kuentzel (66) in a literature review, as well.

The basic concept is applicable to increasing fertility in any body of water. So long as nutrients are biologically assimilable and no toxic materials are present, introduction of nutrients into water will increase nutrient cycling and productivity of the system. Introduction of wastes containing both organic and inorganic nutrients should result in rapid increases of biological activity and concomitant increases in the rate and extent of nutrient cycling. Such increases have been measured in the Flat Creek area of Lake Lanier, Georgia (56).

Data in Figures 2 and 3 show the diel (24 hr) fluctuations in free carbon dioxide and the numbers of bacteria that can be cultured from the Flat Creek area, which is about 2 miles below a secondary treatment plant in Gainesville, Georgia. Data in Figures 4 and 5 (57) reveal increased numbers of bacteria and increased availability of carbon dioxide following the fertilization of a Georgia farm pond with reagent-grade ammonium phosphate, ammonium nitrate, and potassium chloride. The second set of data is included here only to show that addition of organic and inorganic nutrients alone and in combination affects the bacterial population as well as the algal population and that extreme fluctuation in available carbon dioxide occurs in each system.

Complete discussion of the material presented in Figures 2 to 5 and other laboratory and field data pertinent to the concept under discussion can be found in publications by Kerr and co-workers (56,57). The bacterial response to added nitrogen and phosphorus preceded the algal response by some 36 to 48 hr. Antia et al. (38) also observed that during experimentation with plastic spheres, respiration (heterotrophic activity) initially exceeded photosynthesis for approximately 10 days.

Different species of organisms require different absolute amounts of each essential nutrient, so population changes frequently accompany changes in nutrient availability. The amount of any nutrient passing through an organism during its life span is much greater than the amount of that nutrient in the organism at any finite period of time. Secretory, excretory, and respiratory

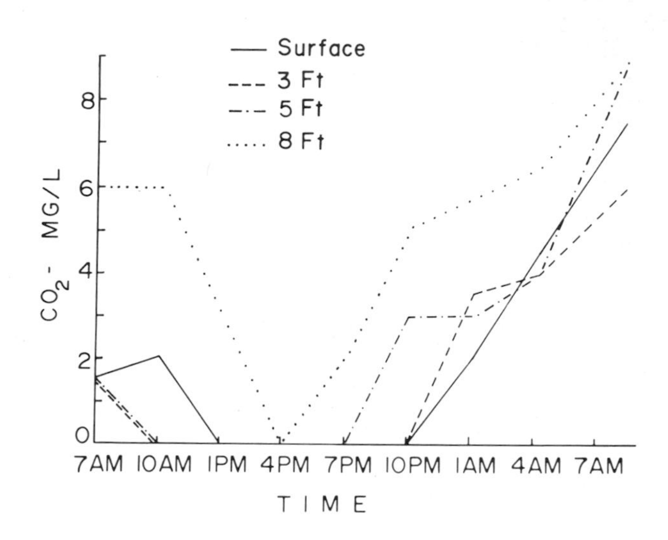

Fig. 2. Diel fluctuation of CO_2 in water from the Flat Creek area of Lake Lanier, Georgia, July 30 and 31, 1969.

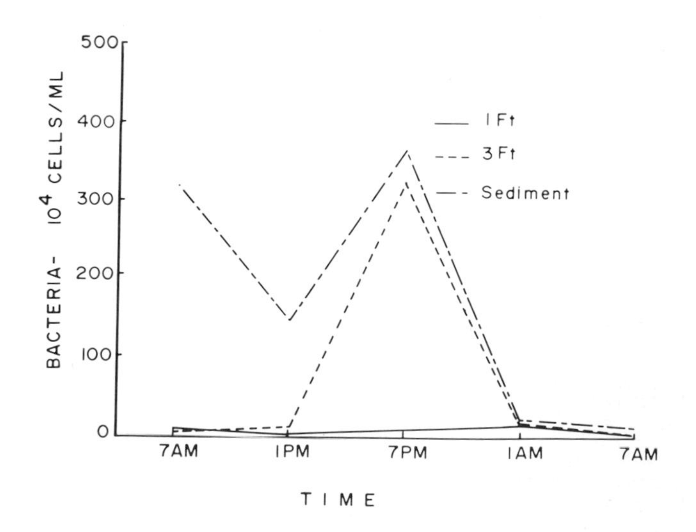

Fig. 3. Diel fluctuation of bacteria in the water and sediments of the Flat Creek area of Lake Lanier, Georgia, July 30 and 31, 1969.

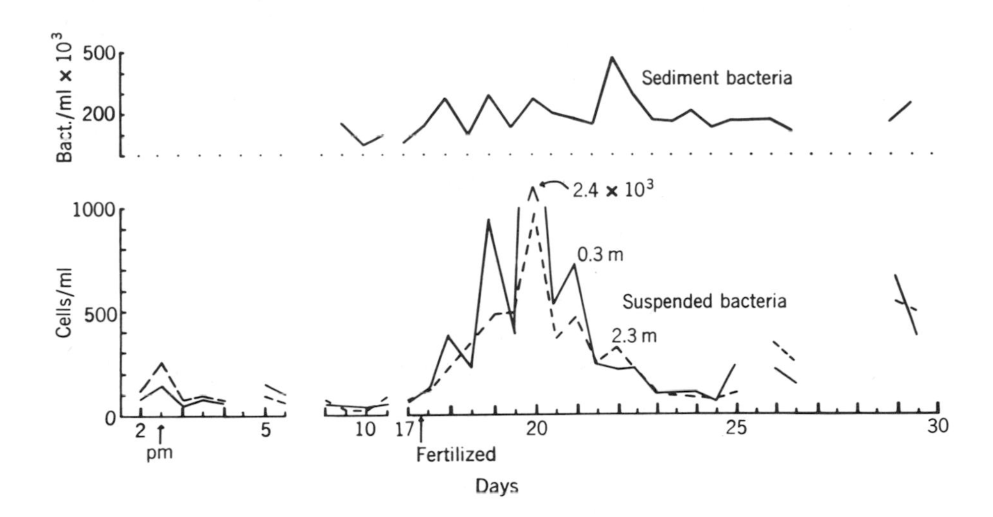

Fig. 4. Bacteria in Shriner's Pond before and after fertilization with reagent-grade nitrogen, phosphorus, and potassium salts. Suspended bacteria were enumerated from the 0.3- and 2.3-m depths. Samples taken at 6 am and 3 pm during June 1970 are plotted by date on the abscissa; only consecutive samples are connected. Reprinted from Ref. 57 with permission of the American Society of Limnology and Oceanography.

losses, as applicable, must be considered in arriving at a true nutrient requirement for any organism. Since these losses are not included in computing the elemental ratios of carbon to nitrogen to phosphorus in organisms, the ratios obtained are very unreliable indicators of the amount of carbon, nitrogen, and phosphorus required for the growth of organisms. In fact, the ratios would be predictive only if the sum of relative rates of uptake, assimilation, and loss were in the same mathematical proportion for each nutrient. Since algae, for instance, can store luxury amounts of phosphorus and continue to divide in the absence of an external source of phosphorus (56,67-71), these ratios obviously will not apply under the conditions named or any other similar conditions in which an organism is either storing nutrients or utilizing stored nutrients to grow.

Nutrient requirements for different kinds of organisms can be filled only by certain specific molecular species. Since algae generally can utilize only CO_2 and/or HCO_3^-

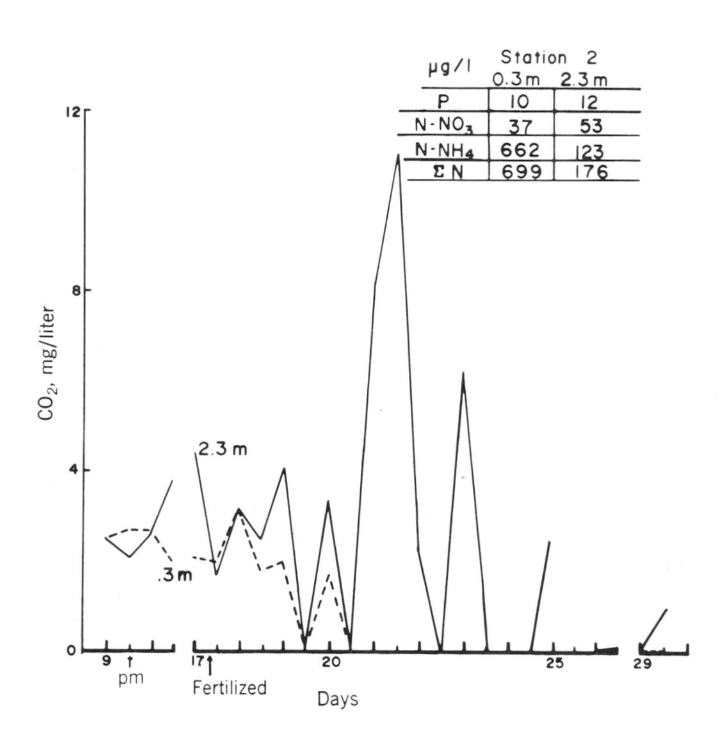

Fig. 5. Carbon dioxide at the 0.3- and 2.3-m depths of Shriner's Pond before and after fertilization with reagent-grade nitrogen, phosphorus, and potassium salts. Samples taken at 6 am and 3 pm during June 1970 are plotted by date on the abscissa; only consecutive samples are connected. Reprinted from Ref. 57 with permission of the American Society of Limnology and Oceanography.

for growth (1,64,72-74), only the amount of these forms of carbon is important to their immediate growth. However, the total amount of carbon in the system and the input, by any means, into the system are important as storage of potential CO_2, HCO_3^-, or both.

It seems that loss of carbon dioxide from the water to the atmosphere is rarely considered in discussions of inorganic carbon availability. Teal (60) has calculated diffusion fluxes for carbon dioxide; his data indicate, as do those of others (1,56,75), that biological production of carbon dioxide can result in diffusion of it from the water into the atmosphere. Teal's (60) data reveal that the pCO_2 of water in Woods Hole varies with seasonal differences in biological activity. The pCO_2 is high during the summer (nearly twice saturation) when the bacteria and other consumers are respiring the organic material produced during the spring and summer. Biological activity decreases in the fall and the concurrent temperature drop and autumnal storms reduce the pCO_2 toward air equilibrium. During the winter algae bloom, pCO_2 decreases to as little as 150 ppm; warming in spring and decomposition of the winter bloom again raise the pCO_2 to summer levels.

Overall, Teal's data indicate a flux of carbon dioxide from the water of Woods Hole to the atmosphere during spring, summer, and early fall. From data reported by Kerr et al., we learn that diel fluctuations of carbon dioxide in waters of Lake Lanier (56) and Shriner's Pond (57) can result in loss of carbon dioxide to the atmosphere during the night and early morning, whereas these waters were unsaturated with carbon dioxide during the late morning and afternoon. Data reported by Takahashi (75) indicate that biological activity can affect the direction of the flux of carbon dioxide into and from the Atlantic Ocean. These data also suggest that diffusion of atmospheric carbon dioxide into water is very slow. This conclusion is further substantiated by the data of Takahashi (75) showing that the greater part of the Atlantic Ocean is undersaturated with respect to atmospheric carbon dioxide. Thus nutrient regulation is determined by the rate of nutrient cycling, not by instantaneous water concentrations.

An organism must be in intimate contant with the specific molecular species before uptake can occur. Molecular and eddy diffusion, convection, and other physical parameters that result in collision of nutrient and organism may well constitute the rate-limiting step in nutrient regulation. Quantitative data indicating that mixing is the rate-limiting step in carbon fixation by algae have been discussed by Krauss (12), Myers (76), and James (77). Button (78) discussed data indicating that mixing is the rate-limiting step in growth of yeast.

Mixing could be important in regulating biological activity in natural systems, since temporal and spatial differences in predominantly autotrophic and heterotrophic activity seem to exist. Since algae are dependent on light as an energy source, they are more metabolically active during daylight hours. The diel periodicity (population maxima during light) in abundance of algae is well known. Unpublished data collected by the authors indicate that bacterial population densities are maximal during the night. Data reported by Birge and Juday (1), Barlow and Bishop (30), and Johannes et al. (79) reveal that protozoa are more numerous in the epilimnion during darkness. These relative population maxima correlate with the metabolic measurements of CO_2, O_2, and pH reported by Verduin (80), Kerr et al. (56,57), and Beyers (81) in which CO_2 is utilized and O_2 generated during the light, and O_2 is utilized and CO_2 generated during the dark.

Seasonal fluctuation in bacteria correlated with supplies of organic material has been discussed by Teal (60) for Woods Hole waters.

Lyford and Phinney (82) described the seasonal photosynthetic-respiratory changes occurring in an estuarine impoundment.

Data from several Wisconsin lakes show peaks of bacterial density in mid-February, mid-August, and early October (55). These bacterial peaks follow the classic winter, spring, and summer algal blooms.

Data from a Texas estuary indicate that the mean (1957 to 1960) respiratory activity exceeded the mean photosynthetic activity from mid-December to mid-May (83). Peak respiratory activity occurred from mid-April to mid-May; peak photosynthetic activity occurred from mid-May to mid-June. Hargrave's (5) data indicate that the bacterial component of the total community respiration was higher in winter than in summer.

Odum (84) reported that certain areas of flowing water are predominantly areas of respiratory activity; others are predominantly areas of photosynthetic activity. Downstream distribution of inorganic nutrients formed in the area of respiratory activity results in downstream areas of photosynthetic activity. Flow of organic materials from the areas of photosynthetic activity results in downstream areas of respiratory activity, and so on.

In lakes and ponds, similar areas of biological activities have been observed from surface to sediment. Photosynthetic activity predominates in the upper, photic zone and respiratory activity predominates in the nonphotic zone, in the sediments, or in both areas (1,4,55-57).

Nonnutritional

Biological activity is regulated by genetic, environmental, and other chemical factors as well as by the availability of nutrients. Endogenous rhythms must be considered in studies of biological activity and regulation. Pamatmat (6), Beyers (81), and Kerr et al. (57) discussed the possible effects of these cycles in natural systems. Light quality and quantity, temperature, and pH are known to regulate biological activity. Oxygen is known to inhibit photosynthesis (85) and to inhibit cytochrome formation and growth of some aerobic bacteria (11; Kennedy, unpublished); carbon dioxide is required in small amounts for the growth of some heterotrophic bacteria (10, 11). These types of biological regulation override nutrient regulation in the classic sense. Therefore, each type of biological regulation--physical, chemical, genetic, and nutritional--is critical and must be considered in attempts to manage our water resources successfully.

Just as any of the essential nutrients can regulate biological activity of each of the various types of organisms at different times and places, any of the four major types of biological regulation can regulate biological activity of each of the various types of organisms at different times and places. In other words, there is no one universal biological regulating agent.

Fortunately, the gene pool is so diverse that some organisms could grow under most any condition. Population selection follows changes in the four types of biological regulation, and it behooves the water manager to ascertain the combination of conditions essential for the maintenance of the mixed populations considered most desirable and appropriate.

OTHER OBSERVATIONS

In this chapter we have emphasized biological carbon cycling; however, data indicate one very interesting interaction between organic material and inorganic carbonates that effectively prevents these carbonate particles from participating in the carbonate system. Chave (86) and Wetzel (87) have noted that suspended particles of carbonate in seawater and marl lakes, respectively, can acquire a coating of organic material and cease to be reactive. Wetzel's data indicate that the organic material is no longer available to bacteria. Thus, this interaction results in removal of both organic and inorganic carbon from the carbon cycle.

Another potentially important interaction between biochemical and inorganic chemical processes in aquatic ecosystems revolves around pH effect on the carbonate system. Algae probably cannot directly utilize carbonate-carbon; so at higher pH values the rate and extent of chemical change to HCO_3^- and CO_2 could regulate the amount of inorganic carbon available for algal utilization. Since organisms produce carbon dioxide in respiration and secrete organic acids, the pH of the water tends to be lowered. Concomitant with a lower pH (9 to 7), more of the total carbonate could exist as HCO_3^- and CO_2. As the pH further decreases, more of the total inorganic carbon can exist as carbon dioxide, and when the concentration of carbon dioxide in the surface water exceeds the equilibrium concentration with the atmosphere, the gas can be lost from the water to the atmosphere.

Carbon is the most abundant element in all known organisms. Carbon compounds are also the energy source for the heterotrophic organisms. Availability and rate of supply of specific molecular forms of carbon (CO_2, HCO_3^-, CO_3^{2-}, and organic) can regulate the extent and rate of biological activity. Natural systems are dynamic rather than equilibrium or steady-state systems. Therefore, in order to comprehend fully the complexities of aquatic ecosystems, it is necessary to ascertain the rates of nutrient cyclings and the genetic, environmental, and chemical factors that affect the rate of these cycles and thus nutrient availability.

REFERENCES

1. E. A. Birge and C. Juday, "The Inland Lakes of Wisconsin. The Dissolved Gases of the Water and Their Biological Significance," Wisc. Geol. Nat. His. Survey Bull., 22, 64-92 (1911).
2. W. Ohle, "The Dynamics of Metabolism of Lakes as a Function of Gas Ebullition from their Mud," Wasser, 25, 127-149 (1958).
3. P. S. Conger, "Ebullition of Gases from Marsh and Lake Waters," State of Maryland Board of Natural Resources Publ. 59, 3-42 (1943).
4. W. Ohle, "Lakes as Victims of the Sewage Calamity," in "Report of the Sewage," Technology Union, Inc. F. Oldenbourg, Munich, 1956, pp. 268-276.
5. B. T. Hargrave, "Epibenthic Algal Production and Community Respiration in Sediments of Marion Lake," J. Fish. Res. Bd. Can., 26, 2003-2026 (1969).
6. M. M. Pamatmat, "Ecology and Metabolism of a Benthic Community on an Intertidal Sandflat," Int. Rev. Ges. Hydrobiol., 53, 211-298 (1968).
7. D. A. Culver and G. J. Brunskill, "Fayetteville Green Lake, New York: V. Studies on Primary Production and Zooplankton in a Meromictic Marl Lake," Limnol. Oceanogr., 14, 862-873 (1969).

8. M. Takahashi and S. Ichimura, "Vertical Distribution and Organic Matter Production of Photosynthetic Sulfur Bacteria in Japanese Lakes," Limnol. Oceanogr., 13, 644-655 (1968).
9. S. I. Kuznetsov, "Recent Studies on the Role of Microorganisms in the Cycling of Substances in Lakes," Limnol. Oceanogr., 13, 211-224 (1968).
10. G. P. Gladstone, P. Fildes, and G. M. Richardson, "Carbon Dioxide as an Essential Factor in the Growth of Bacteria," Brit. J. Exp. Pathol., 16, 335-348 (1935).
11. J. W. T. Wimpenny, "Oxygen and Carbon Dioxide as Regulators of Microbial Growth and Metabolism," Symposium of Society for General Microbiology, Cambridge University Press, 1969, pp. 161-197.
12. R. W. Krauss, "Physiology of the Freshwater Algae," Ann. Rev. Plant Physiol., 9, 207-244 (1958).
13. J. M. Bristow, "The Effects of Carbon Dioxide on the Growth and Development of Amphibious Plants," Can. J. Bot., 47, 1803-1807 (1969).
14. J. B. Martin, Jr., B. N. Bradford, and H. G. Kennedy, "Factors Affecting the Growth of Najas in Pickwick Reservoir," National Fertilizer Development Center, TVA, Muscle Shoals, Ala., 1969, 58 pages.
15. D. L. King, "The Role of Carbon in Eutrophication," J. Water Pollut. Control Fed., 42, 2035-2051 (1970).
16. J. C. Wright and I. K. Mills, "Productivity Studies on the Madison River, Yellowstone National Park," Limnol. Oceanogr., 12, 568-577 (1967).
17. E. C. Cantino, "The Relationship between Biochemical and Morphological Differentiation in Nonfilamentous Aquatic Fungi," in "Microbial Reaction to Environment," 11th Symposium of the Society of General Microbiology, Cambridge, England, Cambridge University Press, 1961, pp. 243-271.
18. H. J. Hatcher and E. L. Schmidt, "Nitrification of Aspartate by Aspergillus flavus," Appl. Microbiol., 21, 181-186 (1971).
19. J. W. Hes, "Function of CO_2 in the Metabolism of Heterotrophic Cells," Nature, 141, 647 (1938).
20. A. A. Held, R. Emerson, M. S. Fuller, and F. H. Gleason, "Blastocladia and Aqualinderella: Fermentative Water Molds with High Carbon Dioxide Optima," Science, 165, 706-708 (1969).
21. R. D. Durbin, "Straight-line Function of Growth of Microorganisms at Toxic Levels of Carbon Dioxide," Science, 121, 734-735 (1955).
22. J. O. Stout, "Reaction of Ciliates to Environmental Factors," Ecology, 37, 178-191 (1956).

23. E. B. Nelson, A. Cenedella, and N. E. Tolbert, "Carbonic Anhydrase Levels in Chlamydomonas," Phytochem., 8, 2305-2306 (1969).
24. E. B. Nelson and N. E. Tolbert, "The Regulation of Glycolate Metabolism in Chlamydomonas reinhardtii," Biochim. Biophys. Acta, 184, 263-270 (1969).
25. N. E. Tolbert and L. P. Zill, "Excretion of Glycolic Acid by Algae during Photosynthesis," J. Biol. Chem., 222, 895-906 (1956).
26. J. A. Hellebust, "Excretion of Some Organic Compounds by Marine Phytoplankton," Limnol. Oceanogr., 10, 192-206 (1965).
27. J. McN. Sieburth and A. Jensen, "Production and Transformation of Extracellular Organic Matter from Littoral Marine Algae: A Resume," in "Symposium on Organic Matter in Natural Waters," D. W. Hood, ed. Institute of Marine Science, University of Alaska, College, 1970, pp. 203-244.
28. H. L. Allen, "Phytoplankton Photosynthesis, Micronutrient Interactions, and Inorganic Carbon Availability in a Soft-Water Vermont Lake," Limnol. Oceanogr., Special Symposium I, 63-80 (1972).
29. Z. Kajak, A. Hillbricht-Ilkowska, and E. Pieczynska, "Production in Several Mazurian Lakes," Preliminary papers for UNESCO-IBP Symposium on Productivity Problems of Fresh Waters, Kazimierz Dolny, Poland, 1, 173-189 (1970).
30. J. P. Barlow and J. W. Bishop, "Phosphate Regeneration by Zooplankton in Cayuga Lake," Limnol. Oceanogr., 10, R15-R25 (1965).
31. E. I. Butler, E. D. S. Corner, and S. M. Marshall, "On the Nutrition and Metabolism of Zooplankton. VI. Feeding Efficiency of Calanus in Terms of Nitrogen and Phosphorus," J. Mar. Biol. Assoc. (U.K.), 49, 977-1001 (1969).
32. B. T. Hargrave and G. H. Geen, "Phosphorus Excretion by Zooplankton," Limnol. Oceanogr., 13, 332-341 (1968).
33. R. E. Johannes, "Uptake and Release of Dissolved Organic Phosphorus by Representatives of a Coastal Marine Ecosystem," Limnol. Oceanogr., 9, 224-234 (1964).
34. R. E. Johannes, "Phosphorus Excretion and Body Size in Marine Animals: Microzooplankton and Nutrient Regeneration," Science, 146, 923-924 (1964).
35. R. E. Johannes, "Nutrient Regeneration in Lakes and Oceans," in "Advances in Microbiology of the Sea," Vol. I, M. R. Droop and E. J. Ferguson, Eds., 1968, pp. 203-213.

36. A. P. Vinogradov, "The Elementary Chemical Composition of Marine Organisms," Mem. Sears Found. Mar. Res., 2, 1-647 (1953).
37. H. L. Golterman, "Studies on the Cycle of Elements in Fresh Water," Acta Bot. Meer, 9, 1-58 (1960).
38. N. J. Antia, C. D. McAllister, T. R. Parsons, K. Stephens, and J. D. Strickland, "Further Measurements of Primary Production Using a Large-Volume Plastic Sphere," Limnol. Oceanogr., 8, 166-183 (1963).
39. E. V. Grill and F. A. Richards, "Nutrient Regeneration from Phytoplankton Decomposing in Sea Water," J. Mar. Res., 22, 51-69 (1964).
40. S. A. Waksman, J. L. Stokes, and M. R. Butler, "Relation of Bacteria to Diatoms in Sea Water," J. Mar. Biol. Assoc. (U.K.), 22, 359-373 (1937).
41. C. E. Renn, "Bacteria and the Phosphorus Cycle in the Sea," Biol. Bull. Biol. Lab. Woods Hole, 72, 190-195 (1937).
42. R. T. Wright, "Glycolic Acid Uptake by Planktonic Bacteria," in "Symposium on Organic Matter in Natural Waters," D. W. Hood, Ed. Institute of Marine Science, University of Alaska, College, 1970, pp. 521-536.
43. R. T. Wright and J. E. Hobbie, "The Use of Glucose and Acetate by Bacteria and Algae in Aquatic Ecosystems," Ecology, 47, 447-464 (1966).
44. R. G. Wetzel, "Dissolved Organic Compounds and their Utilization in Two Marl Lakes," J. Hung. Hydrol. Soc., 47, 298-303 (1967).
45. N. Handa, "Dissolved and Particulate Carbohydrates," in "Symposium on Organic Matter in Natural Waters," D. W. Hood, Ed. Institute of Marine Science, University of Alaska, College, 1970, pp. 129-152.
46. H. L. Allen, "Chemo-Organotrophic Utilization of Dissolved Organic Compounds by Planktonic Algae and Bacteria in a Pond," Int. Rev. Ges. Hydrobiol., 54, 1-33 (1969).
47. W. Ohle, "The Metabolism of Lakes as a Basis for a General Metabolic Dynamism in Water," Kiel. Meeresforsch., 18, 107-120 (1962).
48. T. A. Wojtalik, "Elements of a Program for Carbon-Oxygen Budgets in Reservoirs," in "TVA Activities Related to Study and Control of Eutrophication in the Tennessee Valley," National Fertilizer Development Center, Muscle Shoals, Ala., 1970, pp. 29-33.
49. N. K. Kaushik and H. B. N. Hynes, "Experimental Study on the Role of Autumn-Shed Leaves in Aquatic Environments," J. Ecol., 56, 229-243 (1968).
50. J. W. G. Lund, "Studies on _Asterionella formosa_ Hass II. Nutrient Depletion and the Spring Maximum,"

Ecology, 38, 15-35 (1950).
51. F. J. Triska, "Seasonal Distribution of Aquatic Hyphomycetes in Relation to the Disappearance of Leaf Litter from a Woodland Stream," University Microfilms, Inc. 71-8425, Ann Arbor, Mich., 1970, 189 pages.
52. C. T. Butterfield, "Experimental Studies on Natural Purification in Polluted Waters. III. A Note on the Relation between Food Concentration in Liquid Media and Bacterial Growth," Pub. Health Rep., 44, 2865-2872 (1929).
53. F. J. Fredeen, "Bacteria as Food for Blackfly Larvae Diptera: Simuliidae in Laboratory Cultures and in Natural Streams," Can. J. Zool., 42, 527-548 (1964).
54. W. J. McConnell, "Limnological Effects of Organic Extracts of Litter in a Southwestern Impoundment," Limnol. Oceanogr., 13, 343-349 (1968).
55. B. P. Domogalla, E. B. Fred, and W. H. Peterson, "Seasonal Variation in the Ammonia and Nitrate Content of Lake Waters," J. Am. Water Works Assoc., 15, 369-385 (1926).
56. P. C. Kerr, D. F. Paris, and D. L. Brockway, "The Interrelation of Carbon and Phosphorus in Regulating Heterotrophic and Autotrophic Populations in Aquatic Ecosystems," Water Pollution Control Research Series 16050 FGS, Government Printing Office, Washington, D.C., 1970, 53 pages.
57. P. C. Kerr, D. L. Brockway, D. F. Paris, and J. T. Barnett, Jr., "The Interrelation of Carbon and Phosphorus in Regulating Heterotrophic and Autotrophic Populations in an Aquatic Ecosystem, Shriner's Pond," Limnol. Oceanogr. Special Symposium I, 41-57 (1972).
58. G. W. Minshall, "Role of Allochthonous Detritus in the Trophic Structure of a Woodland Spring Brook Community," Ecology, 48, 139-149 (1967).
59. D. J. Nelson and D. C. Scott, "Role of Detritus in the Productivity of a Rock-Outcrop Community in a Piedmont Stream," Limnol. Oceanogr., 7, 396-413 (1962).
60. J. M. Teal, "Biological Production and Distribution of pCO_2 in Woods Hole Waters," in "Esturaries," G. H. Lauff, Ed. Publ. 83AAAS, Washington, D.C., 1967, pp. 336-340.
61. J. C. Neess, "Development and Status of Pond Fertilization in Central Europe," Trans. Am. Fish. Soc., 76, 335-358 (1946).
62. H. S. Swingle, "Experiments on Pond Fertilization," Bull. 264, Agricultural Experimental Station, Auburn, Ala., 1947.
63. Y. Satomi, "Physiological Significance of Carbon

Sources in Fertilized Fish Ponds," Bull. Fresh Water Fish. Res. Lab., 15, 99-104 (1966).
64. M. B. Allen, "General Features of Algal Growth in Sewage Oxidation Ponds," State Water Pollut. Control Bd. Publ. 13, Sacramento, Calif., 1955.
65. W. Lange, "Effect of Carbohydrates on the Symbiotic Growth of Planktonic Blue-Green Algae with Bacteria," Nature, 215, 1277-1278 (1967).
66. L. E. Kuentzel, "Bacteria, CO_2, and Algal Blooms," J. Water Pollut. Control Fed., 41, 1737-1747 (1969).
67. A. A. Al Kholy, "On the Assimilation of Phosphorus in Chlorella pyrenoidosa," Physiol. Plant, 9, 137-143 (1956).
68. J. C. Batterton and C. van Baalen, "Phosphorus and Phosphate Uptake in the Blue-Green Alga, Anacystis nidulans," Can. J. Microbiol., 14, 341-348 (1968).
69. E. D. Goldberg, T. J. Walker, and E. Whisenand, "Phosphate Utilization by Diatoms," Biol. Bull., 101, 274-284 (1951).
70. B. H. Ketchum, "The Development and Restoration of Deficiencies in the Phosphorus and Nitrogen Composition of Unicellular Plants," J. Cell. Comp. Physiol., 13, 373-381 (1939).
71. F. J. Mackereth, "Phosphorus Utilization by Asterionella formosa Hass," J. Exp. Bot., 4, 296-313 (1953).
72. J. A. Hellebust, "The Uptake and Utilization of Organic Substances by Marine Phytoplankters," in "Symposium on Organic Matter in Natural Waters," D. W. Hood, Ed. Institute of Marine Science, University of Alaska, College, 1970, pp. 225-256.
73. J. Pearce and N. G. Carr, "The Metabolism of Acetate by the Blue-Green Algae, Anabaena variabilis and Anacystis nidulans," J. Gen. Microbiol., 49, 301-313 (1967).
74. A. J. Smith, J. London, and R. Y. Stanier, "Biochemical Basis of Obligate Autotrophy in Blue-Green Algae and Thiobacilli," J. Bact., 94, 972-983 (1967).
75. T. Takahashi, "Carbon Dioxide in the Atmosphere and in Atlantic Ocean Water," J. Geophys. Res., 66, 477-494 (1961).
76. J. Myers, "The Growth of Chlorella pyrenoidosa under Various Culture Conditions," Plant Physiol., 19, 576-589 (1944).
77. W. O. James, "Experimental Researches on Vegetable Assimilation and Respiration. XIX. The Effect of Variations of Carbon Dioxide Supply upon the Rate of Assimilation of Submerged Water Plants," Proc. Roy. Soc., CIII-B, 1-42 (1928).
78. D. K. Button, "Some Factors Influencing Kinetics

Constants for Microbial Growth in Dilute Solution," in "Symposium on Organic Matter in Natural Waters," D. W. Hood, Ed. Institute of Marine Science, University of Alaska, College, 1970, pp. 537-547.

79. R. E. Johannes, S. L. Coles, and N. T. Kuenzel, "The Role of Zooplankton in the Nutrition of some Scleractinian Corals," Limnol. Oceanogr., 15, 579-586 (1970).
80. J. Verduin, "Phytoplankton Communities of Western Lake Erie and the CO_2 and O_2 Changes Associated with Them," Limnol. Oceanogr., 5, 372-380 (1960).
81. R. J. Beyers, "The Metabolism of Twelve Aquatic Laboratory Microecosystems," Ecol. Monogr., 33, 281-306 (1963).
82. J. H. Lyford, Jr. and H. K. Phinney, "Primary Productivity and Community Structure of an Estuarine Impoundment," Ecology, 49, 854-866 (1968).
83. H. T. Odum, "Environment, Power, and Society." Wiley, New York, 1971.
84. H. T. Odum, "Primary Production in Flowing Waters," Limnol. Oceanogr., 1, 102-117 (1956).
85. M. Gibbs, "The Inhibition of Photosynthesis by Oxygen," Am. Sci., 58, 634-640 (1970).
86. K. E. Chave, "Carbonate-Organic Interactions in Sea Water," in "Symposium on Organic Matter in Natural Waters," D. W. Hood, Ed. Institute of Marine Science, University of Alaska, College, 1970, pp. 373-386.
87. R. G. Wetzel, "The Role of Carbon in Hard-Water Lakes," Limnol. Oceanogr. Special Symposium I, 84-96 (1972).

Chapter IV

THE CHEMICAL ANALYSIS OF NUTRIENTS

P. D. Goulden

Water Quality Division, Inland Waters Branch
Department of the Environment
Ottawa, Canada

In considering the effects of nutrients in natural waters, it is obviously an advantage to know the concentrations of particular nutrients at particular points in space and time. The chemical analysis of water samples for nutrients is one way that this information is obtained. The Water Quality Division laboratories of the Inland Waters Branch of the Canadian Department of Fisheries and Forestry are engaged in a program of monitoring water quality across Canada through their National Water Quality Network. They also provide analytical services to other federal and provincial government departments. Because of these commitments to analyze a large number of water samples, the laboratories have placed emphasis on automated techniques for analysis and data handling, with methodology development aimed at lowering the limits of detection.

In this discussion, methods of analysis for the various forms of carbon, nitrogen, phosphorus, and silica are reviewed, and the methods used in the Water Quality Division laboratories are described. Consideration is also given to how the "micronutrients" such as trace metals may be determined. Analysis as discussed here includes:

1. Sampling.
2. Sample preservation.
3. Sample pretreatment.
4. The measurement process.
5. Data handling.

SAMPLING

Sampling may be carried out on a discrete or a continuous basis. Discrete sampling involves water sampled into a container for analysis either immediately or later. In continuous sampling, water is pumped from the water body of interest for either continuous or periodic analysis. In discrete sampling it is important that the sample taken into the container truly represents the water desired for analysis. Since this is difficult to determine before the analyses have been carried out, it is suggested that often an indication of how representative the sample is may be obtained by measurements such as conductivity or perhaps temperature, made with portable equipment at locations around the sampling point.

PRESERVATION

If a sample is taken in a container for later analysis, consideration must be given to ways of preserving the sample so that the results of the measurements made represent the original constitution of the sample. Samples change by adsorption of materials onto the walls of the container, by desorption of materials from the walls, by chemical reactions, and by microbiological action. Biological action can result in rapid loss from solution of such nutrient forms as ammonia, nitrite, nitrate, and orthophosphate. The material of the sample container has an effect on the sample preservation. Linear polyethylene or linear polypropylene are accepted sample-bottle materials.

Several preservation techniques for nitrogen and phosphorus forms have been reported (1); of these, acidification, mercuric chloride, and use of low temperature appear to be the most useful. A technique that will preserve all nutrient forms is not available, although low-temperature storage is the nearest approach to it. Acidification of the sample (to pH 2) will stop microbiological action, but it may remove nitrite from solution (2), and it will cause reversion of polyphosphates to orthophosphates and decomposition of carbonates, with evolution of carbon dioxide. If, however, nitrite does not contribute a significant proportion of the total nitrogen levels and the purpose of the analysis is to determine, for example, total nitrogen, total phosphorus, and organic carbon, then acidification is an acceptable technique. In addition, acidification is probably the best way to preserve the trace metals in solution.

Mercuric chloride at a concentration of 40 mg/liter of mercury combined with low-temperature storage has been reported as a preservative for nitrogen and phosphorus forms (1). However, there is a serious drawback to the use of mercuric chloride: namely, when this compound is used, the analysis for phosphorus cannot be carried out by the direct molybdenum blue reaction. In most of the samples analyzed in the Water Quality laboratories, the molybdenum blue method is the only one available with the necessary sensitivity to measure low levels of phosphate. In the phosphate procedure, mercuric chloride is reduced to mercurous chloride and mercury, which precipitates in the color-forming step and makes measurement for phosphate impossible.

A study has been carried out to determine ways of removing the mercury prior to the phosphate analysis, but no method could be found that did not at the same time remove a significant amount of phosphate. Also, because of the current concern for mercury contamination in the environment, water samples are being analyzed for mercury, with a detection limit of 0.05 μg/liter. The possibility of cross contamination of sample bottles with mercuric chloride is sufficiently high that it was decided in the Water Quality Division not to employ mercuric chloride as a nutrient sample preservative. The technique used for preservation is rapid cooling and prompt analysis.

PRETREATMENT

When a water sample is taken it may consist of water in which are dissolved materials in "true-solution," materials in colloidal solution and particulate matter. In general, the interest is in determining the "total" levels of a constituent (i.e., the level in an untreated sample as taken) and the levels of the "soluble" constituents. For "total" level determinations, the pretreatment may consist of homogenization; for "soluble" levels, the pretreatment is generally filtration. The term "soluble" is one of definition. It is that material which passes through the filter used in the pretreatment. A common definition, and the one that is used in the Water Quality Division laboratories, is that a "soluble" material is one which passes through a 0.45-μ membrane filter.

In continuous sampling and analysis no preservation treatment is needed for the samples, but precise pretreatment is difficult. The sample flow is required to be filtered continuously if a measurement of the "soluble" nutrients is to be made. If at the same time a measure of the "total" levels is required, then either two sets of analysis equipment are needed or the analytical system must be switched alternately between the filtered and unfiltered sample flows. If there is an appreciable amount of solids in the water, the filter becomes coated with solid and the effective pore size is reduced, with consequent change in the "soluble" material levels. In some of the equipment sold for continuous monitoring (3), there are cartridge filters with provision for measuring the pressure drop across them so that the filters can be changed when this pressure drop becomes too high. Unfortunately, when the filter is plugged enough to give a change in pressure, the effective pore size is changed to a considerable degree. It would appear that the most valid type of filter for this purpose is one in which the filter medium is a large strip that is fed through the filter, so that a clean filter is being continually presented to the water sample.

The valid analysis of a sample that contains particulate matter demands filtering a known volume of the sample, analyzing the filtrate, and dissolving and analyzing the solid material that is filtered off. The analysis of the filtrate gives the levels of "soluble" material, and this, together with the analysis of the solid material, gives the "total" levels of the constituents. It is customary in nutrient studies to analyze a well-shaken sample without filtration to determine the "total" levels of the nutrients, and to analyze a filtered sample to determine the "soluble" concentrations. This method is convenient in that it allows experimenters to use the automated procedures available for liquid samples and avoids the need for handling solids and filters in a digestion process. Where manual digestion procedures are followed and large (e.g., 50 ml) volumes of the sample are measured into the digestion vessels, there are no apparent difficulties with this technique. Where automated procedures are used and sample portions of the order of a milliliter or less are taken, equipment is available to keep the sample agitated in the sample cup while the sample volume is drawn off. This is a necessary precaution when measuring "total" parameters, although it has not been established that this is a valid technique for all types of samples.

ANALYTICAL METHODS

Carbon

Carbon may occur in waters as carbonate-bicarbonate, the so-called inorganic carbon, or as a variety of organic compounds. The organic compounds may be dissolved, they may be present as colloidal material, or they may occur as particulate matter. The relative contributions of the colloidal material and the particulate matter to the "soluble" and "total" carbon levels is again dependent on the pore size of the filter used in the pretreatment of the samples.

A measure of the organic material in a sample may be made by determining the amount of oxidizing material consumed in oxidizing the sample. This is the basis for the chemical oxygen demand (COD) measurement in which the sample is boiled with potassium dichromate and the excess dichromate is later determined. Automated equipment may be used (4).

Another method for determining the COD is described by Stenger and Van Hall (5). In this method a small sample (ca. 20 μl) is evaporated in a tube furnace in a stream of carbon dioxide. Some of the carbon dioxide is converted to carbon monoxide by the reducing material in the sample, and the carbon monoxide concentration is measured in an infrared (IR) stream analyzer. The equipment is available as a commercial package. As in the "dichromate" COD test, reducing materials other than organics may interfere and the test may require calibration with known amounts of the interfering substances if a measure of the organic material in a series of samples is to be obtained. The measurement of the amount of oxygen consumed when seawater samples are irradiated with UV light has been described by Armstrong (6) as a means of determining the empirical organic carbon content of water samples. This irradiation also serves to convert the nitrogen and phosphorus in the sample to readily measured forms.

These COD-type measurements are not specific for carbon, since other reducible substances consume oxygen, and since materials such as nitrate make oxygen available in the test. However, the measurements are not affected by the presence of inorganic carbon. In most of the carbon-specific tests to be described, it is necessary to remove the inorganic carbon before the organic carbon can be determined. In some circumstances a COD-type test may be more convenient to use and may give an equally valid result.

A technique that specifically measures the carbon content is conversion of carbon to carbon dioxide and measurement of the amount produced. There are a variety of methods available for the oxidation of organic carbon and for the measurement of the carbon dioxide produced. In the method described by Menzel and Vaccaro (7) for seawater, the sample is oxidized by autoclaving in a sealed tube with potassium persulfate and phosphoric acid. The amount of carbon dioxide produced is measured by sparging it from the solution in a stream of nitrogen and measuring the absorption of this gas stream with an IR analyzer.

Another method described by Van Hall (8) uses a tube furnace packed with catalyst in which is injected a small sample into a stream of oxygen. The organic material is oxidized to carbon dioxide, which is measured in an IR analyzer. The inorganic carbonates decompose under these conditions to yield carbon dioxide. The equipment is available as an "Infra-Red Carbonaceous Analyzer." It has been modified to monitor plant effluents (9) and could be used as a continuous monitor. A similar system, utilizing a fluidized bed reactor for the oxidation, is available as a package for the on-stream monitoring of organic carbon (10).

An alternative means of detection is reduction of the carbon dioxide to methane and measurement of the concentration with a hydrogen flame ionization detector (11). The extreme sensitivity of this detector might make possible a detection limit much lower than is obtainable with other techniques. However, the problems associated with obtaining carbon-free water for reagent preparation and blank determinations appear to be the limiting factor rather than the sensitivity of the detector, so that the detection limit with this technique is the same, about 100 μg/liter of carbon, as with the other techniques described here.

Erhardt (12) has reported an automated method for organic carbon in seawater using Technicon AutoAnalyzer equipment. The samples are mixed with potassium persulfate and pumped through a silica coil irradiated with UV light, where the organic carbon is oxidized. The resulting carbon dioxide is stripped from the solution and after purification is absorbed in sodium hydroxide solution. The conductivity of this solution is measured to determine the carbon dioxide level. Another application of a Technicon AutoAnalyzer system is described by Grasshoff (13) for the determination of inorganic carbon in seawater. The carbon dioxide is liberated by acidification and the levels are detected by the change in color of the sodium-carbonate/phenolphthalein solution used to absorb it.

It is often desirable to distinguish between the inorganic and organic carbon. The inorganic carbon may be removed from the sample by acidification to liberate carbon dioxide and sparging with an inert gas. The carbon dioxide can be measured or the difference in total carbon level with and without the treatment used to calculate the inorganic carbon level. The removal of inorganic carbon has been discussed by Van Hall (14). The carbonate removal will also remove volatile organic materials and a third determination of the total carbon without acidification, but with air sparging may be necessary to determine the magnitude of this effect. In one version of the "Infra-Red Carbonaceous Analyzer," two furnaces are used, one at low temperature and one at high temperature, to differentiate between inorganic and organic carbon.

In the Water Quality Division laboratories, Infra-Red Carbonaceous Analyzers are currently used to determine the total and organic carbon on lake and river water samples. Some difficulty has been experienced in obtaining acceptable precision when unfiltered samples containing a few milligrams per liter of organic carbon are analyzed. This is due to the difficulty of obtaining a representative sample of the particulate matter in a sample as small as 20 µl. None of the homogenization techniques that have been tried have overcome this problem. Thus it has been concluded that, for analysis of these types of samples, separate determinations of carbon on a filtered sample and on the filtered-off particulate matter are more valid procedures. The analysis of the solid material can be carried out in any of the commercial elemental analyzers. The use of the automated Technicon AutoAnalyzer equipment, into which a sample of a few milliliters can be drawn, is being investigated as another way of overcoming this problem.

If the particulate matter is filtered off for separate determination of the carbon content, a filter that can be incinerated in the elemental analyzer must be used. This means using a glass (or metal) filter rather than the membrane filter used in the determination of "soluble" content for other parameters, and hence the term "soluble" for carbon analysis has a different meaning from all the other analyses. It is to avoid this ambiguity that work is being carried out on methods that will give a valid "total carbon" analysis on unfiltered samples. With the exception of this difficulty in handling some unfiltered samples, there seems to be no outstanding characteristic that makes any of the methods just described inherently better than the others. The choice of method depends on the circumstances and on the availability and convenience of the equipment.

The determination and identification of specific organic materials that may be of interest in a nutrients study is now receiving a great deal of attention. Methods are becoming available that will enable particular organic materials to be monitored on a routine basis.

Nitrogen

The forms of nitrogen that occur in water and can be separated analytically are ammonia, nitrite, nitrate, and "organic nitrogen." The classical digestion method, first proposed by Kjeldahl for the determination of "organic nitrogen" and ammonia, has been automated in many forms. The most common forms use the "Technicon" continuous digestor for the reaction with digestion mixtures, and the indophenol blue reaction serves to measure the amount of ammonium ion formed (15,16). With the use of sodium nitroprusside as a color catalyst, the lower limit of detection has been reported as 30 μg/liter of nitrogen. The many reported variations of the automated procedure differ mainly in the method of separating and measuring the ammonia formed in the digestion reaction. Some organic nitrogen compounds resist reduction to ammonia in the Kjeldahl reaction, but these are probably not significant in water analysis.

Experience with the automated Kjeldahl nitrogen systems in the Water Quality laboratories has shown that the methods have high sensitivity and that, at any particular time, levels as low as 20 μg/liter of nitrogen yielded a measurable signal. However, the baseline was noisy and characterized by considerable drift during a few hours; furthermore, with the systems tried it was not possible to measure reliably levels lower than about 200 μg/liter. Low organic nitrogen levels are therefore measured by a UV oxidation method (described below). However, the work carried out in the development of this method showed that a major problem in nitrogen measurement is the absorption of nitrogenous materials from the laboratory atmosphere.

With the advantage of hindsight it is now believed that the problems encountered with the automated Kjeldahl methods were probably due to this absorption. It also appears that with equipment to maintain the digestion, samples, and so on, in a controlled atmosphere, the methods would have given the desired precision and detection limit.

Ammonia can be measured by the indophenol blue reaction, which is the reaction of phenol, ammonia, and hypochlorite in alkaline solution to produce a blue color. The use of catalysts to promote the color development at room temperature has made this method more convenient (17). Limits of detection of about 1 μg/liter of ammonia have been reported (18). This reaction does not work well for seawater, and Grasshoff (13) has used the reaction of ammonia with hypobromite (which gives a blue compound with iodine-starch) to measure the ammonia in seawater. There have been doubts expressed (19) about the specificity for ammonia of this reaction used by Grasshoff. The method used in the Water Quality Division laboratories (20) is that described by Sawyer and Grisley (21), in which the ammonia is chlorinated and determined with o-tolidine. By using metaphosphate to complex the calcium in the sample and by separating the ammonia by dialysis, interference from the common ions in natural waters is eliminated. The limit of detection of the method is about 1 μg/liter of nitrogen.

Nitrite can be measured by the diazotization of sulfanilamide followed by coupling with naphthylethylenediamine (22). This is a very sensitive reaction and will measure nitrite levels to 1 μg/liter of nitrogen. Nitrate can be measured with the same sensitivity if it is first reduced to nitrite. The method described by Brewer and Riley (23), which features a coil packed with cadmium filings for reduction, is now almost universally used. The nitrite-nitrate total is measured by this technique. The sum of the "Kjeldahl including ammonia" determination and the nitrite-nitrate level gives the total nitrogen in the sample.

Another measure of the total nitrogen can be obtained by a microcoulometer measurement. This has been reported to determine nitrogen levels down to 100 μg/liter of nitrogen in waste water (24). In this measurement the nitrogen is reduced to ammonia by reduction with hydrogen over a nickel catalyst and the ammonia is titrated coulometrically.

The use of UV irradiation to convert the nitrogen in a water sample to nitrate has been reported (6), as has the development of a manual method for the analysis of soluble organic nitrogen in seawater (25) by UV irradiation. In the Water Quality Division laboratories, an automated method using UV irradiation serves to determine the soluble nitrogen forms in lake and river waters (26). It was found that the oxidation of different types of organic compounds is pH dependent. Urea, for example, is oxidized rapidly at low pH, but slowly at high pH, whereas ammonium compounds are oxidized rapidly at high pH and slowly at low pH.

Because of this dependence on pH, the irradiation is carried out in two stages. First the sample is irradiated under acidic conditions, then the solution is made alkaline and reirradiated. The nitrate formed is measured by the method of Brewer and Riley (23). The sample taken into the analytical system is split, with portions being analyzed for ammonia and nitrate-without-irradiation, at the same time that the rest of the sample flow is irradiated. The lower detection limit for the differentiation of the forms of nitrogen is 25 μg/liter of nitrogen.

During this work it was found that a cause of lack of precision in nitrogen analysis on Technicon equipment was the absorption of nitrogenous materials from the atmosphere while the water samples were in the cups in the sampling tray. The problem was overcome by covering the sample tray with Saran wrap. The sampler probe is replaced with a sharpened hypodermic needle, which pricks a hole in the Saran wrap when the sample is taken. Because of the need to keep the samples covered until the time they are taken, the equipment for agitating the sample in the sample cups is not used and only soluble nitrogen is measured in this technique.

Phosphorus

The phosphorus forms that are conveniently measured are: soluble orthophosphate, inorganic phosphate, and total phosphorus. The soluble orthophosphate is the phosphate in a filtered sample that reacts in the molybdenum blue reaction; it is also called the reactive phosphate. There have been some claims that the molybdenum blue method gives a result that is in fact higher than the orthophosphate level, and it has been suggested that hydrolysis of organic phosphorus compounds during the determination is responsible for the error (27).

Inorganic phosphate is the phosphate present as orthophosphate, polyphosphate, and insoluble inorganic phosphates. Total phosphorus is the inorganic phosphorus plus the phosphorus that is present as organic material. Determination of these three parameters allows the concentration of the individual forms to be calculated. Orthophosphate is determined by reacting the sample with ammonium molybdate in acid solution to form the molybdophosphoric acid, which is then reduced to a heteropoly blue complex. Many automated systems permitting choice of different acid media and use of different reducing agents are available. Reducing agents such as ascorbic acid, amino-naphthol sulfonic acid (ANSA), and stannous chloride have been used. The sensitivity of the method is dependent on the reducing agent; the lowest limit of detection, which is obtained with stannous chloride, is about 3 μg/liter of phosphate (20). The limit of detection of the method can be lowered by extraction of the heteropoly acid into solvents such as isobutanol (1,28).

Silicon is a potential interference in this test and the normality of the acid medium used must be controlled to avoid this interference. If arsenic is present, it will give a positive interference.

Inorganic phosphate and total phosphorus are measured by conversion to orthophosphate. Inorganic phosphate is converted by digestion with acid, which hydrolyzes the polyphosphates and leaches phosphate from insoluble materials in the sample. Automated methods are available to determine the levels of the different species of condensed phosphate. One method employing Technicon AutoAnalyzer equipment uses an ion-exchange chromatographic column to separate the species, which are then determined by hydrolysis and molybdenum blue color measurement. The organic phosphates are converted by digesting with acid and an oxidizing agent that destroys the organic material and releases the phosphorus as phosphate. These digestions can be carried out manually or in automated systems. The most common systems use the Technicon continuous digestor with an oxidizing agent such as potassium persulfate. In some of the automated systems, the same digestion serves for the Kjeldahl nitrogen determination and also for the conversion of total phosphorus to orthophosphate (29).

In the manual digestion procedures the sample is heated with sulfuric acid and potassium persulfate, and it may be convenient to autoclave the reaction mixture at 120°C for one hour. This has been found to be a satisfactory method for freshwater samples (20). For seawater, evaporation with perchloric acid has been recommended. An advantage of this method is that it removes the arsenic, which otherwise is measured as phosphate.

Ultraviolet irradiation to break down the organic phosphorus compounds has been used by Armstrong (6) in a manual method for the determination of total phosphorus in seawater. Grasshoff has used an automated method for total phosphorus in seawater that also features UV irradiation. Before the samples are analyzed, they are acidified and heated in their storage bottles to hydrolyze any polyphosphate present.

The total phosphorus procedure which the Water Quality Division laboratories finds most convenient involves autoclaving an aliquot of the sample for one hour with persulfate and sulfuric acid. The resulting orthophosphate is then determined in an automated system, using stannous chloride as the reducing agent. This procedure is necessary because not all samples are free from turbidity after the digestion, and the manual operation gives an opportunity for inspection and, if required, filtration before the colorimetric determination. (The importance of this is discussed below.)

When a sample is received in the laboratory, an aliquot is immediately taken and placed in a stoppered conical flask. The flask can be stored for several weeks before the acid and persulfate are added and the autoclaving is carried out, without any change in the total phosphorus level measured. Thus the total phosphorus determination can be performed on an "as-convenient" basis, and this lightens the workload of the analysts, who are concerned with the prompt analysis of highly perishable constituents.

Silicon

Silicon occurs in water as silicic acid and its salts. Silicic acid polymerizes quite readily, and it is probable that only short (i.e., less than three or four silicic acid units), straight-chain polymers are measured by the customary silicomolybdate complex method. For this reason the term "reactive silicate" is used to describe the results of this test. If the sample is boiled with sodium hydroxide, the longer chain polymers are broken down and the silicate measurement then gives the "total silicate" that is measured normally, since it is believed that this is the form of silicon that is readily available to growing plant cells.

Silicic acid reacts with ammonium molybdate to form the yellow silicomolybdate complex, which can be used directly for colorimetric determination. A much more sensitive measurement involves reducing this complex to a blue compound. The reaction is similar to that used in the phosphate determination; the intensity of the blue is dependent on the reducing agent used. However, the sensitivity is greater than for phosphate, and using ANSA as the reducing agent, a limit of detection of 5 μg/liter of silicon dioxide can be obtained (20). Phosphate interference is prevented by using oxalic acid in the reaction solution.

BLANK DETERMINATIONS

In a large number of the analyses for nutrients, the automated systems rely on a colorimeter reading to determine the concentration of the material. If the absorbance measured by the colorimeter is due to the chemical reaction of the reagents, and if the species being measured is the one that was present in the original sample, the result is a valid one. If, however, the absorbance is due to some other colored substance in the water, or to turbidity that either occurred in the original sample or developed in the analytical procedure, then the result is invalid. It has been our experience (30) that a major consideration in carrying out nutrient analysis, particularly phosphorus and silicon in the microgram per liter range, is the assurance that the results are valid ones and have not arisen because of color or turbidity developed in the sample.

The means of overcoming this potential problem include running "sample blanks," filtering treated samples after any digestion process and, preferably, transferring the species being measured from the sample to a clean liquid stream. In the case of volatile materials such as ammonia, the transfer can be made by distillation. The transfer of nonvolatile materials can be made by dialysis. Unfortunately, dialysis normally transfers only a small proportion of the ions of interest, and although it has served in phosphate analysis at high levels, the sensitivities of the phosphate and silicate analyses are not great enough to allow the use of such an inefficient process. Solvent extraction of the heteropoly acids is another way of minimizing the sample blank in these determinations, and manual and automated procedures are reported for these extractions (28,31).

In the analysis of nitrate and nitrite, the color formed by the diazotization and coupling reaction is so intense that the sample can be diluted with a considerable volume of distilled water and adequate sensitivity still is obtained. Under these conditions, the contribution due to the sample blank is minimized. In the Water Quality Division laboratories, methods of high sensitivity are used to minimize the effect of sample blanks even if, on the face of it, the sensitivity is not required for the levels of the species being measured. It is probably a safer procedure to dilute the sample and use an intense-color method than to use an undiluted sample with a color development that can be measured directly.

To be certain that the results obtained in automated colorimetric determinations are valid, it is necessary to run the samples through the system without the color-forming reagent. In the case of the phosphate analysis, for example, the ammonium molybdate-sulfuric acid reagent is replaced with sulfuric acid so that the acid strength in the blank determination is the same as in the actual determination. This is important because the "natural color" in some waters will change with changing pH. If peaks are obtained in the blank determinations, then correction for them must be made in the sample determinations.

In order to run sample blanks in a system that is sampling and analyzing continuously, it is necessary periodically to replace the color-forming reagent with a "blank" reagent. This can be done with a system similar to that used to periodically replace the sample flow with distilled water in determining the "reagent blanks."

Another potential source of error is contamination of the sample in the pretreatment process. Some filter media contain phosphate and trace metals that can contaminate the sample, and they must be washed before the sample is filtered. On the other hand, some filters will adsorb phosphate from the sample. These and other errors can best be identified by a program of preparing standard solutions at the sampling point and treating these standards in exactly the same way that the samples are preserved, shipped, and pretreated.

DATA HANDLING

The development of automated analytical techniques has made possible the collection of an enormous amount of data. Now the automation generally ends with the analytical process. The results of the analyses are presented on charts, which are then read and transcribed manually. This involves considerable effort in terms of man hours and also affords the opportunity for error (particularly when various parameters, such as orthophosphate, inorganic phosphate, and total phosphate are measured in a sample and other parameters, such as organic phosphate are calculated from these results).

Ideally, once an instrument has seen a result, the result should be recorded automatically in a form that can be dealt with directly by a data-handling system, so that subsequent calculations and transformations can be performed automatically. In such a system each sample being analyzed has to be identified by the analyst, but after that the data-handling system can carry out the calculations and present the results in whatever form is desired. If the data are to be entered in an automated data bank, the results can be printed on punched tape or magnetic tape; then the entry can be made directly, and there is very little opportunity for operator error.

Where continuous sampling and analysis is carried out, the results can be periodically digitalized and printed out or stored. A selection of reasonably priced equipment is available to do this. Where discrete samples are analyzed, the majority of automated analytical systems such as the Technicon AutoAnalyzer or the Infra-Red Carbonaceous Analyzer present the data on charts in analog form, from which it is necessary to determine the "peak height" for each sample. The analog signals can be converted to digital form in a Technicon Auto-Analyzer system by running each sample through the analytical system long enough to reach a steady state so that, instead of a sharp peak, there is a plateau that is broad enough for the digitalizing and printing to be synchronized with the maximum reading for the sample (13). A disadvantage of this technique is that it restricts the rate at which samples can be processed.

Another approach is to use a peak-holding device to store the maximum reading, which is later digitalized and printed (32). Other systems use a small computer on-line to decide what the peak heights are and to do the calculations of concentrations, and so on. In multichannel operation these systems become expensive, particularly for the software development.

In the Ottawa laboratory of the Water Quality Division, a system has been developed which digitalizes the peak heights from the six channels of a Technicon CSM 6 and prints them on punched tape. The tape is then processed off-line to produce a print-out of the concentrations in the samples (33). This system, which involves peak-holding devices, is used because it is low in cost and because it will serve for any analytical process that presents the results on a recorder.

In considering automated systems, the so-called discrete sample analyzers should be discussed. In this type of instrument the reagents are added to the sample in a small vessel. The resulting colored solution is drawn into a colorimeter cell for the optical density measurement. Since only one reading is obtained per sample, it is very simple to bring the instrument into a data-handling system; in addition, the throughput in samples per hour is higher than in the flow-through systems.

TRACE METAL ANALYSIS

In recent years atomic absorption spectrophotometry has been the most common method of analysis for trace metals. The technique combines speedy analysis with ease of sample preparation and relative freedom from interferences. In a large number of natural waters the metals of interest are not present in sufficiently high concentrations to be measured directly, and a variety of solvent extraction procedures have been developed in which the metal is complexed and the complex is then extracted into an organic solvent. This gives a concentration and also an enhancement of the response. The combination of solvent extraction and atomic absorption spectrophotometry with conventional flame atomization makes possible the determination of most elements of interest in the microgram per liter range.

In the Water Quality Division laboratories these determinations are carried out by the technique described by Mulford (34): aliquots of the sample are shaken with the complexing agent and solvent in volumetric flasks and, after settling, the solvent layer is aspirated into the flame from the neck of each flask. The sensitivity of the atomic absorption is increased by the use of the sample boat and by the use of the so-called flameless techniques.

In one version of the flameless technique (35), the sample is evaporated in a small boat mounted in an absorption chamber containing an inert gas. The remaining solid is vaporized by quick heating to a high temperature and the absorption is measured with a fast response detector. The great sensitivity of this technique is due to the relatively high concentration of the atoms of interest at the moment vaporization takes place and to the absence of flame "noise," which makes possible the amplification of the signal from the detector.

Atomic fluorescence is a technique that has inherently greater sensitivity for many elements than atomic absorption. Equipment for atomic fluorescence measurement is now available from several manufacturers. The fluorescence intensity is dependent on the intensity of the light source. Using hollow cathode lamps that are pulsed to emit short light bursts of high intensity, very low limits of detection can be obtained for a number of metals. Since the light source is at an angle to the flame-detector axis, the optical system is simple, and the use of a number of light sources with "pulses" synchronized with the detector system makes possible simultaneous multielement analysis. Higher intensity light sources, such as electrodeless discharge lamps and lasers, will lead to lower detection limits. The development of flameless techniques will give even lower detection limits.

X-Ray fluorescence is an analytical technique that is more applicable to solids than to water samples. However, there are methods in which the metal of interest is complexed and extracted with organic solvent. This solvent is evaporated onto a solid support, which is then examined by X-ray fluorescence. X-Ray fluorescence instruments are now available at a price comparable to the price of an atomic absorption spectrophotometer. As X-ray fluorescence instruments become more generally available, it is likely that more methodology will be developed for the analysis of water samples by this technique.

Methods such as atomic absorption, atomic fluorescence, and X-ray fluorescence are normally used as destructive determinations; that is, the purpose of the analysis is to determine the total level of the trace metal in the sample. As more information is gathered on the effects of trace metals on plant growth, it is believed that more emphasis will be placed on the amounts of the trace metals that are available to the living organisms. The distinction made now between reactive silicon and total silicon in a water sample is the type of differentiation that will be required between, for example, the amount of trace metal available as metal ions and the amount complexed by organic materials in the water. In the Water Quality Division it has been concluded that the best technique for accomplishing this is probably differential-cathode-ray polarography, and a program is underway to determine its potentialities.

Differential-cathode (D.C.) polarography of course was used for many years for the analysis of metal ions. The method distinguishes between oxidation states but is of limited sensitivity. It is easy to measure the low current due to the reduction of low levels of metal ions. In the D.C. polarograph, however, the residual current, the currents due to capillary maxima, and impurities in the base electrolyte give such a high background current that it is impossible to detect the small signal current.

In cathode-ray polarography, the voltage sweep is made over a small portion of the life of each mercury drop, and the resulting current-voltage relationship is displayed on the cathode-ray tube. This fast sweep eliminates the fluctuations that occur when each drop grows and falls away. There is also an increase in sensitivity, since at a fast sweep rate the reaction is not controlled by the diffusion rate. The differential mode of operation, with the two cells balanced so that they give identical signals with identical solutions, overcomes the changes in current due to capillary maxima; thus the effects of impurities in the base electrolyte can be balanced out.

In the past, the other objection to polarographic analysis has been that it is time consuming, and therefore fewer samples can be analyzed per man day compared with atomic absorption spectroscopy. To overcome this disadvantage, the operation of the D.C. polarograph in the Water Quality Division laboratory has been automated so that samples can be presented to the instrument at a rate of 10/hr if it is desired to carry out anodic stripping, or 20/hr if stripping is not desired (36). The recording of the trace is currently being done photographically, with the photograph being measured to calculate the result. A data collection system is being installed which will enable the recording and calculation to be carried out automatically.

With the straightforward D.C. polarograph, it is possible to measure trace metals in the microgram per liter range. The free metal ion concentration and the amount of a particular metal complexed by the organic materials in the sample can also be ascertained. In some cases it has been possible to determine the approximate stability constant of the organic complexing agents combined with the metal of interest. With this straightforward technique, solution of 10^{-7}M concentration (which corresponds to 5 to 20 μg/liter depending on the atomic weight of the metal) can be determined easily.

For levels lower than this, some concentration step is necessary. Anodic stripping, one means of doing this, involves "plating" the metal onto the surface of the mercury drop. The enriched layer responds when the voltage sweep is made in the polarographic analysis. Using various voltages and times for the anodic stripping, it is probable that this step can be used as a concentration step for the determination of total metal content in a sample. The same step should also serve to differentiate among the various forms in which the metal exists.

Anodic stripping onto a mercury drop is a very useful technique for some metals, such as zinc, copper, lead, and cadmium; but with other metals, such as cobalt, nickel, manganese, and iron, the method is less valuable because amalgams are formed with mercury that are difficult to oxidize in the polarographic determination. Unfortunately, it is these very metals that are probably of most interest in nutrient studies.

Solvent extraction is currently being investigated as another possible concentration method (37). With the polarographic equipment now available it is possible to perform measurements directly on the organic solvent extracts. The preliminary work accomplished to date indicates that, by the use of various complexing agents with different stability constants with the metals being extracted, it is possible to determine at least the equilibrium concentrations of the metal ions in solution and the concentrations of the metals that would be complexed by the complexing agents in the sample.

The advantage of the polarographic method in this type of work is seen if a complexing agent is added to a sample and the metals complexed by this agent are extracted--it is then possible to identify the extracted metals by their respective reduction potentials. In addition, the use of the differential technique means that other materials that are extracted at the same time cancel out in the two cells and the polarographic trace obtained is due only to the complexing agent that was added to the sample. Considerable work is required to optimize the conditions and to render this technique a useful part of trace metals studies. Preliminary studies, however, indicate that the technique is a viable one.

REFERENCES

1. D. Jenkins, "The Differentiation, Analysis and Preservation of Nitrogen and Phosphorus Forms in Natural Water," in "Trace Inorganics in Water." Advances in Chemistry Series No. 73, American Chemical Society, 1968, pp. 253-280.
2. P. L. Brezonik and G. F. Lee, "Preservation of Water Samples for Inorganic Nitrogen Analysis with Mercuric Chloride," Air and Water Pollut. Int. J., 10, 549-553 (1966).
3. A. Conetta and M. H. Adelman, "The CSM-6: A System for the Multiple Analysis of Water Parameters," in "Technicon Symposia, Advances in Automated Analysis," Vol. II. Mediad Inc., White Plains, N.Y., 1969, pp. 125-128.
4. M. H. Adelman, "Simplified Automated COD Determination: Advanced Procedures," in "Technicon Symposia, Automation in Analytical Chemistry," Vol. 1. Mediad Inc., White Plains, N.Y., 1966, pp. 552-556.
5. V. A. Stenger and C. E. Van Hall, "Rapid Method for Determination of Chemical Oxygen Demand," Anal. Chem., 39, 206-211 (1967).
6. F. A. J. Armstrong and S. Tibitts, "Photochemical Combustion of Organic Matter in Sea Water, for

Nitrogen, Phosphorus and Carbon Determination," J. Mar. Biol. Assoc., (U.K.), 48, 143-152 (1968).
7. D. W. Menzel and R. F. Vaccaro, "The Measurement of Dissolved Organic and Particulate Carbon in Sea-Water," Limnol. Oceanogr., 9, 138-142 (1964).
8. C. E. Van Hall, J. Safranko, and V. A. Stenger, "Rapid Combustion Method for the Determination of Organic Substances in Aqueous Solutions," Anal. Chem., 35, 315-319 (1963).
9. G. Kramig and R. B. Schaffer, "Use of the Total Carbon Analyzer for Pollution Control," in "The Analyzer," Vol. 5. Scientific Process Instruments, Division of Beckman Instruments Inc., 1964, p. 4.
10. Environ. Sci. Technol., 11, 1152 (1969).
11. R. A. Dobbs, R. H. Wise, and R. B. Dean, "Measurement of Organic Carbon in Water using the Hydrogen-Flame Ionization Detector," Anal. Chem., 39, 1255-1258 (1967).
12. M. Erhardt, "A New Method for the Automatic Measurement of Dissolved Organic Carbon in Sea Water," Deep Sea Res., 16, 393-397 (1969).
13. K. Grasshoff, "A simultaneous Multiple-Channel System for Nutrient Analysis in Sea Water with Analog and Digital Data Record," in "Technicon Symposia, Advances in Automated Analysis." Vol. II. Mediad Inc., White Plains, N.Y., 1969, pp. 133-145.
14. C. E. Van Hall, D. Barth, and V. A. Stenger, "Elimination of Carbonates from Aqueous Solutions Prior to Organic Carbon Determinations," Anal. Chem., 37, 769-771 (1965).
15. P. A. Kammerer, M. G. Rodel, R. A. Hughes, and G. F. Lee, "Low-Level Kjeldahl Nitrogen Determinations in the 'Technicon AutoAnalyzer,'" Envir. Sci. Technol., 1, 340-342 (1967).
16. W. H. McDaniel, R. N. Hemphill, and W. T. Donaldson, "Automatic Determination of Total Kjeldahl Nitrogen in Estuarine Waters," in "Technicon Symposia, Automation in Analytical Chemistry." Vol. I. Mediad Inc., White Plains, N. Y., 1967, pp. 363-367.
17. J. E. Harwood and D. J. Huyser, "Automated Analysis of Ammonia in Water," Water Res., 4, 695-704 (1970).
18. R. D. Britt, "Precise Automatic Spectrophotometric Analysis in the Low Parts Per Billion Range," Anal. Chem., 34, 1728-1731 (1962).
19. J. P. Riley, in "Chemical Oceanography," J. P. Riley and G. Skirrow, Eds. Academic Press, New York, 1965, p. 369.
20. W. J. Traversy, "Methods for Chemical Analysis of

Waters and Waste Waters," Tech. Bull., Dept. of Fisheries and Forestry, Ottawa, Canada.
21. R. Sawyer and L. M. Grisley, "The Determination of Ammonia in the ppb Level in Solution," in "Technicon Symposia, Automation in Analytical Chemistry," Vol. I. Mediad Inc., White Plains, N. Y., 1967, pp. 347-350.
22. P. G. Brewer, K. M. Chan, and J. P. Riley, "Automatic Determination of Certain Micronutrients in Sea Water," in "Technicon Symposia, Automation in Analytical Chemistry." Mediad Inc., White Plains, N. Y., 1965, pp. 308-314.
23. P. G. Brewer and J. P. Riley, "The Automatic Determination of Nitrate in Sea Water," Deep Sea Res., 12, 765-772 (1965).
24. R. D. Moore, R. J. Joyce, and M. E. Riddle, "Extension of Microcoulometric Determination of Total Bound Nitrogen in Hydrocarbons and Water from 0.1 ppm to 1% N," presented at the Pittsburgh Conference on Analytical Chemistry and Applied Spectroscopy, March 1970.
25. J. D. H. Strickland and T. R. Parsons, "A Practical Handbook of Sea-Water Analysis," Fisheries Research Board of Canada, Bull. 167, 1968.
26. B. K. Afghan, P. D. Goulden, and J. F. Ryan, "An Automated Method for Determination of Soluble Nitrogen in Natural Waters," presented at the Technicon International Congress, New York, November 1970, paper 72.
27. F. H. Rigler, "Further Observations Inconsistent with the Hypothesis that the Molybdenum Blue Method Measures Orthophosphate in Lake Waters," Limnol. Oceanogr., 13, 7-13 (1968).
28. A. Hendriksen, "An Automatic Method for Determining Low Level Concentrations of Phosphates in Fresh and Saline Waters," Analyst, 90, 30-34 (1965).
29. W. F. Milbury, V. T. Stack, and F. I. Doll, "Simultaneous Determination of Total Phosphorus and Total Kjeldahl Nitrogen in Activated Sludge with the 'Technicon' Continuous Digestor System," presented at the Technicon International Congress, New York, November 1970, paper 73.
30. V. K. Chawla and W. J. Traversy, "Methods of Analyses on Great Lakes Waters," Proc. 11th Conf., Great Lakes Res., 524-530, 1968.
31. K. Stephens, "Determination of Low Phosphate Concentration in Lake and Marine Waters," Limnol. Oceanogr., 7, 484 (1963).
32. R. F. Farr, J. A. Newell, T. P. Whitehead, and G. M.

Widdowson, "Digital Concentration Print-out from a Four-Channel 'AutoAnalyzer' System," in "Technicon Symposia, Automation in Analytical Chemistry," Vol. II. Mediad Inc., White Plains, N. Y., 1966, pp. 225-229.

33. P. D. Goulden and A. DeMayo, "Automated Handling of Data from a Technicon AutoAnalyzer Model CSM-6," presented at the International Symposium on Identification and Measurement of Environmental Pollutants, Ottawa, Canada, June 1971.
34. E. C. Mulford, "Solvent Extraction Technique for Atomic Absorption Spectroscopy," At. Absorption Newsl., 5 (4), 88-90 (1966).
35. H. M. Donega and T. E. Burgess, "Atomic Absorption Analysis by Flameless Atomization in a Controlled Atmosphere," Anal. Chem., 42, 1521-1524 (1970).
36. B. K. Afghan, P. D. Goulden, and J. F. Ryan, "An Automated Method for the Determination of NTA in Natural Water, Detergents and Sewage Samples Using Twin-Cell Oscillographic D.C. Polarography," in preparation, 1971.
37. B. K. Afghan, R. M. Dagnall, and K. C. Thompson, "Analytical Applications of Metal-Acetylacetonates in Organic Solvents," Talanta, 14, 715-720 (1967).

Chapter V

BIOASSAY ANALYSIS OF NUTRIENT AVAILABILITY

George P. Fitzgerald

Water Resources Center, University of Wisconsin
Madison, Wisconsin

Bioassays can be used to evaluate the amount of a particular nutrient or nutrients available to algae or aquatic weeds in a water sample or to assess the nutritional status of in situ plants. Bioassays for any required plant nutrient can be carried out by growth experiments in the laboratory, using selected species of algae. Relatively short-term tests can be performed by measuring changes in certain enzymatic activities or chemical fractions that have been shown to reflect meaningful nutritional changes. An evaluation of the nitrogen or phosphorus nutritional status of in situ algae or aquatic weeds at any particular time can be made by measuring the ammonium-nitrogen absorption rates in the dark, relative amounts of phosphate-phosphorus extracted and alkaline phosphatase activities, or nitrogen-fixation rates by blue-green algae.

Bioassays of water samples demonstrate the level of available nutrients, whereas bioassays with in situ plants demonstrate whether the environment has supplied limited or surplus quantities of nutrients. The latter tests also indicate transitory changes that might have taken place in nutrient sources between sampling dates (e.g., the effects of slugs of nutrients that might not be detected without continuous monitoring).

ANALYSIS OF NUTRIENT CONTENT BY GROWTH

The value of measuring the growth of algae in water samples is that differentiation can be made between the total nutrient content of water samples as obtained by chemical analyses and the nutrients that are available to support the growth of algae under certain circumstances. The concentration of any nutrient required for the growth of algae can be determined by measuring the growth attained by selected algal species in dilutions of the water sample or after suitable spikes of other nutrients are added. Standard techniques for this type of bioassay are being developed (1). Growth experiments often require two or three weeks of incubation, but preliminary results can sometimes be detected after only one or two days. Long incubation periods may allow an original source of a potential nutrient to degrade to an available form (such as a polyphosphate compound degrading to orthophosphate). Therefore it is sometimes desirable to expose the algae to the potential nutrient for relatively short periods of time and then transfer them to another medium to grow on the absorbed nutrient (2).

Some factors that can influence the results of nutritional tests are the type, the source, and the amount of algae used, and the nutrients carried over when the algae are added to the test media. The algae to be used must be readily available and must respond to the nutrients of interest. In addition, the selected species should represent problem-causing species, since not all algae will respond similarly to the same environment. There is frequent controversy over whether the natural flora of a water sample should be used for nutrient bioassays or whether a standardized culture, which may not be related to a particular flora but which has been shown to respond similarly to various nutrients, should be used instead.

As pointed out, the alga to be used must be available when the water is to be tested; if the tests are to be run on samples taken at spring overturn of a lake, the in situ algae will probably not be the same as those which cause problems in midsummer. If the samples must be collected and stored until a convenient time for the bioassays, it is probable that the original flora will no longer be present. Several workers have noted that the algal species composition of in situ tests frequently changed when samples are confined for a few days time, as seen in the replacement of phytoplankton by epiphytic types of algae that had become attached to the walls of the container. Therefore, except for certain specific studies, the use of selected and tested species of algae is preferred over the use of in situ algae for growth bioassays.

Algae are able to concentrate certain nutrients in excess of their present needs when they are grown in media with surplus nutrients and this factor must be taken into account in selecting culture media and determining the amount of algae used. The media effect on nutrition tests was demonstrated by Gerloff and Skoog (3). They showed that the planktonic blue-green alga, *Microcystis aeruginosa* (Wis. 1036), cultured in a medium with low nitrogen (6.8 mg N/liter) would not grow when transferred to a medium lacking nitrogen, but *Microcystis* from a medium with surplus nitrogen for maximum growth (27 mg N/liter) was able to increase twofold in a medium lacking nitrogen.

Gerloff and Skoog (3) also showed that *Microcystis* from a medium with excessive phosphorus (1.8 mg P/liter) could increase fourfold when transferred to a medium lacking phosphorus. They performed a series of 30 experiments using the green algae *Chlorella pyrenoidosa* (Wis. 2005) and *Selenastrum capricornutum* (PAAP), and the blue-green algae *Microcystis* and *Anabaena flos aquae* (Ind. 1444). The results revealed that there was no significant further growth in media lacking nitrogen or phosphorus with algal inoculation levels of from 10,000 to 1 million cells/ml from the relatively dilute PAAP medium (4.7 mg N/liter and 0.2 mg P/liter).

Thus if algae are cultured in relatively dilute media, the amount of growth in subsequent media will be dependent on the nitrogen or the phosphorus of the latter media, regardless of inoculum size. However, if algae are precultured in more concentrated media, such as Allen's (4) (178 mg N/liter and 45 mg P/liter) or Gorham's (5) (82 mg N/liter and 7 mg P/liter), *Selenastrum* could increase two- to threefold in nitrogen-free media and fourfold in phosphorus-free media. The growth that occurs in media free of nitrogen or phosphorus is due to surplus nutrients (luxury consumption) inside the algal cells.

Therefore, the amount of algae added to test media from more concentrated media should be low enough that the excess nutrients in the cells would be insignificant compared with the nutrients in the test media. Statistical analysis has shown that the lowest concentration of algae that could be readily measured by cell counts using a hemocytometer was 100,000 cells/ml (100 cells, ± 20, in 175 microscope fields); thus inoculations at levels less than 100,000 cells/ml should be employed when cells from relatively concentrated media are used.

The carryover of extracellular nutrients from the

preculture medium can be minimized by washing cells in media lacking the nutrients of interest (e.g., by centrifuging and resuspending in distilled water containing 50 mg/liter $NaHCO_3$). The sensitivity of growth tests carried out with these precautions can be evaluated by the minimum detectable concentrations of the nutrients of interest. By washing the green alga Selenastrum from any nutrient medium and inoculating to a concentration of 50,000 cells/ml, according to the measurements used, concentrations of various nutrients as low as those presented in Table 1 will consistently yield increased growth. Measurements more sensitive than those indicated will make it possible to detect lower concentrations of nutrients.

Table 1. Minimum Detectable Concentrations, mg/liter, of Nutrients by Growth of Selenastrum when Algal Growth is Measured by Different Techniques

Nutrient	Absorbance, 1 cm, 750 mμ	Cell counts, by hemocytometer	Chlorophyll a, by fluorometry
N	0.2	0.1	0.1
P	0.02	0.01	0.01
Mg	0.02	0.01	0.01
S	0.02	0.01	0.01
Fe	0.002	0.001	0.001

By using data on the relationship between nutrients and the amount of algae produced, the amount of nutrients required to produce a heavy bloom of algae in a eutrophic lake can be calculated. During the summers from 1915 through 1957, the more eutrophic lakes in the Madison, Wisconsin, area (Lakes Waubesa and Kegonsa) usually contained blooms of algae such that a 20-cm white Secchi disc would disappear at depths of 50 to 100 cm, and the dry weight of such blooms amounted to 5 to 20 mg/liter. If we assume that algae absorb their nutrients and grow in the layers of water in which they are found, the concentrations of nutrients in Table 2 would be approximately those required to support the growth of a 20 mg/liter bloom of algae.

In order to evaluate the reproducibility of data on the relation between algal growth measurements and nutrient levels in natural waters, the amount of growth attained by Selenastrum was measured by three to five

Table 2. Concentrations of Nutrients Required for the Development of a "Bloom" of Algae[a] Based on the Amount of Growth Attained Experimentally per Milligram of Nutrient

Nutrients	Concentration required, mg/liter
N	0.3[b]
P	0.02[b]
S	0.05
Mg	0.02
Fe	0.002

[a]20 mg/liter (dry weight); Secchi disc reading of less than 50 cm.

[b]The values for N and P are 1.5 and 0.15 mg/liter, respectively, when calculated from chemical composition data on normal bloom algae.

tests in each of eight samples of autoclaved water taken from three lakes in the Madison area. Concentrations of available nitrogen in five separate tests of a Lake Mendota surface water sample, collected and preserved on August 17, 1970, and assayed during December 1970, were 0.15, 0.17, 0.20, 0.20, and 0.25 mg N/liter. This degree of reproducibility is typical for bioassays. The level of available nitrogen in six surface water samples collected in midsummer 1970 was about 0.2 mg N/liter, whereas a hypolimnion sample contained 1.2 mg N/liter, and a surface sample collected in December contained 0.5 mg N/liter. These values correlate well with levels of $NH_4^+ + NO_3^- - N$ obtained by chemical analyses of similar samples.

The availability of different forms of nutrients can also be evaluated by growth experiments. Certain relatively insoluble sources of nutrients (e.g., iron-phosphorus compounds and teeth for phosphorus, hair for nitrogen, pyrite for iron, and marble for carbon) were readily used by algae, whereas other sources, including the nitrogen or phosphorus of aerobic lake muds or the nitrogen or phosphorus contained in other live plants, were relatively unavailable.

EXTRACTIVE AND ENZYMATIC ANALYSES FOR PHOSPHORUS

An extractive procedure can be used to differentiate between algae that have surplus or stored phosphorus and those that are phosphorus limited. Algae and aquatic weeds containing adequate phosphorus will release more than 0.08 mg PO_4-P/100 mg (dry weight) of plant material when extracted in a boiling water bath for 1 hr. For example, after three weeks of culture in a sample of Lake Wingra water, the extractable PO_4-P of the blue-green alga _Microcystis_ sp. was 0.036 mg PO_4-P/100 mg of algae, whereas in a lake sample fortified with additional PO_4-P (7.2 mg P/liter) the extracts of _Microcystis_ contained 0.32 mg PO_4-P/100 mg. Therefore, this extractive procedure can be used to measure the phosphorus-nutritional status of algae or aquatic weeds and to follow the effects of environmental changes that might influence the phosphorus nutrition of plants (6).

We also know that phosphorus-limited plants will have 25 times as much alkaline phosphatase activity as plants grown with surplus phosphorus. Analysis of alkaline phosphatase activity can thus be used to confirm that plants with low extractable PO_4-P levels are alive but limited in phosphorus. The two procedures have been used together to detect long-term nutritional changes, such as seasonal changes in the availability of phosphorus in lake waters or recent additions of available phosphorus. Plants that have only recently been exposed to increased supplies of available phosphorus have higher extractable PO_4-P but also have relatively high alkaline phosphatase activities, because the alkaline phosphomonoesterase content is only lowered by dilution through growth of the cell under adequate phosphorus conditions.

During the period of July 22 to 31, 1968, the _Cladophora_ sp. along the shore of Lake Mendota appeared to be phosphorus limited in that very little PO_4-P could be extracted (0.03 to 0.07 mg/100 mg of algae) and relatively high alkaline phosphatase activities were recorded (ca. 1000 units/mg of algae). There was a sudden increase in extractable PO_4-P from the _Cladophora_ during the period of August 5 to 7, 1968 (probably associated with the 1.4 in. of rain occurring then). The values rose to 0.12 mg PO_4-P/100 mg on August 6, although the algae still had 1500 units of alkaline phosphatase per milligram. With growth under surplus phosphorus conditions (probably associated with a lack of competition because there were few phytoplankton present; Secchi depth of 2.5 m), by August 29 the _Cladophora_ had 0.17 mg of extractable PO_4-P/100 mg, but only 90 units of alkaline phosphatase per milligram.

The use of the extractive procedure in laboratory studies of the availability of phosphorus sources has indicated that 10 mg of either the green alga _Cladophora_ sp. or leaves of the aquatic weed _Myriophyllum_ sp. could detect concentrations as low as 0.04 mg P/liter. When phosphorus-limited _Cladophora_ sp. from Lake Wingra was tested under aerobic conditions in the laboratory, it was shown that 1% or less of the phosphorus of various lake muds was available to the algae (7).

RATE OF AMMONIA ABSORPTION FOR NITROGEN NUTRITION

Plants that are limited by the supply of available nitrogen are able to absorb ammonia (NH_4-N) in the dark 4 to 5 times more rapidly than plants with adequate or surplus nitrogen (8). The effect of changes in the environmental supply of nitrogen to in situ algae or aquatic weeds can be followed, and different sources of nitrogen can be evaluated in laboratory experiments by adding NH_4-N to water in the dark and analyzing for NH_4-N after a 1-hr incubation using 10 to 20 mg of plant material.

Examples of the results of this test are presented in Table 3, which shows the influence of rain on the nitrogen nutrition of the green algae _Spirogyra_ sp. and _Cladophora_ sp. from different lakes in Wisconsin. Related data have shown that 4 to 7 days of sunny weather during midsummer can cause the algae growing in the surface waters of the Madison, Wisconsin, lakes to become nitrogen limited; however, rain will provide enough available nitrogen to cause their rate of NH_4-N absorption to decrease and their colors to change from yellow to bright green.

Table 3. Effect of Rain on Nitrogen Nutrition of Algae

Algae	Source	Date	Micrograms of NH_4-N absorbed/(10 mg) (hr)	
			Before Rain	After Rain
Spirogyra sp.	Salmo Pond	June 7, 1967	16	8
Cladophora sp.	Lake Mendota	June 7, 1969	18	6
Cladophora sp.	Lake Mendota	June 12, 1970	24	7
Cladophora sp.	Monona Bay	July 22, 1970	14	2

An interesting correlation has been made between the nitrogen nutrition of filamentous algae or aquatic weeds and the growth of epiphytic algae. Observations in the field and in controlled laboratory tests have indicated that plants growing for a week or more in the presence of surplus nitrogen usually become coated with epiphytic algae. Therefore, the presence of a dense coating of epiphytes on algae is an indication that the algae have had surplus nitrogen available in their immediate history (9).

NITROGEN-FIXATION RATES RELATED TO SOURCES OF N AND P

It has recently been shown that the capacity of blue-green algae and other plants to fix nitrogen could be followed by measuring the rate of acetylene (C_2H_2) reduction by the same nitrogenase enzymes used to fix the nitrogen (10). The rate of reduction of C_2H_2 to ethylene (C_2H_4) can be measured easily with gas chromatography, and this measurement is a useful tool in limnology. Studies on nitrogen fixation by blue-green algae in lakes have revealed that the rate of fixation per milligram of algae varies during the summer. This is to be expected, since the physiological state of the algae will influence the activity of the plants. Two factors that influence the rate of nitrogen fixation (or C_2H_2 reduction) by algae are their supplies of nitrogen and phosphorus.

Algae grown in an environment containing adequately fixed nitrogen (NH_4 or NO_3) do not fix N_2 without a preliminary starvation period during which the nitrogenase enzymes can develop (11). Therefore, when blue-green algal blooms occurring after the fall, 1970, turnover of Lake Mendota were found to consist mainly of nitrogen-fixing species which had almost no capability for fixing N_2, it was thought that lack of a preliminary starvation period was the controlling factor. This hypothesis was correlated with the knowledge that filaments of these algae did not contain hererocysts (clear cells in which the nitrogenase enzymes seem to be located), whereas during earlier periods of the year, when these species of algae had relatively high fixation rates, there were many heterocysts present. Specifically, on July 6, 1970, 35 filaments from a bloom of *Aphanizomenon* sp. contained 9 heterocysts and these algae reduced C_2H_2 at the rate of 45 nmoles/mg of algae/30 min, whereas on October 21, 1970, 35 filaments from an *Aphanizomenon* sp. bloom had no heterocysts and these algae only reduced C_2H_2 at the rate of 3 nmoles/mg/30 min.

If nitrogen-fixing algae are limited in phosphorus, they will respond after only 0.5-hr incubation in the presence of available phosphorus by having higher C_2H_2 reduction rates (12). Since other factors can affect the rate of C_2H_2 reduction, only a response to added phosphorus should be used as an indication of phosphorus limitation.

In mid-June 1970, samples of blooms of *Aphanizomenon* sp. from Lake Mendota and Monona Bay were compared. The *Aphanizomenon* from Lake Mendota appeared to have adequate phosphorus for the amount of algae present: their extractable PO_4-P amounted to 0.22 mg PO_4-P/100 mg of algae, and after incubation in lake water or lake water containing an additional 0.04 mg of PO_4-P/liter, there was little increase in their C_2H_2 reduction rate (47 to 53 nmoles/mg/30 min). In contrast, the *Aphanizomenon* from Monona Bay had only 0.06 mg of PO_4-P extracted from 100 mg of algae, and the rate of C_2H_2 reduction increased to more than 150% of the original value on incubation in lake water supplemented with 0.04 mg of PO_4-P/liter (91 to 140 nmoles/mg/30 min).

Thus this test can be used to indicate whether nitrogen-fixing species of algae have developed with adequate or limiting quantities of available phosphorus and to demonstrate that the nutritional status of the same bloom-producing species can vary with the season, perhaps being different in lakes less than 1 mile apart when sampled on the same day.

By using laboratory cultures of *Anabaena flos aquae* (Ind. 1444) grown under phosphorus-limiting conditions, the availability of the phosphorus in various lake waters has been measured using the C_2H_2 reduction rate. The rate of reduction of C_2H_2 by such phosphorus-limited algae will be doubled by the addition of as little as 0.010 mg of P/liter to phosphorus-free medium or lake waters with low phosphorus contents.

Thus, by comparing the rates of C_2H_2 reduction by *Anabaena* in control media having different levels of phosphorus with the rates obtained when lake waters are used, the concentration of available phosphorus in the lake waters can be predicted. It was found that from June to mid-July, 1970, the surface waters of Lake Mendota had surplus phosphorus; there was no increase in C_2H_2 reduction rates when PO_4-P was added to *Anabaena* in surface water samples. However, from mid-July through August, 1970, logical responses to added PO_4-P were obtained: the addition of 0.02 mg P/liter to phosphorus-free medium or a sample of surface water obtained on

August 8, 1970, caused the same response (rate of C_2H_2 reduction for increasing a 30 min incubation from 2.9 to 5.6 nmoles/mg). On this date and at other times, phosphorus-limited Anabaena incubated in samples of water from below the thermocline had relatively high C_2H_2 reduction rates and did not respond to added PO_4-P.

OTHER BIOASSAY METHODS

Other methods for the bioassay of available nutrients have been reviewed by various authors. These include both chemical analyses and measurements of reactions related to the nutritional status of algae or aquatic weeds. Laboratory evaluations of the availability of any nutrient under specific environmental conditions can be accomplished using chemical analyses before and after exposure of a water sample to nutrient-limited algae. Chamberlain and Shapiro (13) have employed this technique to evaluate various chemical methods for the analysis of phosphorus. It also would be applicable to any other nutrient that could be readily separated from the algae after a suitable incubation period. Results can be obtained in a relatively short period of time by this method, but growth experiments should follow to show that the loss of a particular nutrient from solution results in increased growth of the test organism.

As more sensitive methods of analysis for particular compounds or reactions of algae are developed, they are soon applied to nutritional bioassays. The response of nutrient-limited algae to the addition of various nutrients as measured fluorometrically by an increase in chlorophyll is under investigation. The stimulation of the photosynthesis of in situ phytoplankton by the addition of either organic or inorganic nutrients has been followed by ^{14}C assimilation studies (14,15). These methods are usually used as alternatives to growth studies with in situ phytoplankton.

Increases in rates of photosynthesis after different nutrients or combinations of nutrients are added are believed to indicate which nutrient factors are limiting further development of the algae. As pointed out by Wetzel (15), when dealing only with stimulatory responses without a direct increase in growth of an organism, a great deal of caution should be exercised in interpreting the results. The short incubation times required for responses (a few hours to several days) make these tests valuable in predicting results to be obtained with growth studies, but there is a lack of basic data correlating the stimulation of photosynthesis by various nutrient additions with ultimate growth responses. Also, as in all nutrient assays, the degree of normal variability to be expected in results should be well understood; this will permit critical evaluation of coincidental values that happen to match the researchers' theories. However, these methods do give limnologists a tool to predict responses that could be obtained by further growth tests without the relatively long incubation periods required for algae to respond by increases in growth.

Total nutrient analysis for the diagnosis of the nutritional status of plants has been an important tool for analyzing economic plants growing in soil. The concentration of an element in an organism varies over a wide range in response to the availability of the element in the environment. Plant content of an element below a critical concentration indicates that growth is being limited by the supply of that element.

Various workers have used the concentrations of phosphorus per cell to follow the utilization of phosphorus under different environmental conditions and to show whether algae had surplus or limiting quantities of the element (16,17). However, the utilization of such data is limited to comparing algae of similar sizes. Gerloff and Fishbeck (18) have reported critical cell concentrations of several elements on a dry weight basis, and this type of calculation may be more practical. Yet Gerloff (19) has pointed out that it might be necessary to use some correction factors if the algae analyzed had different amounts of sheath materials. Therefore, additional knowledge of the ratio of protoplasmic to nonprotoplasmic constituents may be needed for the application of the tissue analysis technique.

DISCUSSION

Growth tests with algae in water samples can determine the relative ability of different water samples to support the growth of algae, which nutrient may be limiting the growth of algae, and how algae will respond to an increase in a nutrient or a combination of nutrients. Suppose that the objective is to determine the relative level of eutrophication or the ability of water samples to support the growth of algae by comparing either different lakes or different areas or depths in one lake. The first step, then is to determine the reproducibility of the tests to be applied. If results from five tests of preserved lake water indicate that there was a range of obtained values for the available nitrogen between 0.15 to 0.25 mg N/liter, then a difference of 50% between two lake water samples having values near these concentrations should not be considered significant.

The value of bioassays is the record of responses of algae to growth conditions, and responses with some variability can be expected in any survey. Of course it is first necessary to determine the amount of difference among results that constitutes significance, so that minor differences that may reflect the researcher's own opinions will not be taken seriously. Once critical evaluations of the reproducibility of growth bioassays are established for different concentration ranges of the essential nutrients, the growth potential of different waters would be a valuable tool: the relative eutrophication of lakes could be established, sources of nutrients that have meaningful effects on a lake could be identified, and tests could be made of nutrient removal processes that might have a practical effect on the lacustrine environment.

Another function of growth bioassays is to determine how the algae in a lake environment would respond to increases in certain nutrients, thereby indicating which nutrient(s) might be considered most likely as the limiting factor for the growth of algae in that water sample. An increase in growth due to the addition of a nutrient (nutrient spiking) is the most logical response to determine but, as has been pointed out, a week or more of incubation might be required. Faster responses can be obtained by following preliminary changes that might take place, such as an increase in chlorophyll as measured by fluorometry (1 to 3 days for response) or an increase in photosynthesis as measured by the ^{14}C technique (a few hours for response).

Caution must be used with such short cuts until basic data have shown that positive responses by either of the tests are followed by the ultimate response, growth. These tests should not replace growth tests; they should merely be used as guides to make better decisions on the amount and type of nutrient to use for the more time-consuming growth tests. Here again, the normal variability of results must be known so that significant results will not be confused with lucky coincidences.

It is obvious that if increasing quantities of a limiting nutrient are added to a water sample, eventually another nutrient is forced to become the limiting factor. In order to keep some perspective on the addition of nutrient spikes, a response to an added nutrient should be sought that is significantly different from the normal variations in bioassays. The quantity of a nutrient that causes a two- to fivefold increase in growth might be appropriate. For instance, if growth in oligotrophic waters is to be measured by cell counts or fluorometry measurements, the lower level of a series of N spikes could be 0.2 mg N/liter. If a sample water is relatively fertile (e.g., hypolimnion waters containing 1 mg N/liter), the response to a spike of 0.2 mg N/liter could be lost in the normal variability to be expected, and it would take a spike of about 1 mg N/liter or more to give a significant increase.

When dealing with lakes of known nutrient content, it might be possible to select spikes that give significant response the first time they are tried. However, if samples of unknown nutrient content are being tested, a relatively wide range of spikes is necessary to lessen the chance of missing the concentration that could produce a significant response. It is here that short-term informative tests, such as fluorometry or ^{14}C, can be especially valuable.

Since growth of algae in lake water samples can be limited by two or more nutrients in close succession, spikes of combinations of nutrients that might be significant must also be tested. The most obvious nutrients to test are nitrogen, phosphorus, and iron (also silicon, if diatoms are important in the lake flora). It is thus evident that a great deal of preliminary basic research should be carried out in order to limit the ranges of combinations of nutrients to be tested to those most frequently giving significant results. Because the possible combinations of concentrations and nutrients in an unknown sample of lake water that could be significant may be very large, bioassays should be approached with the view that one test may not answer all the questions. Indeed, it might be better to be prepared to run several tests, each based on results of the previous one. The results of these tests will give information that can be used to develop a rational framework for dealing with practical problems.

Descriptions have been presented of short-term bioassays that measure changes in certain enzymatic activities or chemical fractions that reflect meaningful nutrient changes. A great deal of information about significant sources of nutrients and the comparative nutritional status of plants from different environments can be obtained from such tests. However, as in growth tests, caution must be used in the interpretations placed on the data collected. If algae attached to a substrate in an environment, such as _Cladophora_ sp. along a lake shore, are used it can be reasonably assumed that changes in the nutritional status of the test algae were due to changes in the environment. If phytoplankton are tested, it is very difficult to be sure that changes recorded are due to environmental changes, since it is possible that different phytoplankton populations have been sampled.

It has been pointed out that the same species of algae collected at different depths in a lake could have different nutritional characteristics: *Gloeotrichia* sp. from subsurface samples of Lake Mendota contained surplus phosphorus (0.36 mg of PO_4-P extracted/100 mg of algae), whereas *Gloeotrichia* from surface samples were phosphorus limited (0.12 mg of PO_4-P/100 mg). Two interpretations of such results would be that:

1. Algae in the surface water had exhausted the supply of available phosphorus at a time when subsurface waters had not yet been stripped of theirs, probably because fewer algae are present in subsurface waters.
2. The light intensity at subsurface environments was low enough that algae from these depths had not yet been forced to use their surplus stored nutrients, whereas the light intensity at the surface forced higher metabolic rates to be maintained in order to overcome the effects of high light intensities.

One assumption that can be made is that algae growing in surface waters, such as *Cladophora* sp., change in significant ways when the surface waters cease to be able to provide adequate nitrogen or phosphorus for the growth of algae in the high light intensities of the surface waters. Algae at lower depths and lower light intensities will require longer periods between nutrient supplies to show a lack of adequate nitrogen or phosphorus in their environment. When a bloom of *Microcystis* sp. which is not limited in nitrogen or in phosphorus suddenly appears at the surface of Lake Mendota, it may be assumed that the surface waters contain adequate nitrogen or phosphorus to support this amount of algal growth. In fact, however, these algae have only recently ascended to the surface, and the nutrient supply of the surface waters is not related to their nutritional status. Furthermore, if bioassays are carried out with mixtures of nitrogen-fixing and nonnitrogen-fixing phytoplankton, it might be difficult to interpret the results, since the nitrogen-fixing species could be phosphorus limited and the nonfixing algae could be nitrogen limited but have surplus phosphorus. Thus tests with in situ algae must be frequent enough so that trends can be followed and careful scrutiny given to the species composition of samples tested at different times.

The evaluation of nutrient sources, such as rain as a source of nitrogen in in situ tests or the lack of availability of the nitrogen or phosphorus from aerobic lake muds in field or laboratory tests, leads to judgments about the relative importance for algae of the different nutrient sources. It must be borne in mind, however, that only certain environmental conditions are being evaluated with each test. Rains in certain portions of the United States contain little or no available nitrogen compounds, whereas the Midwest area annually obtains about 10 lb of nitrogen per acre (20) from this source.

Surface aerobic lake muds have not revealed significant nitrogen and phosphorus sources when tested by several bioassay techniques. However, in nature these muds are stratified and are anaerobic not too far from the mud-water interface. The interstitial waters of such muds do contain available nitrogen and phosphorus compounds. Therefore, by disturbing in situ anaerobic muds it would be possible to release certain amounts of available nutrients that would never be determined by tests of aerobic muds alone. Here again, it is always important to check results for significance, and to check for the significance of the results to the aquatic environment.

SUMMARY

In order to determine the biological availability of algal nutrients in a sample of water and to measure the response to changes in the growth-limiting nutrient, the following tests can be used:

1. Growth attained by selected algae can be measured in spiked and untreated samples (as much as 2 to 3 weeks of incubation required).
2. Available nitrogen of the sample can be calculated from NH_4-N absorption rates after incubation with nitrogen-limited algae (1 or 2 days incubation required).
3. Available phosphorus of the sample can be calculated from increases in extractable PO_4-P after incubation with phosphorus-limited algae (1 or 2 days incubation required) or increases in the rate of C_2H_2 reduction by phosphorus-limited, nitrogen-fixing algae (0.5 hr incubation required).

To determine if an environment or a source of nutrient has supplied adequate or limiting amounts of nitrogen or phosphorus for in situ algae or aquatic weeds, the following information can be used:

1. Nitrogen-limited algae or aquatic weeds will absorb NH_4-N in the dark at rates greater than 15 μg NH_4-N/(10 mg)(hr).

2. Nitrogen-fixing algae grown in environments with surplus available fixed nitrogen (NH_4 + NO_3) will have relatively few heterocysts and low nitrogen fixation or C_2H_2 reduction rates.

3. Phosphorus-limited algae or aquatic weeds will have less than 0.08 mg of extractable PO_4-P per 100 mg of plant material and will have relatively high alkaline phosphatase activities (800 to 1600 units/mg).

4. Phosphorus-limited nitrogen-fixing algae will respond to incubation with added PO_4-P with increased rates of C_2H_2 reduction.

The following is a summary of limnological facts recently corroborated by bioassay analyses:

1. The nutritional status of certain species of algae can vary from lake to lake, or even from different areas or depths in the same lake, on the same sampling date; subsurface samples of planktonic algae have been shown to have surplus nitrogen or phosphorus at times when the same species in surface waters were nitrogen or phosphorus limited.

2. Lake Mendota algae contain surplus nitrogen and phosphorus in spring, can become nitrogen or phosphorus limited (at the same time or independently) during summer, and yet have surplus nitrogen and phosphorus again after the fall overturn. This pattern may represent the changes that take place in similar eutrophic lakes with spring and fall overturns.

3. Rain can be a significant source of available nitrogen to algae in surface waters in the Madison lakes. Less dramatic increases in available phosphorus were also associated with certain rains.

4. Filamentous green algae, such as *Cladophora* sp., and aquatic weeds that have been in an environment containing surplus nitrogen for a week or more are usually visibly coated with epiphytes (i.e., have a brown appearance).

5. In certain laboratory environments, solubility is not a limiting factor in the nutrition of algae, since the equilibrium between soluble and insoluble nutrients allows algae to compete successfully with insoluble forms of nutrients.

6. Factors other than insolubility prevent the

nitrogen or phosphorus of certain samples of aerobic lake muds from being readily available for the growth of algae; phosphorus-limited *Spirogyra* sp. have been found growing through layers of muds with 0.1% total phosphorus content.

7. The nutrients of live algae and aquatic weeds are not effectively available to other plants even when nutrient-limited plants are mixed with plants containing surplus nutrients. When plants containing surplus nutrients are killed, however, their nutrients become available for nutrient-limited plants. Therefore, there is an obvious ecological preference for the physical harvesting of obnoxious growths of aquatic weeds that, if killed by chemical treatment, could release some of their nutrients to the lake water in forms available for the growth of algae.

ACKNOWLEDGMENTS

This research was supported by Grant 16010 EHR from the Federal Water Quality Administration. The technical aid of Mrs. S. L. Faust and the chemical analyses by Mrs. M. Torrey are gratefully acknowledged.

REFERENCES

1. T. E. Maloney (in charge), "Provisional Algal Assay Procedure Evaluation," National Eutrophication Research Program, Pacific Northwest Water Laboratory, Corvallis, Ore., 1970.
2. G. P. Fitzgerald, "Evaluations of the Availability of Sources of Nitrogen and Phosphorus for Algae," J. Phycol., 6, 239-247 (1970).
3. G. C. Gerloff and F. Skoog, "Cell Contents of Nitrogen and Phosphorus as a Measure of Their Availability for Growth of *Microcystis Aeruginosa*," Ecology, 35, 348-353 (1954).
4. M. B. Allen, "The Cultivation of *Myxophyceae*," Arch. Mikrobiol., 17, 34-53 (1952).
5. E. P. Hughes, P. R. Gorham, and A. Zehnder, "Toxicity of a Unialgal Culture of *Microcystis Aeruginosa*," Can. J. Microbiol., 4, 225-236 (1958).
6. G. P. Fitzgerald and T. C. Nelson, "Extractive and Enzymatic Analyses for Limiting or Surplus Phosphorus in Algae," J. Phycol., 2, 32-37 (1966).
7. G. P. Fitzgerald, "Aerobic Lake Muds for the Removal of Phosphorus from Lake Waters," Limnol. Oceanogr.,

15, 550-555 (1970).
8. G. P. Fitzgerald, "Detection of Limiting or Surplus Nitrogen in Algae and Aquatic Weeds," J. Phycol., 4, 121-126 (1968).
9. G. P. Fitzgerald, "Some Factors in the Competition or Antagonism Among Bacteria, Algae, and Aquatic Weeds," J. Phycol., 4, 351-359 (1969).
10. W. D. P. Stewart, G. P. Fitzgerald, and R. H. Burris, "In Situ Studies on Nitrogen Fixation Using the Acetylene Reduction Technique," Proc. Nat. Acad. Sci. (U.S.), 58, 2071-2078 (1967).
11. W. D. P. Stewart, G. P. Fitzgerald, and R. H. Burris, "Acetylene Reduction by Nitrogen-fixing Blue-green Algae," Arch. Mikrobiol., 62, 336-348 (1968).
12. W. D. P. Stewart, G. P. Fitzgerald, and R. H. Burris, "Acetylene Reduction Assay for Determination of Phosphorus Availability in Wisconsin Lakes," Proc. Nat. Acad. Sci. (U.S.), 66, 1104-1111 (1970).
13. W. Chamberlain and J. Shapiro. "On the Biological Significance of Phosphate Analysis; Comparison of Standard and New Methods with a Bioassay," Limnol. Oceanogr., 14, 921-927 (1969).
14. C. R. Goldman, "Micronutrient Limiting Factors and their Detection in Natural Phytoplankton Populations," in "Primary Productivity in Aquatic Environments," C. R. Goldman, Ed. Mem. 1st. Ital. Idrobiol., 18 Suppl. Berkeley, University of California Press, 1965, pp. 121-155.
15. R. G. Wetzel, "Nutritional Aspects of Algal Productivity in Marl Lakes with Particular Reference to Enrichment Bioassays and their Interpretation," in "Primary Productivity in Aquatic Environments," C. R. Goldman, Ed. Mem. 1st Ital. Idrobiol., 18 Suppl., Berkeley, University of California Press, 1965, pp. 137-157.
16. F. J. Mackereth, "Phosphorus Utilization by Asterionella Formosa Hass," J. Exp. Bot., 4, 293-313 (1953).
17. E. J. Kuenzler and B. H. Ketchum, "Rate of Phosphorus Uptake by Phaeodactylum Tricornutum," Biol., Bull., 123, 134-145 (1962).
18. G. C. Gerloff and K. A. Fishbeck, "Quantitative Cation Requirements of Several Blue-green Algae," J. Phycol., 5, 109-114 (1969).
19. G. C. Gerloff, "Evaluating Nutrient Supplies for the Growth of Aquatic Plants in Natural Waters," in "Eutrophication: Causes, Consequences, Correctives." National Academy of Sciences, Washington, D.C., 1965, pp. 537-555.

20. G. F. Lee (Chairman), "Report on the Nutrient Sources of Lake Mendota," Tech. Rep. Lake Mendota Problem Committee, Jan. 3, 1966 (mimeo), 41 pages.

APPENDIX

Analysis of Nutrient Content by Growth

Details of the Standardized Algal Assay Procedure can be obtained from Dr. T. E. Maloney, Environmental Protection Agency, Pacific Northwest Water Laboratory, Corvallis, Oregon 97330.

Extractive Analysis for Phosphorus (P) Nutrition

Materials

Algae or leaves of aquatic weeds--10 to 80 mg samples, in triplicate.
Gorham's medium (- N-P) (Hughes et al., 1968).
Boiling water bath.
Ortho-PO_4-P analysis (APHA, 1965).
Dry weight oven (110°C).

Procedure

Place 10 to 80 mg samples of washed [Gorham's (- N-P) medium] plant material in 40 ml of Gorham's (- N-P) medium. Extract in boiling water bath for 60 min. Analyze supernatant for ortho-PO_4-P. Collect algal debris and measure dry weight.

Calculations

Report ortho-PO_4-P extracted as milligrams of PO_4-P extracted from 100 mg of plant material (dry weight basis).

Limits

In situ phosphorus-limited algal samples or phosphorus-limited algal cultures exposed to potential sources of phosphorus can be used to detect as low as 0.04 mg P/liter using 10-mg samples of algae.

Definition

Extracted algae that give less than 0.08 mg PO_4-P/100 mg algae are considered to be phosphorus-limited.

Enzymatic Analysis for Phosphorus (P) Nutrition

Materials

Algae or leaves of aquatic weeds--1 to 20 mg samples, in triplicate.
Gorham's medium (- N-P).
Buffer solution--1M tris, 0.01M $MgCl_2$, adjusted to pH 8.5 with acetic acid.
p-Nitrophenyl phosphate solution--30 mg/100 ml.
Incubator--35 to 37°C.
Colorimeter--395 mμ.

Procedure

Place 1 to 20 mg of washed [Gorham's (- N-P) medium] plant material in 32 ml of Gorham's (- N-P) medium. Add 4 ml of buffer and 4 ml of p-nitrophenylphosphate solution. Incubate with occasional mixing for 15 to 60 min at 35 to 37°C. Analyze supernatant for nitrophenol (395 mμ). Collect plant material for dry weight analysis (110°C).

Calculations

Report results as units of enzyme per milligram (dry weight) of plant material. One unit of alkaline phosphatase is defined as the amount of enzyme liberating 1 mμmole of nitrophenol per hour.

Limits

Phosphorus-limited algae will have as much as 25 times more alkaline phosphatase activity than algae with surplus phosphorus. However, the alkaline phosphatase of phosphorus-limited algae only decreases with growth under adequate phosphorus conditions, so this test is most appropriate for in situ analyses rather than for evaluating nutrient sources in short-term laboratory tests.

Rate of Ammonia Absorption for Nitrogen (N) Nutrition

Materials

Algae or aquatic weed leaves--5 to 20 mg samples, in triplicate.
Gorham's (- N-P) medium.
NH_4-N stock--0.1 mg of NH_4-N/ml.
NH_4-N analysis by Nesslerization using Rochelle salt (APHA, 1965).

Procedure

Place 5 to 20 mg samples of washed [Gorham's (- N-P) medium] plant material in 30 ml of Gorham's (- N-P) medium. Add 1 ml of NH_4-N stock (0.1 mg N). Incubate in dark at room temperature with occasional mixing for 1 hr. Analyze supernatant for NH_4-N. Collect plant material for dry weight analysis.

Calculations

Report results as micrograms of NH_4-N absorbed per 10 mg (dry weight) × hour.

Limits

Nitrogen-limited algae absorb NH_4-N in the dark 4 to 5 times faster than algae with adequate nitrogen. This test can be used for detection of potential sources of nitrogen by using 1 day or more incubation periods with nitrogen-limited algae. Lower limits of detection have not been established, but 10 mg of algae can be used to detect as little as 0.05 mg of nitrogen.

Definition

Algae with ammonia absorption rates of more than 15 μg N/10 mg × hour are considered to be nitrogen-limited.

Acetylene Reduction Rates for Sources of Phosphorus (P)

Materials

Phosphorus-limited nitrogen-fixing blue-green algae, such as _Anabaena flos aquae_ (Ind. 1444), triplicate 1-mg samples per test source of phosphorus.

5 to 7 ml stoppered serum bottles and 1 and 2.5 ml gas syringes.
Acetylene (purified grade).
Ethylene (purified grade).
Gas chromatography analysis (Stewart et al., 1970).
Total phosphorus analysis (Gales et al., 1966).

Procedure

Place 1-mg samples of phosphorus-limited algae in 25 to 500 ml of sample to be tested for available phosphorus. Incubate in light (100 to 400 footcandles, room temperature) for 1 hr. Centrifuge algae, make volume to 1 ml, and add to 5 to 7 ml serum bottle. Stopper bottle. Add 1.4 ml of acetylene and mix. Restore pressure to atmospheric by pricking the serum stopper. Incubate in light (100 to 400 footcandles, room temperature) for 30 min. Poison system by injecting 0.2 ml of 5N H_2SO_4. Analyze ethylene formed by gas chromatography. Control samples containing 0, 0.01, 0.025, and 0.05 mg P/liter should be used to compare responses of phosphorus-limited algae to known phosphorus sources and test samples. Additional controls made up of test sample + 0.05 mg PO_4-P/liter should be used to test for toxicity in the test sample. Use standards of ethylene to convert chromatograph peak heights to nmoles of ethylene.

Calculations

Report results as available milligrams of phosphorus per liter versus total milligrams of phosphorus per liter in samples. Response of algae as nmoles of ethylene produced with standards of PO_4-P are used to calculate milligrams of available phosphorus per liter.

Limits

Phosphorus-limited *Anabaena* will respond to 0.01 mg P/liter with a 100% increase in ethylene produced, when compared with samples containing no phosphorus.

ADDITIONAL REFERENCES

1. American Public Health Association, "Standard Methods--Water and Wastewater," 12th ed. New York, 1965, 769 pages.
2. M. E. Gales, Jr., E. C. Julian, and R. C. Kroner,

"Method for Quantitative Determination of Total Phosphorus in Water," Am. Water Works Assoc., 58, 1363-1368 (1966).

Chapter VI

NUTRIENT SUBMODELS AND SIMULATION MODELS OF PHYTOPLANKTON PRODUCTION IN THE SEA*

John J. Walsh and Richard C. Dugdale

Department of Oceanography
University of Washington
Seattle, Washington

Prediction of the effect of added nutrients on a natural marine ecosystem is an important goal; it appears, however, that quantitative prediction is hardly possible at present. This situation is the result of inadequacies of both theory and data. These inadequacies in turn probably are attributable to a failure to adopt a systems approach to the prediction problem. The main tool of systems analysis is a simulation model used in conjunction with field and experimental studies. Such models, varying widely in nature, are used for the common and urgent problem of describing and understanding relatively complex phenomena, and phytoplankton productivity is certainly one of these.

Simulation models of marine ecosystems are being developed in our laboratory so that the rapid flood of data from contemporary continuous data acquisition systems can be interpreted, synthesized, and understood by workers in the field. In contrast to results generated from inherently noncausal statistical models such as multiple regression analysis (1), output of our simulation models stems from measured rate data incorporating

*Contribution No. 655, Department of Oceanography, University of Washington, Seattle, Washington.

causal relationships between trophic levels. Our models consist of sets of coupled nonlinear partial differential equations describing the temporal and spatial distribution of nutrients and phytoplankton. More often than not, this degree of nonlinear and spatial complexity makes it difficult or impossible to describe the system in mathematical form sufficiently simple to allow an analytical solution. Modern high-speed computers must be used for numerical solution of the equations, and the fast pace of current developments in simulation modeling is a direct result of the spectacular rate of development of powerful, high-speed electronic computers.

No attempt is made here to survey the literature on nutrient-based phytoplankton simulation models; for the most part it is nonexistent, and the most significant information is not available in the open literature. Instead, the modeling approach employed in our laboratory to study nutrient and phytoplankton processes in two systems, upwelling and marine outfalls, is presented.

UPWELLING ECOSYSTEMS

The complexities of coastal upwelling ecosystems are being unraveled in a project sponsored by the U.S. Component of the International Biological Program. In Figure 1 our ideas of the role of models in research on total ecosystems are illustrated. Initial observations are required to obtain some idea of the system under study. Following data analysis, an elementary simulation model is constructed. The cycle is completed by carrying out experimentation in the laboratory and designing new cruise plans in light of the preliminary results of the modeling.

In spring of 1969 the R/V Thomas G. Thompson made a major cruise to Peru, where coastal upwelling processes form the basis for the world's largest single fishery. The annual catch of anchoveta (*Engraulis ringens*) exceeds 10 million metric tons. An area near San Juan, Peru, about 20 × 50 km in size, was studied in detail (2). Through a variety of methods including discrete station sampling and an automated system for mapping surface parameters, a picture of the distribution of certain variables was obtained. In Figure 2, for example, cold water can be seen rising near the shore and moving away from it in a plume. Plume structures occur in all the major upwelling areas (3), and this environmental heterogeneity may be a key to the evolutionary development of the short, productive diatomaceous phytoplankton-clupeiform fish food chains found in these areas. Distributions

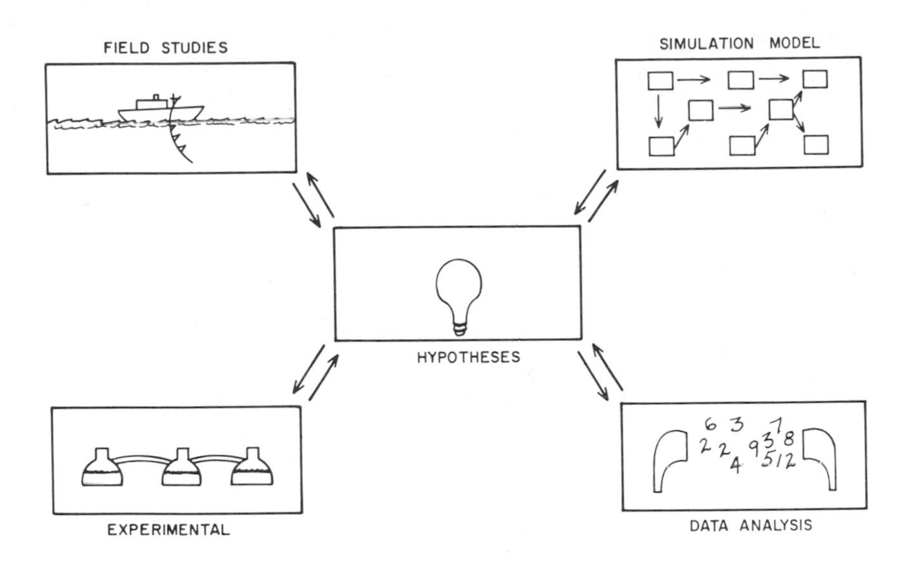

Fig. 1. The cyclic nature of a systems approach to the study of aquatic ecosystems.

of pilchard (4), herring (5), sardine (6), and anchoveta (7) have been observed in association with temperature gradients, and this type of herbivorous fish may use the cold water plumes characteristic of upwelling areas to locate, feed, and stay with the phytoplankton blooms contained within the upwelled water.

The distribution of phosphate (Figure 3) follows the plumelike structure of temperature; the values of phosphates are high near shore and low offshore. Silicate and nitrate distributions correspond to that of phosphate as well. In contrast, the distribution of chlorophyll (Figure 4) shows a minimum near shore and increases as the water moves offshore. The increase along the chlorophyll plume reflects growth of phytoplankton populations and depletion of the nutrients.

Our first model of the Peru upwelling system is based on a one-dimentional, two-layered series of spatial blocks aligned along the axes of the plumes (Figure 5). The model is designed to run on our shipboard IBM 1130 computer, and the spatial resolution of the simulation is thus restricted by the available core storage of this machine. Water circulation and the advective transfer

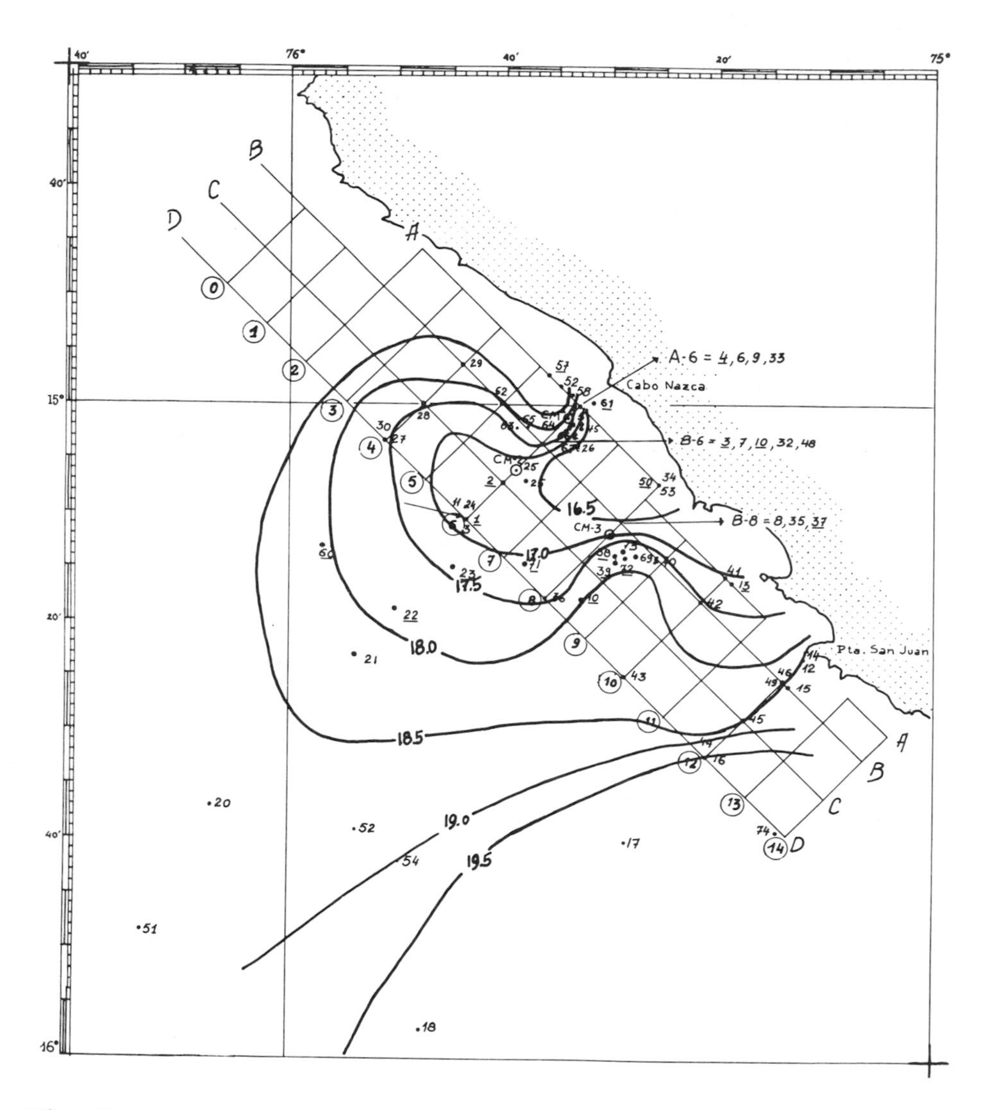

Fig. 2. The distribution of temperature, °C at 3 m depth, off Pt. San Juan, Peru.

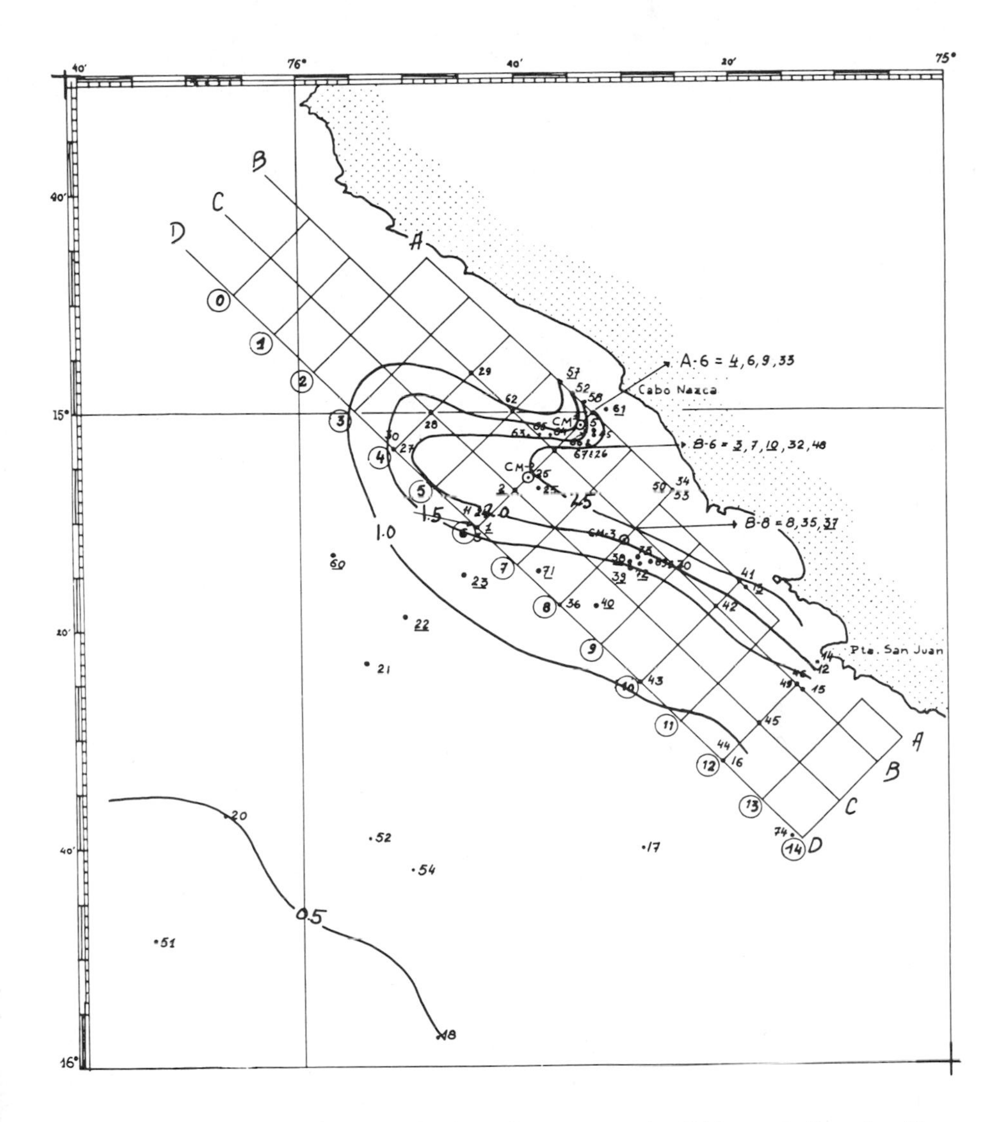

Fig. 3. The distribution of phosphate, μg-at./liter at 3 m depth, off Pt. San Juan, Peru.

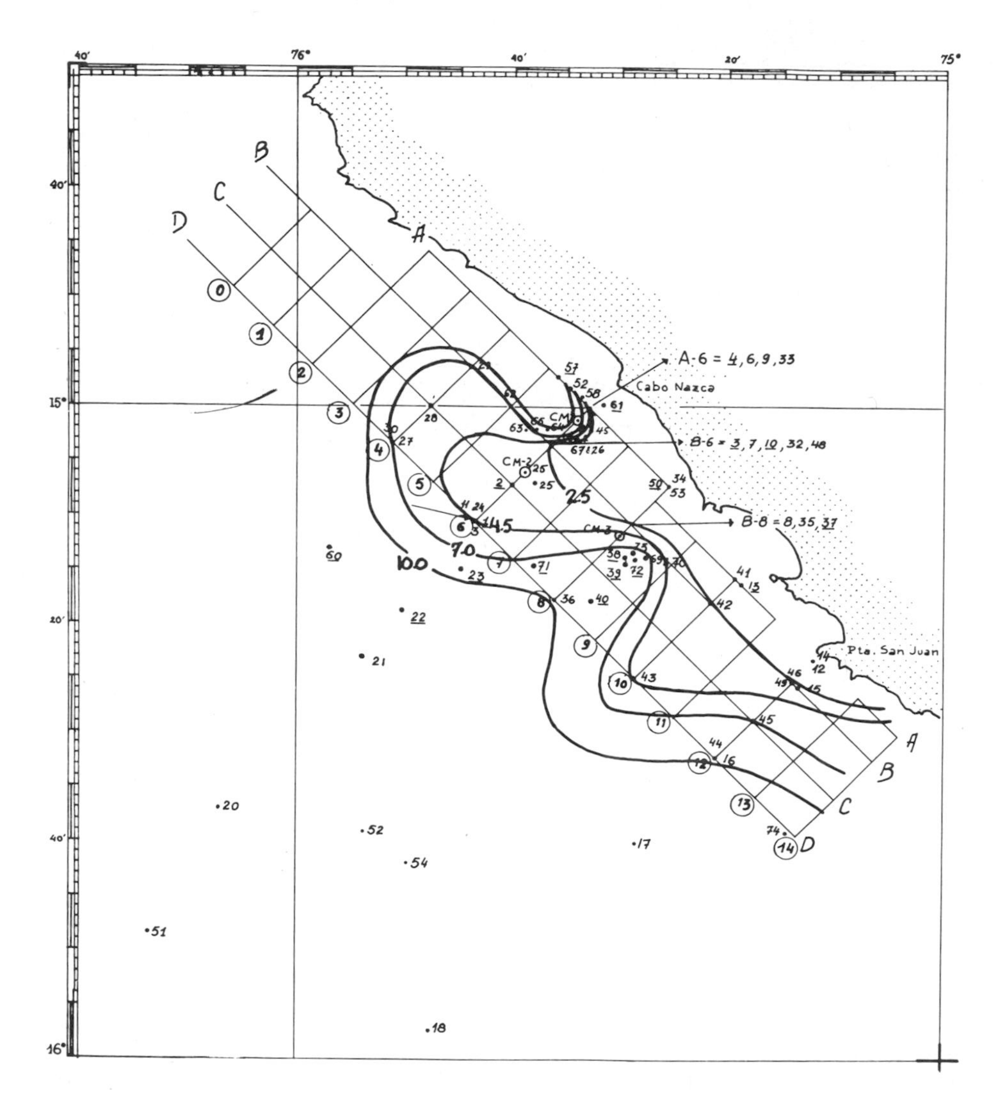

Fig. 4. The distribution of chlorophyll, μg/liter at 3 m depth, off Pt. San Juan, Peru.

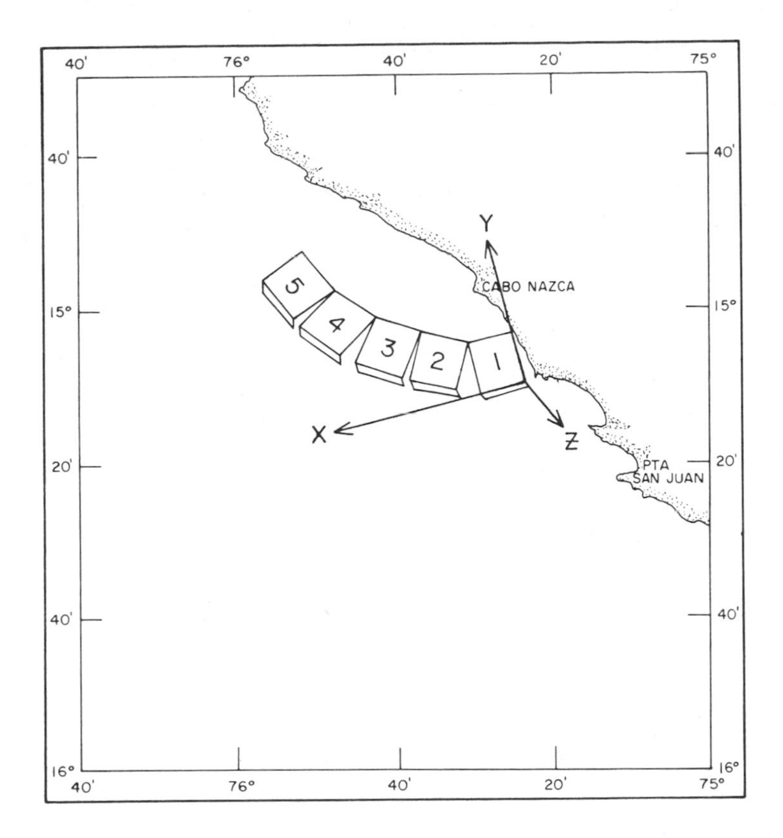

Fig. 5. The spatial resolution of a simulation model representing the plume of upwelled water off Pt. San Juan, Peru.

of materials between blocks is treated as shown in Figure 6. This is essentially an input-output model of nitrogen flux, and all materials in circulation must be accounted for by continuity considerations. In Figure 7, which describes the biological interaction within any one block, some indication of the true complexity of the system is revealed, although important variables have been omitted for visual clarity. Further details can be found in Walsh and Dugdale (8).

To obtain solutions for the state equations of the model, appropriate rate constants are inserted in finite difference approximations of the differential equations for phytoplankton and the nutrients (some rates are quite well known, others are best guesses), initial and boundary

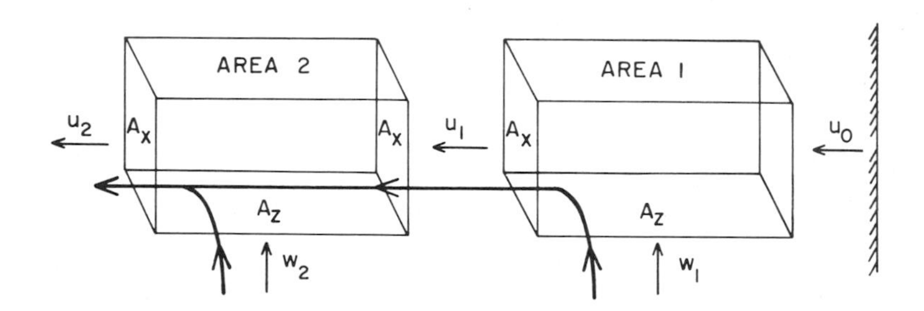

Fig. 6. Conservation of water mass and nutrient flow in the upwelling (w velocity and bottom face Az) and downstream (u velocity and horizontal face Ax) transports of the simulation model.

conditions are stated, and sequential computations of the transfer of material are made for each increment of simulated time, usually one hour. Equation 1 is an example of such a finite difference expression for the temporal and spatial distribution of the phytoplankton:

Change in Phytoplankton — Upwelled Phytoplankton

$$\frac{dP}{dt} \cong - \frac{(w)\, P_{i,j,k} - P_{i,k,l-1}}{\Delta z}$$

Downstream Transport — Lateral Diffusion and Mixing

$$- \frac{(u) P_{i+1,j,k} - P_{i-1,j,k}}{2\Delta x} + \frac{(K_y)(P_{i,j+1,k} - 2P_{i,j,k} + P_{i,j-1,k})}{(\Delta y)(\Delta y)}$$

Sinking — Nutrient Uptake

$$- \frac{(w_s) P_{i,j,k}}{\Delta z} + \frac{(V_{max})(P_{i,j,k})(N_{i,j,k})}{K_t + N_{i,j,k}}$$

Grazing

$$- \frac{(G_{max})(H_{i,j,k})(P_{i,j,k} - P^*)}{K_p + (P_{i,j,k} - P^*)} \qquad (1)$$

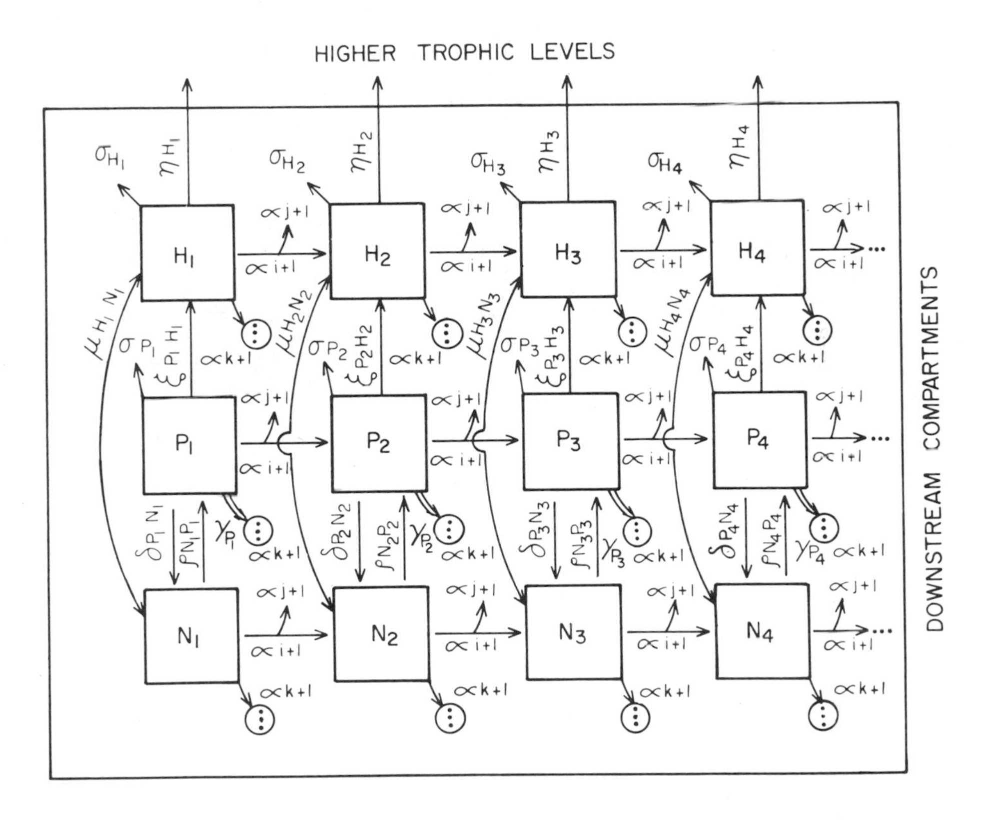

Fig. 7. The simulated fluxes of nutrient uptake ρ, grazing ζ, excretion of phytoplankton δ, and herbivores μ, respiration σ, predation ν, of sinking γ, and downstream, lateral, and vertical advection and diffusion $\alpha_{i,j,k}$ between the nutrients N_i, phytoplankton P_i, and herbivores H_i within each spatial area of the simulation model.

Table 1 lists the symbol definitions for each term of the equation, and a technical report (9) has been prepared on the methods of numerical solution, finite difference approximation, and implementation of the simulation program.

Constant winds are applied in the model to an ocean initially at rest; after 10 days of simulated spin-up time, steady-state solutions are obtained. Through trial-and-error manipulation within reasonable ranges of estimates for missing data, the results presented in Figure 8

Table 1. Definition of the Symbols in Equation 1

Symbol	Definition
$\frac{dP}{dt}$	Change in phytoplankton with time
w	Vertical velocity as a function of wind stress, depth, and distance offshore
$P_{i,j,k}$	Phytoplankton concentration in the surface layer with respect to the offshore x-axis, the alongshore y-axis, and the vertical z-axis
$P_{i,j,k-1}$	Phytoplankton concentration in the subsurface layer
Δz	Depth of the surface layer (i.e., of the spatial block)
u	Downstream velocity as a function of w
$P_{i+1,j,k}$	Phytoplankton concentration to east of surface layer
$P_{i-1,j,k}$	Phytoplankton concentration to west of surface layer
Δx	Length of the spatial block
K_y	Lateral eddy coefficient as a function of distance offshore and scale length
$P_{i,j+1,k}$	Phytoplankton concentration to north of surface layer
$P_{i,j-1,k}$	Phytoplankton concentration to south of surface layer
Δy	Width of the spatial block
w_s	Sinking velocity of the phytoplankton as a function of population age
V_{max}	Maximum nutrient uptake rate under nutrient-saturated conditions; this value may be adjusted to express interaction with other nutrients, inhibitors, or light
$N_{i,j,k}$	Concentration of nutrient in the surface layer with respect to the x, y, z-axes
K_t	Half-saturation constant or the amount of nutrient at which nutrient uptake is half the maximal value
G_{max}	Maximum grazing rate of the herbivores
$H_{i,j,k}$	Concentration of herbivores in the surface layer with respect to the x, y, z-axes
P^*	Grazing threshold, the phytoplankton concentrations below which the herbivores no longer feed
K_p	Concentration of phytoplankton at which grazing occurs at half the maximal value

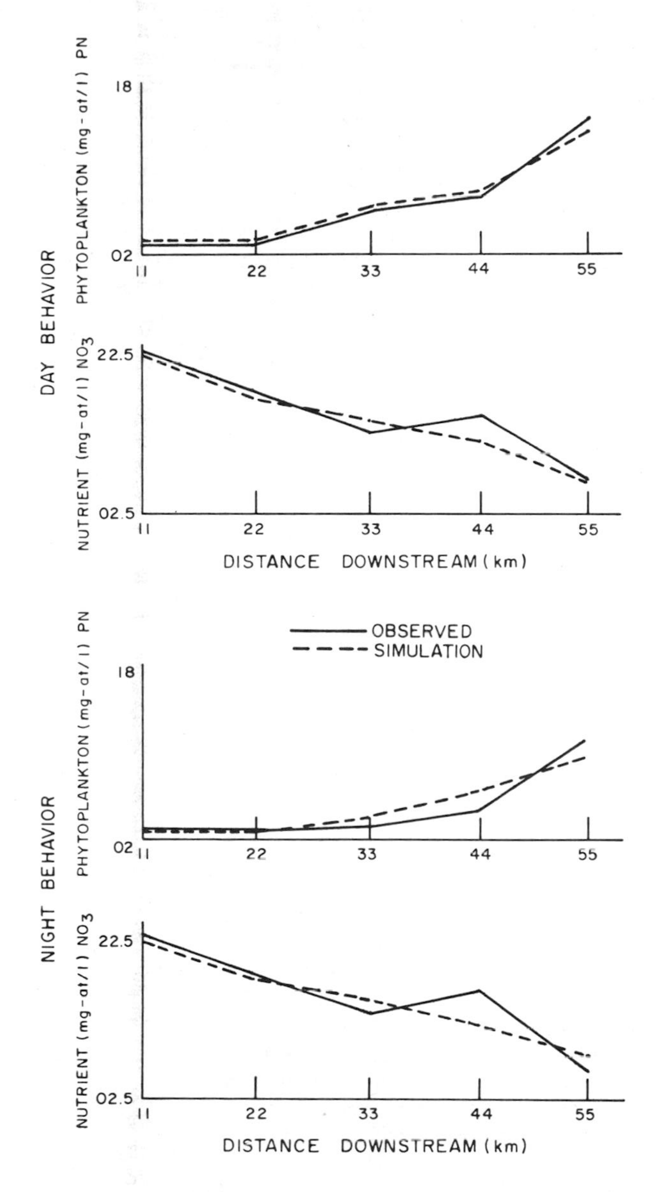

Fig. 8. A comparison of the simulated and observed diel distribution of nitrate and phytoplankton down the upwelling plume off Pt. San Juan, Peru.

were obtained. The agreement between observations and theory is really quite good, although many variables are not known in enough detail to claim any degree of rigorous validation for the preliminary model.

However, it has already been demonstrated from the results that the effect of fish and zooplankton in regenerating primary nutrients cannot be ignored without gross distortion, and our attention has been directed to this much-neglected area (10,11). With incorporation of explicit grazing and excretion terms, and with increased spatial resolution over our present upwelling model, it should be possible to predict the approximate surface phytoplankton and nutrient fields along the axis of a plume of upwelled water such as that observed near Pt. San Juan, Peru.

MARINE OUTFALL ECOSYSTEMS

Ultimately three-dimensional spatial models can be constructed which will help us understand nutrient-phytoplankton processes more fully. A step in this direction was made by developing a two-dimensional, one-layered model of the Hyperion outfall at Los Angeles (12). The downstream distribution of phytoplankton and nutrients in the upper 15 m of the water column was represented with the grid points of Figure 8. The actual dimensions of the diffuser (13) were used, and the limiting nutrient was defined as ammonium with a concentration after dilution, of 3 μg-at./liter. In the model, initial conditions of an ammonium concentration of 1 μg-at./liter, a phytoplankton population of 0.5 μg-at./liter as particulate nitrogen, and a current of 0.2 kn were assumed for the water flowing past the diffuser.

Using a maximum specific uptake rate of 1.0/day inserted in finite difference expressions similar to equation 1, the steady-state phytoplankton and nutrient distributions downstream from the diffuser after 2 days are shown in Figure 9. Interaction occurs between the increase of phytoplankton due to ammonium from the diffuser and the loss due to downstream transport and diffusion by the current, and this generates contours of phytoplankton standing crop intersecting those of ammonium concentration at approximate right angles, giving the appearance of a downstream bloom detached from the sewage field. Although this model is unverified at present, the results could be of use in planning programs in the field. Our recent studies (14) off the Keratsini outfall of Athens, Greece, suggest that ongoing simulation models are invaluable in

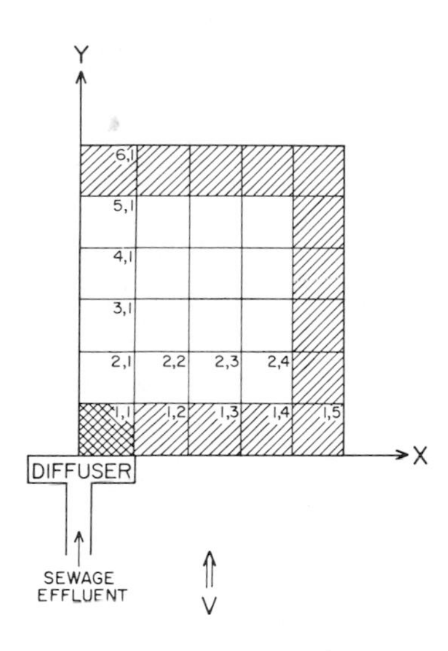

Fig. 9. The spatial resolution of a simulation model representing the Hyperion Sewage outfall off Los Angeles, California.

producing preliminary insight into the dynamics of marine ecosystems.

The results in Figure 10 were obtained from a Fortran program run on an IBM 1130. In a new program for the CDC 6400, the running time has been decreased substantially, even though the spatial resolution was increased from 30 to 300 blocks. However, the results of the CDC simulations appear to be indistinguishable from those obtained with the smaller 1130 version. The sewage model is far below a level of sophistication that could predict the distribution of phytoplankton in an outfall plume. However, as in the case of the upwelling model, a number of improvements are being carried out to meet this objective. The diffusion terms are being revised, terms for toxicity and inhibition of nutrient uptake must be provided, and the present nutrient submodels will be expanded to allow for interaction with the other primary nutrients.

SUBMODELS

The transfer of materials in a complex system model is described through a series of submodels such as the

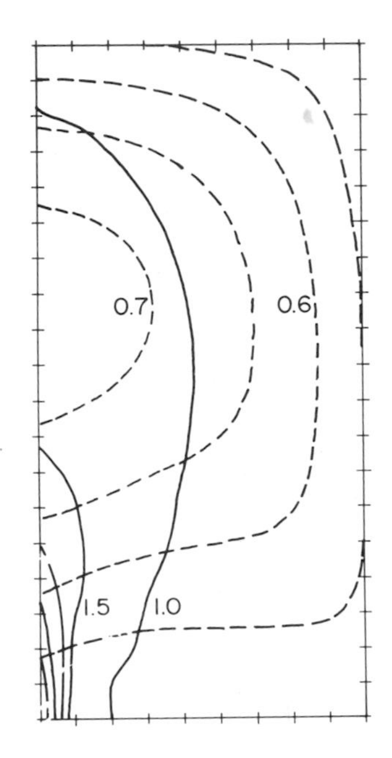

Fig. 10. The simulated downstream distribution of ammonium (solid curves) and phytoplankton (dotted curves) off a sewage outfall.

terms of equation 1. Successful operation of the system model depends on the soundness of all the submodels. Nutrient uptake submodels drive the rest of the trophic submodels constituting either the upwelling or marine outfall system models, and the extent to which the nutrient submodels correctly describe nutrient interactions determines the minimum predictive usefulness of the larger system models. The upwelling ecosystem and the marine outfall models are fundamentally nitrogen-flow models, and substantial progress is being made in the series of submodels required to describe the flow of various species of nitrogen illustrated in Figure 11.

The accumulated knowledge of factors controlling the uptake of inorganic nitrogen is sufficiently detailed to be of great use in the submodels. At any given depth of the water column, uptake of nitrate is limited, usually by the ambient light intensity or by the concentration of nitrate. In general, either form of limitation can be described in terms of Michaelis-Menten kinetics (15).

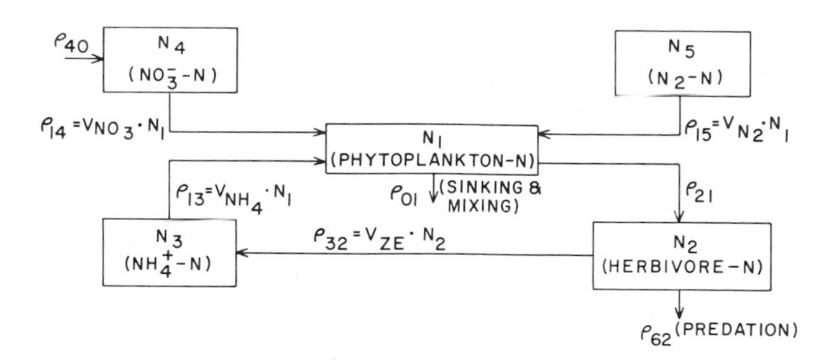

Fig. 11. A flow diagram of the nitrogen species N_i involved in a submodel of nutrient uptake.

The equation for the nutrient uptake curve, a rectangular hyperbola (Figure 12) is

$$V = V_{max} \frac{N}{K_t + N} \tag{2}$$

where V is the specific uptake rate, V_{max} is the maximum specific uptake rate, N is the concentration of the ambient limiting nutrient, and K_t is the nutrient concentration at which the specific uptake rate V is one-half the maximum uptake rate V_{max}.

In addition to consideration of light intensity and nitrate concentrations, laboratory and field data on ammonium and nitrate interaction (14,16) show that ammonium ion inhibits nitrate uptake (Figure 13), apparently by reducing $V_{NO_3 max}$.

These factors have been combined into a submodel that predicts daily uptake of nitrate (17). The effect of ammonium inhibition is considered first in the submodel by calculating a value of $V_{NO_3 max}$ for each depth, both as a function of the nitrate and ammonium concentration at that depth and as a function of the mean value of $V_{NO_3 max}$ obtained from ^{15}N uptake experiments in the study area. The dependence of the specific uptake of nitrate V_{NO_3} on light or nitrate concentration is then considered in a two-step calculation.

Using the $V_{NO_3 max}$ determined previously and equation 2, V_{NO_3} is first calculated as a function of light intensity at that depth and then as a function of nitrate concentration at that depth. The lower of the two V_{NO_3}

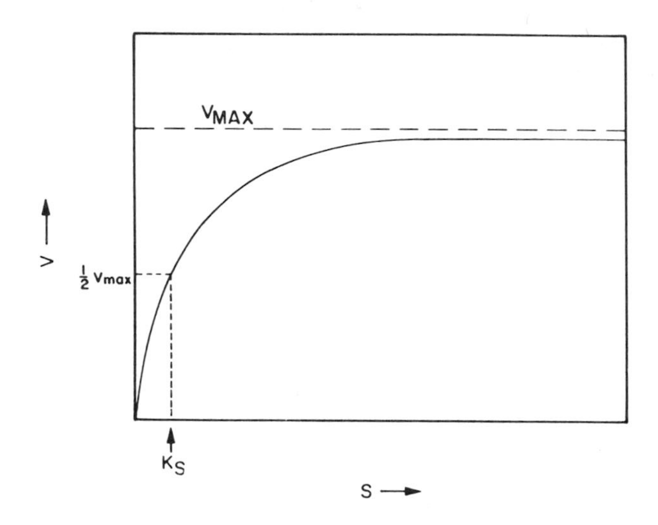

Fig. 12. A Michaelis-Menten description of nutrient limitation with S as the limiting nutrient concentration, K_s as the half-saturation constant, V_{max} as the maximum specific uptake velocity, and V as the specific uptake velocity.

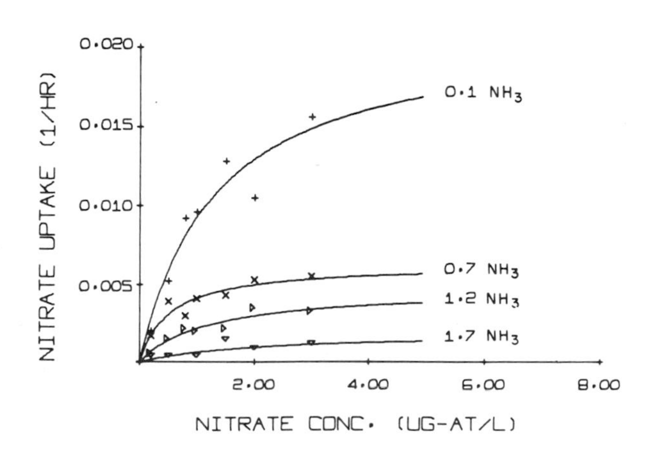

Fig. 13. The inhibition of nitrate uptake by increasing concentrations of ammonium.

calculated values is used in the submodel as the prevailing uptake rate at depth for that time step. The process is repeated for each hour at each depth. Figure 14 illustrates the fit between observed nitrate uptake with depth as measured by ^{15}N incubation experiments and the calculated nitrate uptake with depth generated from the submodel. It should be borne in mind that a form of equation 2 and the complex nitrogen interactions considered in its implementation constitute only one term of equation 1.

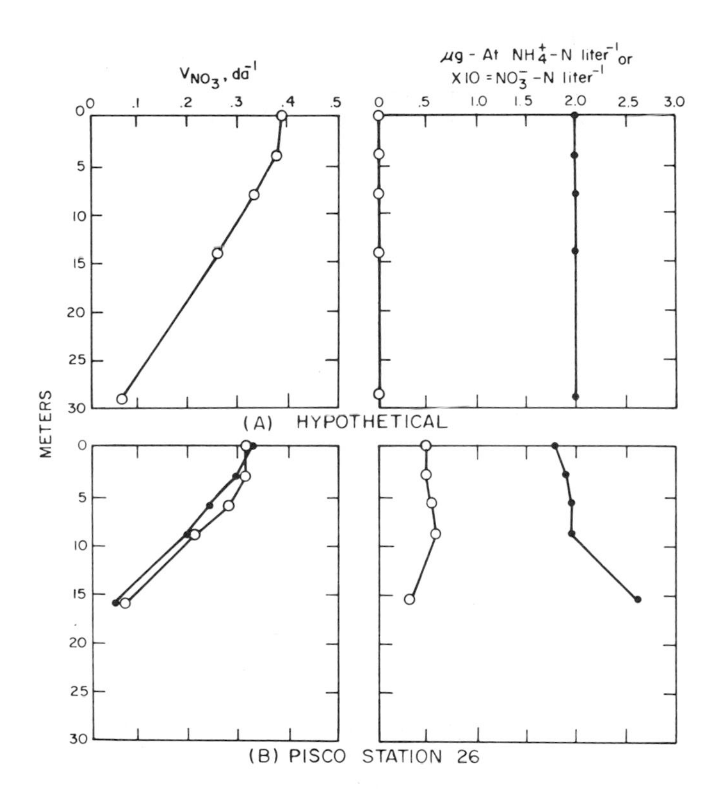

Fig. 14. A comparison of the simulated nitrate uptake and ammonium concentration (open circles) and the observed nitrate uptake and nitrate concentration (solid circles): (a) Hypothetical case with no ammonium present at any depth and 20 μg-at./liter of nitrate at all depths, (b) real case for a station off Pt. San Juan, Peru.

Terms for the remaining primary nutrients, silicate and phosphate, must be incorporated as nutrient uptake submodels into the upwelling and marine outfall system models. The information required for these nutrient terms is being developed in our automated algal chemostat laboratory (16) and in other laboratories. Chemostat experiments also should be useful for developing photosynthesis submodels. Predicting the effect of either pollutant toxicity or available chelators on nutrient uptake and photosynthesis is more complicated. However, algal chemostats should be useful in assessing these interactions as well.

THE IMMEDIATE FUTURE

Simulation models of phytoplankton production are not yet in a sufficiently advanced state for use in the management of marine fisheries or of waste disposal. Considerable work remains to be done in the area of system model structure and function (i.e., the way in which submodels are linked and the way in which nutrient or energy transfer computations are performed). Problems of interfacing between submodels are perhaps more serious than is commonly thought, especially since the appropriate iterative time steps of a simulation may differ greatly between submodels. For example, linkage between fishery submodels and phytoplankton submodels is difficult because the smallest appropriate time step is in the order of days or weeks for the former, and minutes or hours for the latter.

Currently, economic considerations make it impractical to use the output of a phytoplankton simulation as an input to the fishery submodel in the same computer run. Present models are based on the solution of differential equations by numerical integration at each evenly spaced time step of the simulation. Costs of computer usage limit the minimum length of a time step over a fixed interval of simulated time, since the smaller steps require more calculation and more computer processor time. Very large time steps generate unstable temporal behavior of the models; thus in a dynamic system model, submodels with incompatible time steps cannot be used over long or short fixed iteration periods.

Progress (18,19) is being made with these "stiff" systems of fast and slow time constants, however, and there are now two approaches to the problem. Integrated output of a phytoplankton submodel from a previous simulation run can be used as input to a fish submodel (i.e., the fast part of the system appears as a steady-state condition to the slow part of the system). Some of the power of a dynamic, interacting system model is lost with this approach, however, and the alternative is to eventually settle on a compromise variable time step for concurrent submodels based on either error or sensitivity analysis.

A related problem is the step-down transfer to small shipboard computer facilities of simulation models run on large land-based systems. Since relatively large investments in computer programming are being made, it is highly desirable that a substantial measure of compatibility between models and computer systems be achieved. For this purpose, one of the most fundamental requirements is that a common programming language be used. Fortunately, it appears that the trend is away from special simulation languages and toward the use of Fortran by most groups modeling large biological systems. One significant consequence of Fortran usage and the accessibility of modern digital computers has been the tendency of more biologists to attempt to construct simulation models. It seems likely that significant breakthroughs will be made by these biological modelers, who are more concerned with biological reality than with the elegance of equations.

Most of the commonly constructed ecosystem models are designed for describing and predicting events at a single geographical location; they are not particularly useful for aquatic systems. Spatial modeling is likely to receive increased attention by biologists in the near future. Such models have been built and run successfully by physicists to describe the wind-driven circulation of surface water in the North Pacific Ocean (20) and the Indian Ocean (21). However, large numbers of spatial blocks and variables are required, and the most powerful computers available must be used.

In our experience, small spatial models can be run successfully on small computers such as the IBM 1130. Restrictions in computer memory and speed encourage programmers to be more efficient; thus a small computer, if available for long periods at low cost, may often perform quite satisfactorily. Validation of both our submodels and system models is considered as important as building them, if not more so. Testing system models, or at least sequentially testing their submodels at sea is vital to our research program on total ecosystems, and many limitations of small shipboard computers are compensated for by the time available to us for running simulations at sea. An assessment of background variability in comparison with model output while at sea is an essential prerequisite for the ability to predict and observe perturbation responses of marine systems.

REFERENCES

1. J. J. Walsh, "The Relative Importance of Habitat Variables in Predicting the Distribution of Phytoplankton at the Ecotone of the Antarctic Upwelling Ecosystem," Ecol. Monogr., 41 (4), in press.
2. J. J. Walsh, J. C. Kelley, R. C. Dugdale, and B. W. Frost, "Gross Features of the Peruvian Upwelling System with Special Reference to Possible Diel Variation," Inv. Pesq., 35, 25-42 (1971).
3. J. J. Walsh, "Simulation Analysis of Trophic Interaction in an Upwelling Ecosystem," Proc. 1971. Summer Simulation Conf., Boston, 1971, pp. 874-878.
4. G. H. Stander, "Temperature: Its Annual Cycles and Relation to Wind and Spawning," The Pilchard of South West Africa, 9, 1-15 (1963).
5. J. H. Steele, "The Environment of a Herring Fishery," Scotl. Mar. Res., 6, 1-18 (1961).
6. H. Kasahara, "Fishery Development on the West Coast of Africa," Stud. Trop. Oceanogr. (Miami), 5, 423-436 (1967).
7. R. Jordan, "Distribution of Anchoveta (*Engraulis ringens* J.) in Relation to the Environment," Inv. Pesq., 35 (1), 113-126 (1971).
8. J. J. Walsh and R. C. Dugdale, "A Simulation Model of the Nitrogen Flow in the Peruvian Upwelling System," Inv. Pesq., 35, 330 (1971).
9. J. J. Walsh and P. B. Bass, "OCEANS, a Seagoing Simulation Program--A User's Guide to the University of Washington's IBM 1130 Spatial Version of COMSYS1," Tech. Rep., Department of Oceanography, University of Washington, Seattle, 1971.
10. J. J. Walsh, "Implications of a Systems Approach to Oceanography," presented at the I.B.P./P.M. Working Conference, Rome, October 4-8, 1971.
11. R. C. Dugdale, "Chemical Oceanography and Primary Production in Upwelling Ecosystems," Geoforum, in press.
12. R. C. Dugdale and T. E. Whitledge, "Computer Simulation of Phytoplankton Growth near a Marine Sewage Outfall," Rev. Intern. Oceanogr. Med., 17, 201-210 (1970).
13. Hyperion Engineers, "Ocean Outfall Design," Final report to the City of Los Angeles, 1957.
14. R. C. Dugdale, J. C. Kelly, and T. Becacos-Kontos, "The Effects of Effluent Discharge on the Concentration of Nutrients in the Saronikos Gulf," presented at the FAO Technical Conference on Marine Pollution, Rome, December 1970.

15. R. C. Dugdale, "Nutrient Limitation in the Sea: Dynamics, Identification, and Significance," Limnol. Oceanogr., 12, 685-695 (1967).
16. H. L. Conway, P. Harrison, C. Davis, and S. Pavlou, "Potential Uses of the Algal Chemostat in Pollution Studies," Presented at the Pacific Northwest Oceanographers' Conference, Victoria, B.C., February 1971.
17. R. C. Dugdale, and J. J. MacIsaac, "A Computation Model for the Uptake of Nitrate in the Peru Upwelling Region," Inv. Pesq., 35, 299-308 (1971).
18. R. N. Nilsen, "A Study of Adaptive Step Size Control in Numerical Integration Using Discrete Sensitivity Analysis," Proc. 1971 Summer Computer Simulation Conf., Boston, 1971, pp. 119-122.
19. J. Creighton, "Role of Numerical Stability in Computer Solutions of Chemical Chain Reactions," Proc. 1971 Summer Computer Simulation Conf., Boston, 1971, pp. 440-446.
20. J. J. O'Brien, "A Two-Dimensional Model of the Wind-Driven North Pacific," Inv. Pesq., 35, 331-349 (1971).
21. M. D. Cox, "A Mathematical Model of the Indian Ocean," Deep-Sea Res., 17, 47-76 (1970).

Chapter VII

OXYGEN-NUTRIENT RELATIONSHIPS WITHIN THE CENTRAL BASIN OF LAKE ERIE*

N. M. Burns

Canada Centre for Inland Waters
Burlington, Ontario, Canada

and

C. Ross

Region V, U.S. Environmental Protection Agency
Fairview Park, Ohio

During the winter of 1969 a joint Canadian-American study was initiated by the Chemical Limnology Section, Canada Centre for Inland Waters (CCIW), and the U.S. Environmental Protection Agency, Water Quality Office, Fairview Park, Ohio. This study, named "Project Hypo," had as its main aim the investigation of the pattern and causes of oxygen depletion in the hypolimnion of the central basin of Lake Erie; the secondary aim was to outline, if possible, the pattern of nutrient relationships in the basin and to relate these to changing oxygen conditions. Since many factors affect the oxygen levels in a lake, an effort was made to investigate simultaneously the physical, biological, bacteriological, and chemical conditions of the basin.

During the planning stages of Project Hypo, the decision was made to carry out the chemical investigation as quantitatively as possible. It was considered necessary to move beyond the level of the understanding of the system where correlations were observed between simultaneous changes in variables, to the level of trying to establish cause-and-effect relationships. Thus efforts were made to plan the sampling and analysis programs such that chemical budget calculations were possible on the components within the hypolimnion. It was decided that knowledge only of changes in the concentrations of the chemical components with time would probably be insufficient to elucidate many of the complex mechanisms that would occur, and that further knowledge of the masses of water involved in the various changes was also necessary.

METHODS AND PROCEDURES

It is no simple matter to be quantitative about the hypolimnion of the central basin of Lake Erie. It is a thin sheet of cold water having an average thickness of 2.5 m extending almost 6000 miles^2 and lying under an average depth of 17.5 m of thermocline and surface water. However, the implementation of the study was greatly aided by use of an electronic bathythermograph (EBT), which gave a continuous readout of temperature versus depth on an x-y recorder. The instrument had an absolute accuracy of approximately 0.1°C and 0.3 m depth, but relative changes in depth of 0.1 m could be detected. This instrument was strapped to a submersible pump that was used to sample the water. By means of the combination of the two instruments it was possible to have accurate knowledge of the temperature and depth of the water being sampled. By moving the pump up and down it was even possible to sample the middle of the thermocline while it was undergoing internal wave activity. The necessity of the instrument combination in sampling the hypolimnion was only realized when it broke down at the end of survey 2. The conventional method of oceanographic-type sampling with Van Dorn bottles was used during survey 3, but it was found to be difficult to obtain hypolimnion samples that were not contaminated with thermocline water. Hence the results of survey 3 have not been used in the budget calculations presented here.

Twenty-five sampling stations were set up across the central basin of Lake Erie (Figure 1), and seven surveys of the basin were conducted in the space of 28 days. Each survey occupied between two and three days, and the

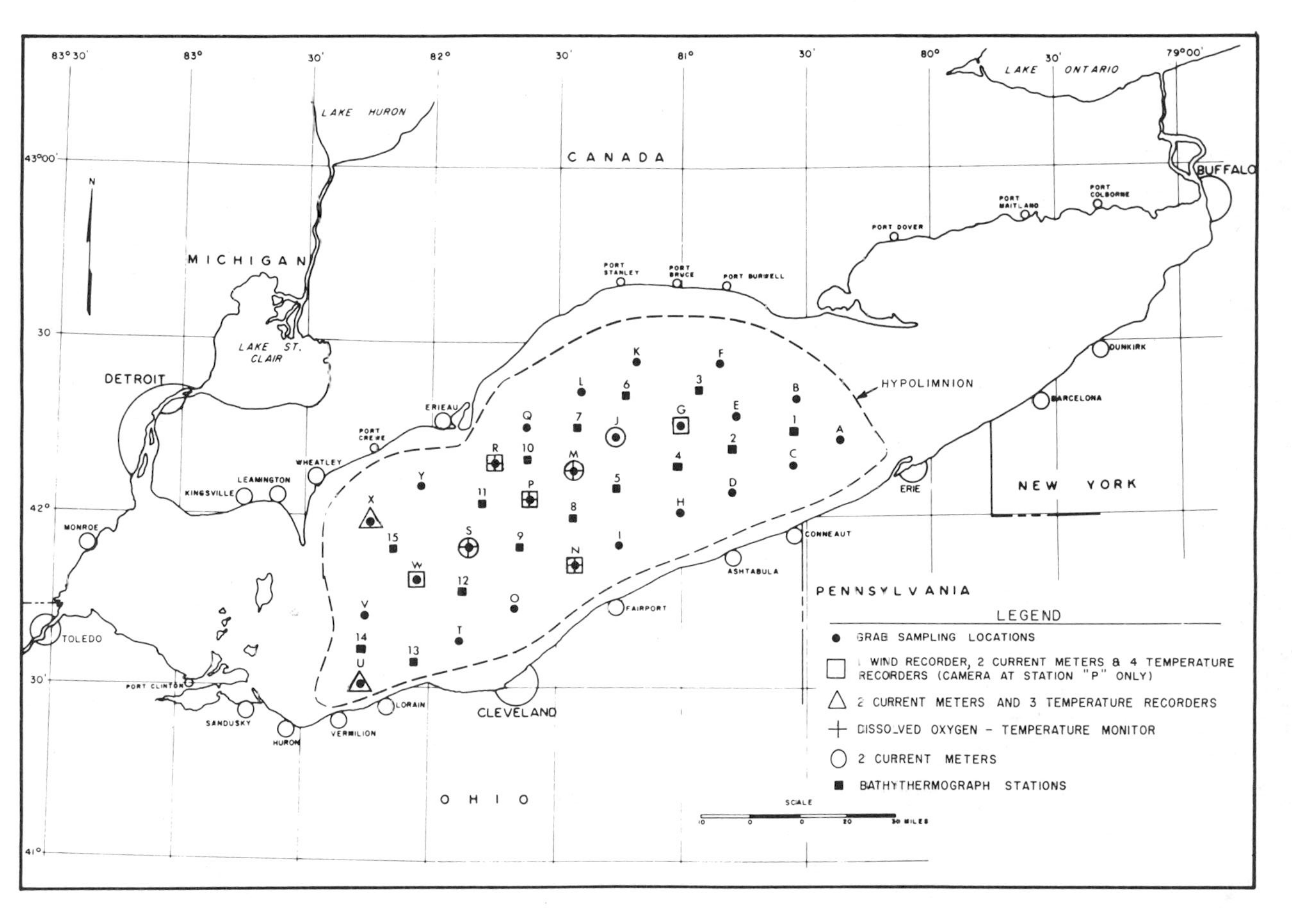

Fig. 1. Position of sampling stations and equipment during Project Hypo.

surveys were spaced approximately four days apart. Water samples were taken from the leeward side of the ship. The EBT pump was lowered to within 1.0 m of the bottom, thereby measuring the thermal structure of the water column. The hypolimnion was then sampled moving upward, at one-meter intervals. Next the pump was raised through the thermocline and two surface water samples were taken, the first at 1.0 m above the thermocline (mesolimnion) and the second at 1.0 m from the surface. At the five major stations (M,N,P,R,S) midthermocline samples were taken when possible.

The list of parameters measured at all the stations was fairly extensive; it included dissolved oxygen, pH, Eh, suspended mineral, total CO_2, NH_3, NO_3, NO_2, and total hardness. In addition, samples were also taken at the five major stations for particulate organic carbon and nitrogen, filtered and unfiltered total phosphorus, filtered and unfiltered total nitrogen, calcium hardness, total iron, total manganese, and total sulfate.

The ship used for this work was the CSS Limnos of the Canada Centre for Inland Waters. The laboratory on the ship was adequate and manned by both Canadian and American scientific workers. Most analyses were done immediately on the ship, except for the analyses on the particulate organic materials and those being performed on the total quantities of components. These analyses were done at the shore laboratories in Cleveland and Burlington.

ANALYTICAL PROCEDURES

The water from the pump was passed over two oxygen probes. When the probes gave a steady reading (ca. 2 to 5 min), water from the pumping system was led into 8.0-l holding bottles; a second small sample was drawn simultaneously and the pH and Eh were measured.

The oxygen probes were calibrated daily by immersing the probes in stirred, air-saturated water, and the correction was made for the barometric pressure. The probes were then calibrated at the low end of the scale by placing them into water that was being vigorously purged with nitrogen gas. The percentage saturation values were converted to micromoles of oxygen per liter from the values in "Standard Methods" (1). In general the duplicate probes gave values for a sample which agreed to within 1% of an oxygen saturation value.

The suspended mineral values were obtained by filtering water through a precombusted, weighted, glass fiber GF/c filter paper and weighing again after combustion had removed the organic materials. This method has a standard deviation of 0.05 mg/liter.

The pH was measured on an unstirred sample at approximately the temperature of the pumped sample, and values are accurate to 0.1 of a pH unit. The Eh was measured simultaneously with pH, using a bright platinum electrode and a standard reference electrode. The platinum electrode was cleaned each day and calibrated against a standard solution.

Total carbon dioxide was determined by purging the CO_2 from an acidified sample and passing the gas mixture through a gas chromatograph. The method had an accuracy of 0.05 mmoles CO_2/liter.

Filtered and unfiltered alkalinity was measured using an AutoAnalyzer method which has been tested against the normal titration method. Methyl orange, buffered by a solution of potassium hydrogen phthalate, was used as an indicator and standardized against sodium carbonate and bicarbonate solutions. The method has a sensitivity of ±1.0 μmoles $CaCO_3$/liter for filtered lake water, but it appears to undergo significant interference with unfiltered lake water.

At the end of the study, the total CO_2 content and the alkalinity of an anaerobic solution of 1.00 mM solution of Na_2CO_3 were measured by the equipment and methods used during the study. The alkalinity method yielded values that were 4% higher than those given by the gas chromatograph, where in fact the alkalinity value expressed as millimoles of calcium carbonate should have been about 1% lower than the total CO_2 value expressed as millimoles of carbon dioxide. Since the major use of both the CO_2 and alkalinity values was in the calculation of budgets, with the initial quantities of each subtracted from the final quantities, it was decided that any errors introduced were small; therefore, no changes were made in the data.

The methods used in the nutrient analyses have already been reported by Chawla and Traversy (2). The sensitivities of the various methods are as follows:

	Detection limit
Phosphorus	0.01 μmoles P/liter
Dissolved silica	0.01 μmoles Si/liter
Ammonia	0.1 μmoles N/liter
Nitrate or nitrite	0.03 μmoles N/liter

Total iron, total manganese, and dissolved sulfate were determined according to the "Standard Methods for Examination of Water and Wastewater," (1) with the following sensitivities:

	Detection limit
Total iron	0.5 μmoles Fe/liter
Total manganese	1.0 μmoles Mn/liter
Dissolved sulfate	20 μmoles SO_4/liter

The particulate organic carbon and particulate organic nitrogen values were obtained by filtering lake water onto GF/c glass fiber filter papers after the samples had been dryed in a desiccator, they were combusted in a Perkin-Elmer CHN Analyser.

RESULTS AND DISCUSSION

General

Our approach in this study has been to monitor a set of five sequential environmental reactions and then sum the five reactions into a net result for the period of observation. The first reaction, labeled R_{12} represents the changes that occurred in the hypolimnion during the period from the end of the first survey to the end of the second survey. The difference then between the dissolved oxygen calculated to be present in the hypolimnion at the end of the first and second surveys represented the dissolved oxygen that had disappeared into other chemical forms during the time interval; its rate of disappearance was also calculated. Also, by noting which other oxygen-containing components increased in quantity during the time interval, it was possible to estimate the extent of the various chemical transformations involving oxygen.

The first and most obvious result of the investigation was that the hypolimnion volume increased by almost 100% during the course of the study. This phenomenon (schematically illustrated in Figure 2) involved thinning and elevation in depth of the thermocline. The hypolimnion volume increase was most unexpected and caused much concern about the possibility of valid budget calculations. The concentration of the various materials in the hypolimnion would have remained essentially unchanged with a loss of volume; however, with increasing hypolimnion volumes, water having a quite different concentration of

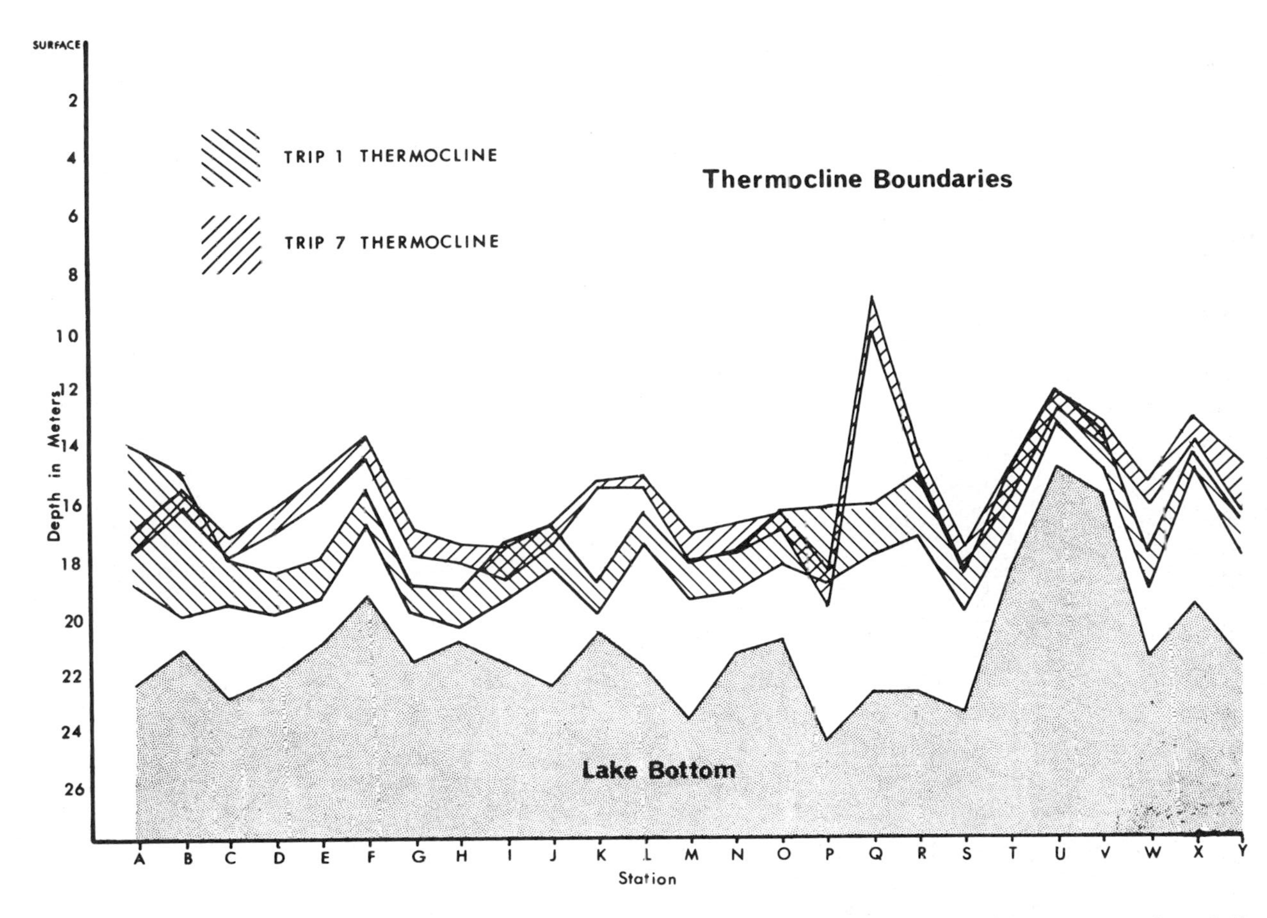

Fig. 2. Thermocline depths and thickness at the various stations during surveys 1 and 7.

the reactants was introduced into the hypolimnion from the epilimnion.

The validity of the results presented here depends, to a large extent, on the accuracy of the estimates of materials that were added to the hypolimnion, because in many cases the quantities of the materials added during a reaction period were greater than the changes in the quantities that occurred in the hypolimnion due to chemical and biological action.

One method of analyzing the situation of increasing hypolimnion volumes is to develop a model of the whole system, with the model predicting the behavior of a certain conservative parameter; then the calculated levels of this parameter would be compared with the levels that were actually observed. If the model is reasonably accurate in predicting the observed level of the chosen conservative parameter, the model, with its patterns of mixing can be used to estimate the movement of nonconservative quantities within the water mass.

In this study, heat was chosen as the conservative parameter within the hypolimnion, and different models were set up to test which was the most reliable in predicting the observed heat budget. The model that made satisfactory heat budget calculations possible was used for the calculation of the budgets of the individual chemical components. This information then made it possible to demonstrate the probable chemical pathways that were followed by the major components.

Heat Budget Model

Of the various models examined, the one here termed as the "sequential mesolimnion erosion model" gave values that showed good agreement between estimated and measured hypolimnion heat contents and also between estimated and observed average hypolimnion temperature. In this model, the structure of the thermocline during a particular survey is established. If in the next survey, the volume of the hypolimnion is found to have increased by Q km^3, the volume of Q km^3 is taken off the bottom of the mesolimnion (as defined in the first survey) in a slice of constant thickness across the whole basin. This volume is then added to the hypolimnion as it was observed in the first survey. The mixture now represents the hypolimnion of the second survey, and predictions of the hypolimnion heat content and average temperature of the hypolimnion during the second survey can be made.

The process can now be applied again, with the

calculated heat change from survey 2 to survey 3 being added onto the calculated heat content of survey 2, not onto the measured heat content of survey 2. The procedure, which is followed whenever there is a volume increase in the hypolimnion, is shown diagrammatically in Figure 3. In the one case where there was a hypolimnion volume decrease, a slice was taken off the top of the hypolimnion and added to the mesolimnion.

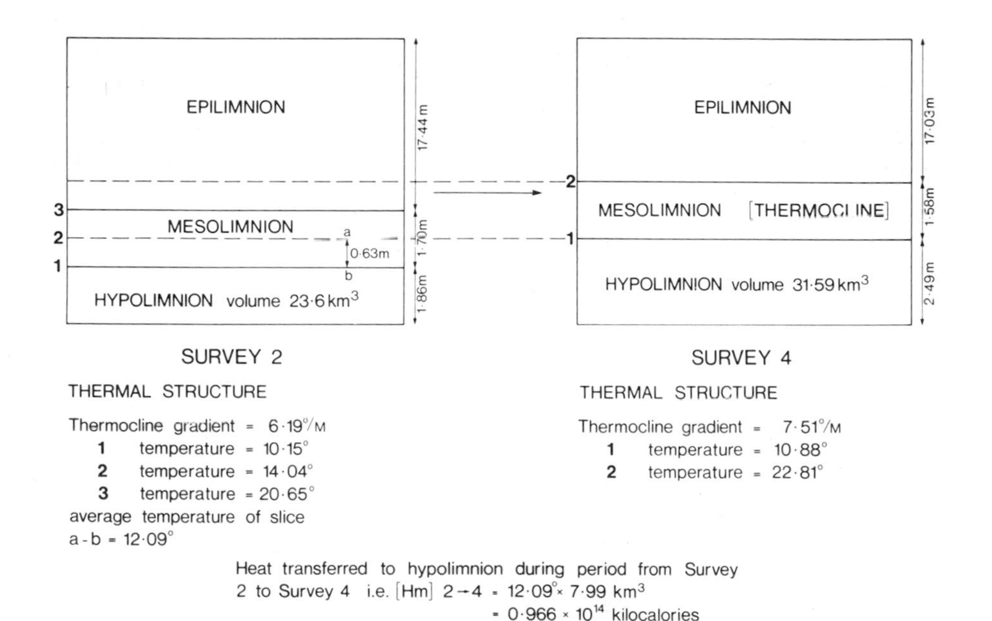

Fig. 3. Example of sequential mesolimnion erosion model, showing heat and volume transfers which occurred between surveys 2 and 4, (i.e., during the reaction period R_{24}).

The effective heat budget equation for the hypolimnion would be

$$H_T = H_i + H_m + H_{adv} + H_{cond} - H_{sed}$$

where H_T = total heat within the hypolimnion

H_i = initial heat within hypolimnion

H_m = heat entrained downward from the mesolimnion (thermocline)

H_{adv} = heat advected into the hypolimnion by horizontal currents

H_{cond} = heat conducted into the hypolimnion by thermal conduction downward

H_{sed} = heat lost to the sediments.

The method of calculating H_m has been explained. The western basin of Lake Erie has no hypolimnion, and during the study the eastern basin hypolimnion appears to have been effectively cut off from the central basin hypolimnion by the ridge separating the two basins (3); thus the assumption was made that $H_{adv} = 0$. The heat conducted down during a period has been calculated by taking the average of the thermocline gradients calculated for surveys before and after the period, using the equation

$$H_{cond} = 0.0014 \frac{\text{cal}}{\text{cm/(sec.)(°C)}} \times \text{gradient (°C/cm)}$$

An estimation of the sediment heat uptake had to be made because the thermal structure of the sediment was not measured during the study; the necessity of this measurement not being realized at the time. Hutchinson (4, p. 506) reported a study by Birge, Juday, and March in which the annual sediment heat budget in Lake Mendota at depths varying between 18.0 and 23.5 m was estimated to be 1100 cal/cm^2. It appears that the bottom water changed in temperature from approximately 1 to 12°C in the course of the year. This would indicate that the sediment heat budget of Lake Mendota was 100 cal/cm^2 for each degree change in the bottom water temperature. The deep part of Lake Mendota appears to be similar to the central basin of Lake Erie in that both lakes have an average depth of approximately 20.0 m and an annual bottom water temperature change of about 12°C. However, they differ quite markedly in their interface characteristics. Lake Mendota has a very ill-defined sediment-water interface (5), whereas during the study the Lake Erie sediment had a very firm, well-defined sediment-water interface, consisting of a layer of tightly matted, organic material approximately 1 cm thick. The assumption is made here that there is a smaller amount of sediment-water interaction in Lake Erie than in Lake Mendota and that the thermal conductivity of the sediment-water interface is lower in Lake Erie. A value for the sediment heat budget of 70 cal/cm^2 for each degree change in bottom water temperature is thus assumed. Since the change in the average bottom water temperature during the study was 1.83°C and the study duration was 26.28 days, the sediment heat flux was calculated from the equation

$$H_{sed} = \frac{70\ \text{cal}/(\text{cm}^2)(°\text{C}) \times 1.83°\text{C}}{26.28\ \text{days}}$$

$$= 4.86\ \text{cal}/(\text{cm}^2)(\text{day})$$

However, measured volumes have to be used in the calculation of the heat budget and, since the measurements of the volumes are themselves subject to error, it is difficult to obtain an objectively correct heat budget. In this regard, the prediction of the average temperature is independent of systematic volume measurement errors because, in the calculation of the volume-weighted average temperature, the same volumes are used in both the numerator and denominator.

A heat budget for the hypolimnion has been calculated and average temperatures have been estimated according to the methods outlined previously (Table 1). The measured volumes of the hypolimnion are shown in Figure 4.

However, examination of Table 1 and Figure 5 reveals poor agreement between predicted and observed temperatures for surveys 2 and 6 (standard deviation of ±0.33°C). During both these cruises a strong wind blew which could have seriously affected any hypolimnion volume measurement. Thus calculations were done to find the volumes of surveys 2 and 6 that gave the best agreement between estimated and observed temperatures (standard deviation of ±0.11°C). A second heat budget has been calculated with two of the six survey volumes adjusted. The results of these calculations appear in Table 2.

Basically, there are two sources of error in a material or energy transfer estimate. The first error is due to incorrect measurement of the hypolimnion volume. This error is difficult to estimate because it is highly variable; the error can be small if the hypolimnion is calm, but it can be large if the hypolimnion is undergoing extensive seiching activity during measurement. The second error is the result of the inability of the mesolimnion erosion model to reproduce the real situation.

In Table 3 the measured transferred heat quantities are compared with the estimated quantities. Because the measured quantities are themselves uncertain, the estimated uncertainty in a quantity during a single reaction

Table 1. Hypolimnion Heat Budget, Measured Volumes: A Comparison of a Theoretical Hypolimnion Heat Budget (Calculated by Means of Sequential Mesolimnion Erosion Model and Measured Hypolimnion Volumes) with the Measured Hypolimnion Heat Budget

Survey		Estimated heat content, $\times 10^{14}$ kcal	Measured heat content, $\times 10^{14}$ kcal	Estimated temperature, °C	Observed temperature, °C
1	H_i	2.276	2.276	10.07	10.07
	(H_m) 1→2	0.605			
	$(H_{cond} - H_{sed})$ 1→2	0.018			
2		2.899	2.827	10.40	10.14
	(H_m) 2→4	0.410			
	$(H_{cond} - H_{sed})$ 2→4	0.041			
4		3.350	3.438	10.60	10.88
	(H_m) 4→5	-0.152			
	$(H_{cond} - H_{sed})$ 4→5	0.018			
5		3.216	3.283	10.66	10.89
	(H_m) 5→6	0.308			
	$(H_{cond} - H_{sed})$ 5→6	0.020			
6		3.544	3.713	10.82	11.34
	(H_m) 6→7	1.249			
	$(H_{cond} - H_{sed})$ 6→7	0.057			
7	H_T	4.850	4.910	11.74	11.90

Table 2. Hypolimnion Heat Budget, Volumes 2 and 6 Adjusted: A Comparison of a Theoretical Heat Budget (Calculated by Means of the Sequential Mesolimnion Erosion Model with the Hypolimnion Volumes of Surveys 2 and 6 Adjusted) with the Measured Hypolimnion Heat Budget

Survey		Estimated heat content, $\times 10^{14}$ kcal	Measured heat content, $\times 10^{14}$ kcal	Estimated temperature, °C	Observed temperature, °C
1	H_i	2.276	2.276	10.07	10.07
	(H_m) 1→2	0.103			
	$(H_{cond} - H_{sed})$ 1→2	0.018			
2		2.397	2.393	10.16	10.14
	(H_m) 2→4	0.966			
	$(H_{cond} - H_{sed})$ 2→4	0.041			
4		3.404	3.438	10.78	10.88
	(H_m) 4→5	-0.152			
	$(H_{cond} - H_{sed})$ 4→5	0.018			
5		3.270	3.283	10.88	10.89
	(H_m) 5→6	0.606			
	$(H_{cond} - H_{sed})$ 5→6	0.020			
6		3.896	3.969	11.13	11.34
	(H_m) 6→7	0.873			
	$(H_{cond} - H_{sed})$ 6→7	0.057			
7	H_T	4.826	4.910	11.69	11.90

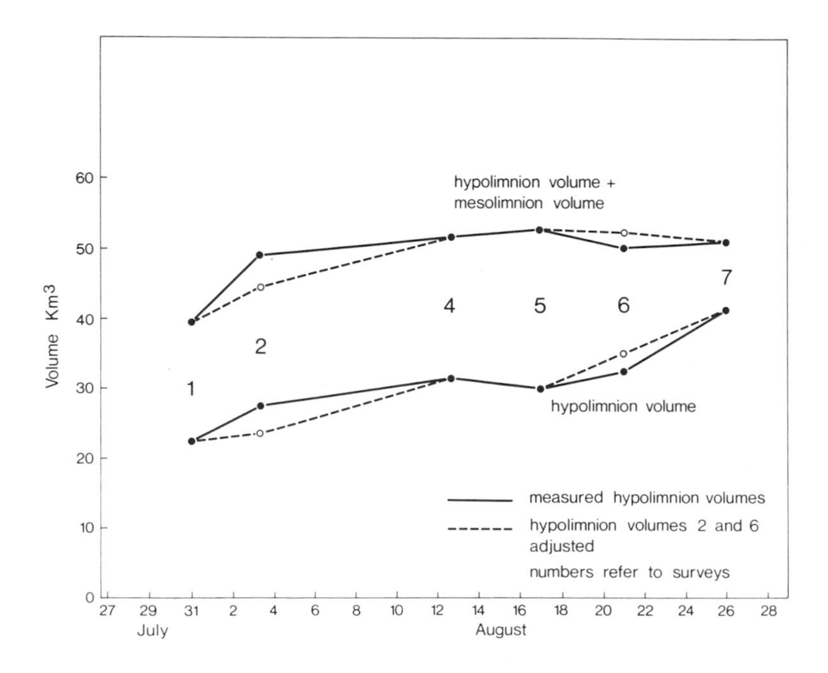

Fig. 4. Hypolimnion and mesolimnion volumes measured during Project Hypo showing adjusted volumes for surveys 2 and 6.

period is raised from 6.3 to 10.0%. The error in the total transferred quantity (R_{17}) is less than the sum of the error in the individual quantities because these errors are self-compensating to some degree. The uncertainty in the net transferred quantity is raised from 2.8 to 5.0% because of volume measurement uncertainties.

The temperature distribution one meter from the bottom is displayed in Figure 6. From the comparison of all the bottom temperature maps, it appears that appreciably greater downward entrainment of mesolimnion water occurred at stations A, B, G, F, K, T, U, and V. If a weighting factor were applied in the model calculations so that slightly greater than average entrainment occurred at these stations and slightly less than average entrainment occurred at the other stations, the hypolimnion heat uptake would be estimated to be a little greater and the estimated average temperatures would agree even more closely with the observed average temperatures. This correction would make the temperature agreement much better, but it is not

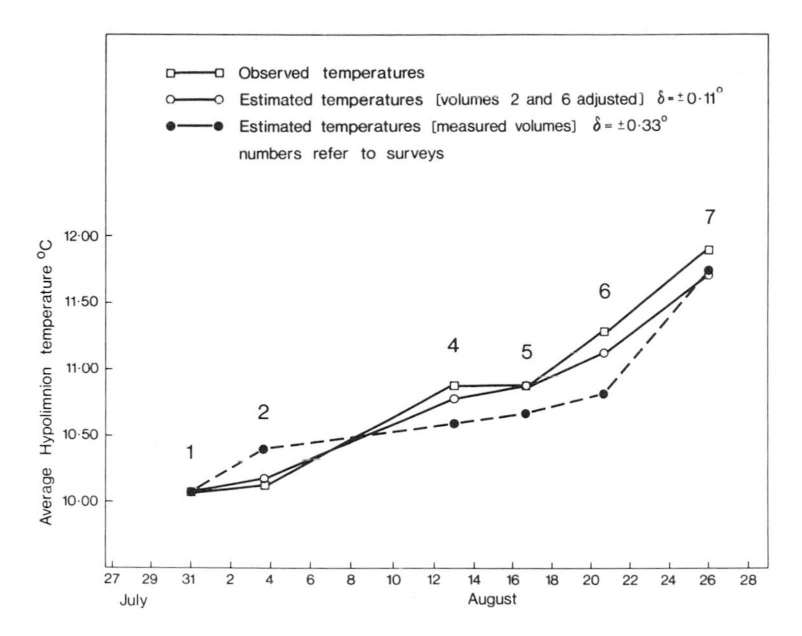

Fig. 5. Comparison of observed average temperatures with average temperatures estimated by means of the sequential mesolimnion erosion model.

Table 3. Comparison of Measured and Estimated Heat Transfers: A Comparison of Measured and Estimated Heat Transfers that Occurred between Surveys (i.e., during respective reaction periods); Hypolimnion Volumes of Surveys 2 and 6 Adjusted

Reaction	Measured net heat transfer, $\times 10^{14}$ kcal	Estimated net heat transfer, $\times 10^{14}$ kcal	Difference, $\times 10^{14}$ kcal	Difference, %
R_{12}	+0.117	+0.121	+0.004	+3.4
R_{24}	+1.045	+1.007	-0.038	-3.6
R_{45}	-0.155	-0.134	+0.021	+13.5
R_{56}	+0.686	+0.626	-0.060	-9.6
R_{67}	+0.941	+0.930	-0.011	-1.2
R_{17}	+2.634	+2.550	-0.074	-2.8

Average % difference R_{12} ,..., R_{67} = 6.3%

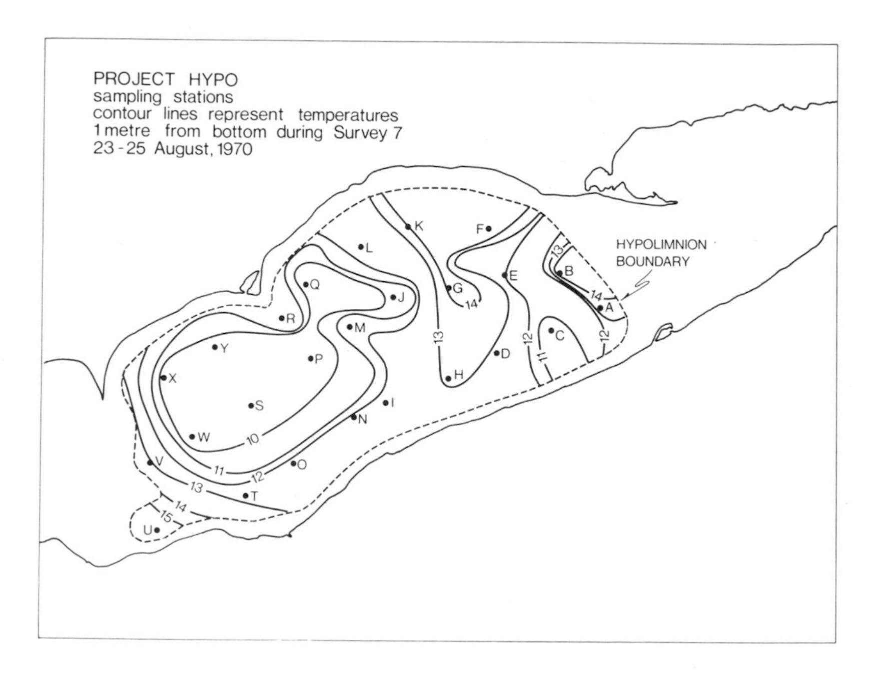

Fig. 6. Temperature pattern in °C one meter from the lake bottom in Lake Erie Central Basin.

considered essential because it would only affect the heat budget by approximately 2% and the chemical budgets by approximately 1% of observed change in quantities.

An interesting aside is that, if greater than average hypolimnion volume entrainment occurred at the suggested localities, hypolimnion currents would be set up away from these localities. A generalized flow pattern can be deduced, and this pattern appears to be in agreement with that outlined by Hartley (6), who used seabed drifters.

When calculating the heat contents of the various hypolimnion segments, the only temperature values used were those which corresponded to the water that was sampled for chemical analysis. The information of the full bathythermograph trace was not used. This was done purposely so that the model developed previously, together with the error estimates, would be directly applicable to the chemical budget studies.

Mesolimnion or Thermocline

Just after a lake has stratified, the only real difference between the epilimnion and the hypolimnion is the temperature; but as the stratification persists, the differences in the chemical concentrations of the two zones increase. A region of chemical transition between these two zones then comes into existence and is commonly named the thermocline. It is suggested that this zone of transition should preferably be referred to as the mesolimnion, not the thermocline, since the chemocline gradient may be much more significant than the thermocline gradient. The temperature may change by 10°C or so on passing through the mesolimnion, but many of the concentrations of the chemical constituents may change by one or two orders of magnitude.

It is known from the bathythermograph traces that the thermocline through the mesolimnion was usually linear with depth. But was the chemocline similarly linear? It was necessary to have some idea of the answer to this question before materials transfer calculations could be done using the mesolimnion erosion model. Mesolimnion data from the five major stations have been processed to obtain some idea of chemical conditions through the mesolimnion. Mesolimnion depth and concentration values have been normalized for comparison because the mesolimnion varied in thickness and concentration gradient. The mesolimnion has been considered to be of unit thickness, with the bottom of the epilimnion having a height of 1.0 above the top of the hypolimnion, which had a height of 0. Similarly, the concentration difference between the top and bottom of the mesolimnion has been taken as 1.0. This is best illustrated with the following example.

Depth from surface to top of mesolimnion (bottom of epilimnion) = 17.4 m NO_3 conc. = 1.2 μmolar

Mesolimnion sample depth = 18.1 m NO_3 conc. = 2.0 μmolar

Depth to top of hypolimnion = 18.6 m NO_3 conc. = 5.5 μmolar

Relative height of sample above hypolimnion $= \dfrac{18.6 - 18.1}{18.6 - 17.4} = \dfrac{0.5}{1.2} = 0.42$

Relative sample concentration $= \dfrac{2.0 - 1.2}{5.5 - 1.2} \quad \dfrac{0.8}{4.3} = 0.19$

These values can now be plotted as a point on the graph. The results of the least-squares second-degree equation were plotted, and these results are shown in Figure 7 (line b). It can be seen that CO_2 and NO_2 have a linear chemocline, and that O_2, pH, filtered alkalinity, NO_3, NH_3 and SiO_2, have a nonlinear relationship. The latter parameters all have lower values than would be expected from a linear chemical gradient (line a) except

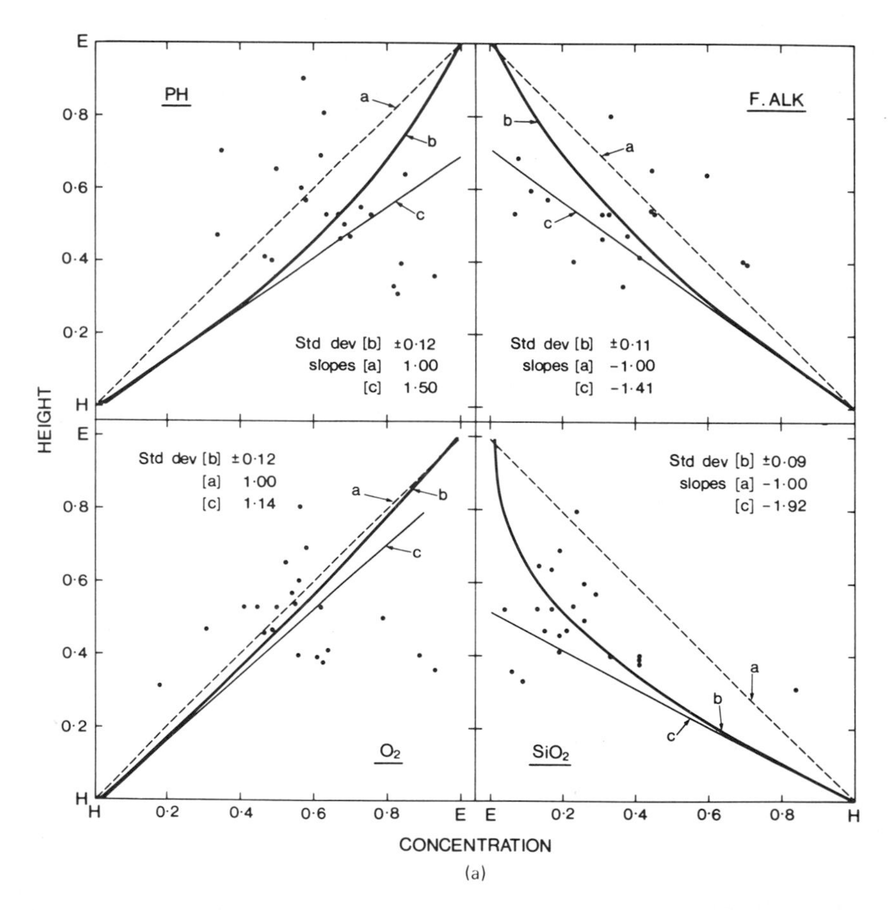

Fig. 7. Chemical concentration gradients through the mesolimnion normalized for different mesolimnion thicknesses--concentration values: H = hypolimnion value, E = epilimnion value; depth values: H = hypolimnion value = 0, E = epilimnion value = 1.0.

for O_2 and pH, which have higher values. These nonlinear features probably occur because water has been eroded from the bottom of the mesolimnion, tending to bring epilimnion values deeper into the mesolimnion; this was very much the situation in the case of PO_4, where the mesolimnion value was always close to the epilimnion value. However, this is not the case with NO_2 and CO_2 and the explanation is not immediately obvious.

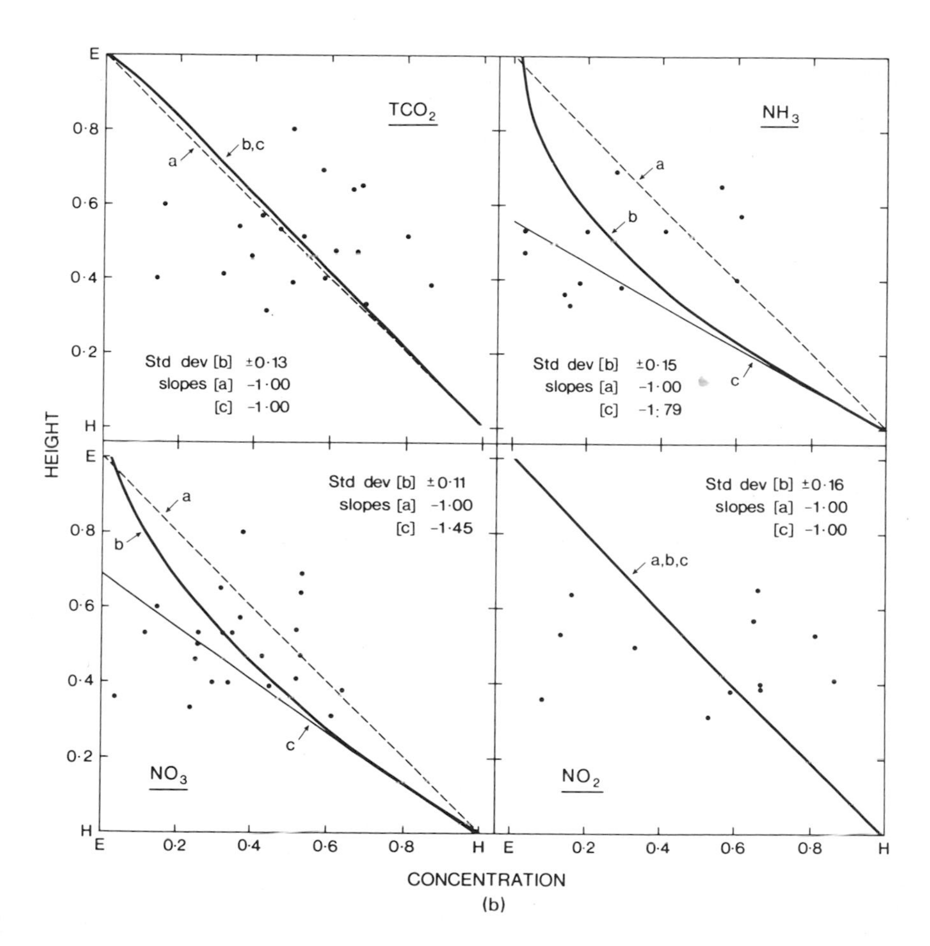

Fig. 7 (continued).

In all cases, the thickness of the layer that was eroded from the bottom of the mesolimnion never exceeded one-third of the thickness of the mesolimnion, so that a straight line (line c) was approximated to the curve, as in Figure 7, and this correction to an assumed linear gradient was used in calculating the average concentration of materials transferred into the hypolimnion. The transfer of materials downward by molecular diffusion was considered negligible in comparison to the entrainment processes.

Budget Calculations

A highly condensed summary of the chemical data obtained during the study is provided in Tables 4 to 6. Table 7 presents two budgets worked out in detail, PO_4 (total) and PO_4 (anoxic); the PO_4 (oxic) here is the difference between the other two. The O_2, NO_3, NO_2, NH_3, SiO_2, and CO_2 total budgets were worked out in the same manner as the total PO_4 budget.

Table 4. Volumes of Various Lake Water Masses*

	Volumes, km^3				
Survey	Total epi-limnion	Total meso-limnion	Total hypo-limnion	Oxic hypo-limnion	Anoxic hypo-limnion
1	229.76	16.84	22.70	22.60	0
2	220.23	21.20	23.60	23.60	0
4	217.65	20.06	31.59	29.03	2.56
5	216.59	22.55	30.16	26.18	3.99
6	219.24	17.32	35.00	28.25	6.76
7	218.32	9.69	41.29	29.08	12.20

*Volumes of various lake water masses observed during Project Hypo in Lake Erie Central Basin, Surveys 2 and 6 adjusted.

The anoxic budget calculations were made on a basis different from the total budgets because the volumes of specific stations varied markedly from survey to survey. When a station changed from the oxic to anoxic condition, the change in the concentration of the parameter under investigation was noted; this value was then multiplied by the volume of the station when it was anoxic, to obtain the quantity which was generated by the change. Once a station became anoxic, it was kept in the calculations of anoxic budgets from survey to survey. The letters in the center column of Table 7 show which stations were found to be anoxic in each survey. The sulfate, iron, and manganese budgets were calculated on a different basis, since these parameters were only sampled at the five major stations that remained oxic throughout the study, except for station R on survey 7. However, in addition on survey 7, all the anoxic stations were sampled as if they were major stations in order to have a better understanding of the chemistry at the anoxic stations. From the five major stations it was possible to estimate an oxic budget for the basin for sulfate, manganese, and iron, and from the anoxic stations samples, the excess quantities generated by anoxic conditions were calculated. Knowledge of the number of days during which certain stations were anoxic, together with their associated surface areas, has made possible the calculation of the anoxic generation rates of various chemical compounds. All these values are summarized in Tables 8 to 10.

External Influences

The physical cause-and-effect relationships of wind, hypolimnion currents, and mesolimnion entrainment are reported on by Blanton and Winklhofer (3). The wind caused almost complete vertical mixing in the thin hypolimnion (significant chemical gradients were seldom noticed), and the downward mesolimnion entrainment was also the partial cause of "algal rains" into the hypolimnion. The algal conditions during Project Hypo were closely followed and are described by Braidech et al. (7). Frequently, high algal concentrations were found in the region of the mesolimnion, and these algae would rain down into the hypolimnion periodically because of wind effects on the mesolimnion. The rains were documented by underwater photography (3). One of the more unexpected findings of the whole study was that the sedimented algae would remain alive and would photosynthesize for a period after they fell to the lake bottom, even at depths of 24 m (80 ft) (8).

The sequence of events associated with algal rains is shown schematically in Figure 8 along with dissolved oxygen and nitrate changes. When the algae first rained down in early August, they formed a fluffy green layer on the bottom, 2 to 3 cm thick (7); but within a week the algae had turned brown and matted down to a thickness of

Table 5. Volume-Weighted Average Concentrations of the Components

Survey	Temperature °C	O_2 μg-at./liter	NH_3 μmolar	NO_2 μmolar	NO_3 μmolar	PO_4 μmolar
				Volume-weighted average		
1	21.61	581.5	1.96	0.34	3.08	0.038
2	20.63	554.0	2.48	0.42	2.21	0.049
4	22.74	539.4	2.52	0.38	1.20	0.059
5	23.58	548.5	2.41	0.28	0.81	0.061
6	23.64	525.0	2.19	0.66	0.66	0.081
7	23.09	530.9	1.85	0.82	0.35	0.105
				Volume-weighted average		
1	10.07	265.7	6.53	1.07	11.08	0.054
2	10.14	225.7	8.15	1.53	12.09	0.095
4	10.88	159.2	8.35	0.63	11.47	0.359
5	10.89	85.4	9.08	0.70	11.19	0.349
6	11.34	70.3	9.95	1.49	10.44	0.353
7	11.90	53.4	4.85	1.68	10.23	0.488
				Volume-weighted average		
1	--	--	6.53	1.07	11.08	0.057
2	--	--	8.15	1.53	12.09	0.095
4	--	--	6.13	0.67	12.46	0.116
5	--	--	5.90	0.67	12.51	0.119
6	--	--	5.68	1.15	13.25	0.140
7	--	--	5.10	1.77	12.05	0.156
				Volume-weighted average		
1	--	--	--	--	--	--
2	--	--	--	--	--	--
4	--	--	33.56	0.02	1.43	3.106
5	--	--	30.93	0.09	2.51	1.853
6	--	--	28.91	0.30	2.38	1.194
7	--	--	15.18	0.15	6.42	1.293

Measured at All Sampling Stations

SiO_2 µmolar	Total CO_2 mmolar	pH	Eh mV	Filtered alkalinity µmolar $CaCO_3$	Suspended mineral mg/liter
epilimnion concentrations					
2.64	1.67	8.33	--	922	0.28
3.03	1.71	8.60	--	929	0.43
2.40	1.69	8.68	408	934	0.14
2.59	1.71	8.54	384	920	0.13
2.38	1.70	8.76	406	917	0.31
2.61	1.69	8.71	421	915	0.16
hypolimnion concentrations					
12.40	1.94	7.38	--	947	0.66
15.25	1.99	7.29	--	966	0.77
15.93	2.01	7.46	374	974	0.73
18.50	2.08	7.39	356	970	0.60
16.90	2.06	7.41	424	967	0.66
16.38	2.06	7.33	410	972	0.64
hypolimnion oxic concentrations					
12.40	1.94	7.38	--	947	--
15.25	1.99	7.29	--	966	--
14.28	2.01	7.45	400	972	--
16.45	2.07	7.38	381	967	--
14.14	2.04	7.41	417	962	--
14.16	2.04	7.32	470	974	--
hypolimnion anoxic concentrations					
--	--	--	--	--	--
--	--	--	--	--	--
34.61	2.11	7.52	71	989	--
28.91	2.16	7.44	192	993	--
29.43	2.11	7.40	298	988	--
21.67	2.10	7.33	267	973	--

Table 6. Average Concentrations of Components[a] at Major Station: Unweighted Average Concentrations, μmoles/liter, of the Components Sampled at Stations M, N, P, R, S Only; Anoxic Water was Sampled at Stations O, R, T, U, V, W, X on Survey 7

Survey	PON	Total SO_4	POC	PP	Total P	SOP	TFP	SRP	Total Fe	Total Mn
		Unweighted average epilimnion concentrations (major stations)								
1	3.43	235	30.14	0.21	0.32	0.07	0.11	0.03	0.13	0.10
2	3.65	230	28.36	0.20	0.35	0.10	0.15	0.06	0.48	0.07
4	3.58	232	28.13	0.20	0.37	0.11	0.17	0.08	0.27	0
5	3.53	245	22.06	0.21	0.38	0.11	0.17	0.08	0.27	0
6	2.57	248	20.77	0.21	0.37	0.08	0.16	0.05	0.48	0
7	3.48	234	24.86	0.22	0.39	0.07	0.17	0.10	0.33	0
		Unweighted average hypolimnion oxic concentrations (major stations)								
1	3.18	248	29.44	0.21	0.37	0.11	0.16	0.05	0.42	1.87
2	4.05	243	28.71	0.21	0.40	0.13	0.19	0.06	1.00	3.59
4	3.81	234	19.72	0.30	0.58	0.17	0.28	0.11	2.07	4.50
5	3.14	246	17.82	0.39	0.71	0.19	0.32	0.13	2.83	5.55
6	3.12	248	18.03	0.43	0.73	0.14	0.30	0.16	3.59	4.58
7	3.92	235	20.67	0.48	0.88	0.19	0.36	0.17	2.72	4.17
		Unweighted average hypolimnion anoxic concentrations								
1	--	--	--	--	--	--	--	--	--	--
2	--	--	--	--	--	--	--	--	--	--
4	--	--	--	--	--	--	--	--	--	--
5	--	--	--	--	--	--	--	--	--	--
6	--	--	--	--	--	--	--	--	--	--
7	3.87	211	18.76	0.84	3.07	0.92	2.23	1.31	6.21	7.27

[a]Abbreviations are as follows: PON = particulate organic nitrogen, POC = particulate organic carbon, PP = particulate phosphorus, SOP = soluble organic phosphorus, TFP = total filtered phosphorus, SRP = soluble reactive phosphorus.

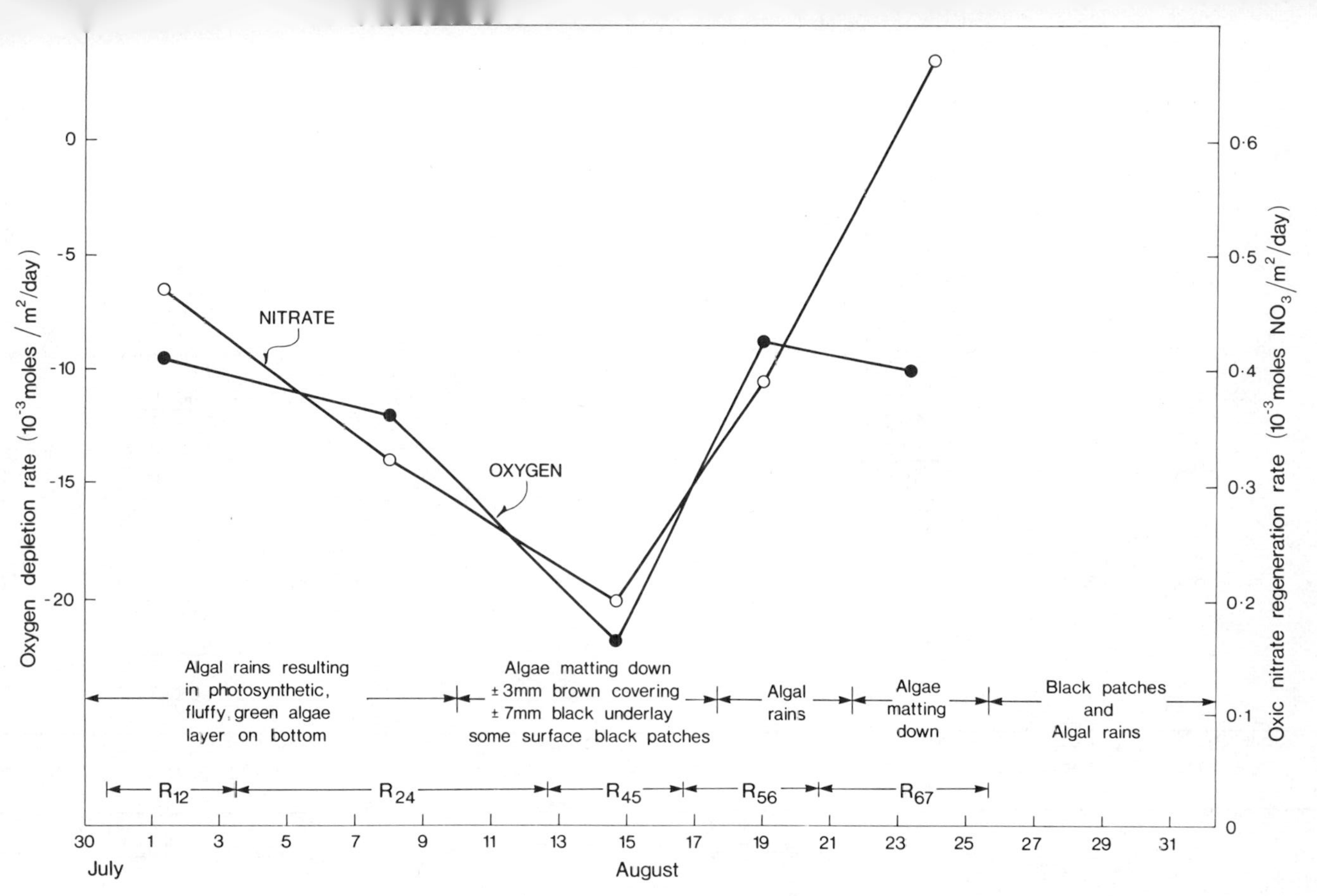

Fig. 8. Sequence of events associated with algal rains, showing a decreased oxygen demand and increased nitrate production when the algae were photosynthetic and a larger oxygen demand and smaller nitrate production when the algae were dead. The duration of the different reaction periods also appears.

Table 7. Detailed Budget for Total Anoxic and Oxic Phosphate[a]

Survey or Reaction[b]	Total PO_4 × 10^6 moles		Survey or Reaction (anoxic)	Anoxic PO_4 × 10^6 moles	Reaction (oxic)	Oxic PO_4 × 10^6 moles
1	1.225					
1→2 T	0.052					
2 E	1.277					
2 M	2.243					
R_{12} PO_4 G	0.966	±0.005	R_{12}	0	R_{12} gain	0.966
			2(U,V)[c] conc., μmoles/liter	0.0233		
2→4 T	0.051		4(U,V) conc., μmoles/liter	3.105		
4 E	2.294		(4-2) conc. diff.	2.872		
4 M	11.340		4(U,V) anoxic vol., km^3	2.559		
R_{24} PO_4 G	9.05	±0.005	R_{24} [(4-2) conc. × (4) vol.]	7.35	R_{24} gain	1.70
			4(T,U,V) conc., μmoles/liter	2.408		
4→5 T	-0.168		5(T,U,V) conc., μmoles/liter	1.853		
5 E	11.172		(5-4) conc. diff.	-0.555		
5 M	10.520		5(T,U,V) anoxic vol., km^3	3.786		
R_{45} PO_4 G	-0.652	±0.017	R_{45} [(5-4) conc. × (5) vol.]	-2.10	R_{45} gain	1.45

			5(T,U,V,O,I) conc., μmoles/liter	1.386		
5→6 T	0.537		6(T,U,V,O,I) conc., μmoles/liter	1.194		
6 E	11.057		(6-5) conc. diff.	-0.192		
6 M	12.370		6(T,U,V,O,I) anoxic vol., km^3	6.752		
R_{56} PO_4 G	1.31	±0.054	R_{56} [(6-5) conc. × (6) vol.]	-1.29	R_{56} gain	2.60
			6(T,U,V,O,I,W,R,X) conc., μmoles/liter	0.654		
6→7 T	0.855		7(T,U,V,O,I,W,R,X) conc., μmoles/liter	1.293		
7 E	13.225		(7-6) conc. diff.	0.639		
7 M	20.610		7(T,U,V,O,I,W,R,X) anoxic vol., km^3	12.204		
R_{67} PO_4 G	7.39	±0.086	R_{67} [(7-6) conc. × (7) vol.]	7.80	R_{67} loss	-0.41
R_{17} total PO_4 G	18.06	±0.084	R_{17} anoxic PO_4 gained	11.76	R_{17} gain	6.31

[a] Detailed budget for soluble reactive PO_4, showing the total PO_4 budget for the complete hypolimnion and the budget calculated for the stations where anoxic water was encountered; finally the oxic PO_4 budget is calculated from the other two budgets. The volumes from Table 4 and the concentrations from Table 5 were used for the total budget. The values for the anoxic calculations are shown in the table.

[b] T = transferred, E = estimated, M = measured, PO_4 G = PO_4 gain.

[c] Letters in parentheses refer to stations where anoxic water was encountered.

Table 8. Total Quantities:[a] Changes Calculated in the Total Quantities of the Various Components in the Hypolimnion which were due to Chemical and Biological Action Only, Including Error Estimates Based on Heat Budget Comparisons

Reaction	O_2	NO_3	NO_2	NH_3	PO_4	SiO_2	Total hardness
R_{12}	-49.1 ±1.67	2.37 ±0.11	1.10 ±0.01	3.88 ±0.06	0.097 ±0.0005	6.93 ±0.11	-15.8 ±13.5
R_{24}	-138.8 ±12.43	0.93 ±0.71	-2.59 ±0.10	2.53 ±0.44	0.905 ±0.0005	6.65 ±0.72	14.8 ±106.3
R_{45}	-110.8 ±1.19	-0.15 ±0.17	0.21 ±0.01	2.84 ±0.11	-0.065 ±0.0017	8.21 ±0.23	72.9 ±19.7
R_{56}	-43.4 ±3.76	-1.49 ±0.43	2.75 ±0.04	4.35 ±0.27	0.131 ±0.0054	-2.77 ±0.60	-300.8 ±67.3
R_{67}	-66.8 ±5.40	1.02 ±0.47	0.69 ±0.10	-7.09 ±0.26	0.739 ±0.0086	2.11 ±0.64	89.7 ±82.3
R_{17}	-408.9 ±12.2	2.68 ±0.945	2.16 ±0.13	6.52 ±0.57	1.807 ±0.0084	21.13 ±1.15	-139.2 ±144.6

Reaction	Total CO_2	Total CO_2 - filtered alkalinity	Filtered alkalinity	Eh $\times 10^{-3}$ V	pH	Suspended minerals mg/liter
R_{12}	104.0 ±19.5	58.6 ±10.0	45.4 ±9.46	-- --	-0.09 --	0.010 --
R_{24}	142.0 ±154.8	111.4 ±78.7	30.6 ±76.5	-- --	0.17 --	-0.041 --
R_{45}	177.0 ±28.7	186.1 ±14.8	-9.1 ±13.89	-17.74 --	-0.07 --	-0.131 --
R_{56}	-43.0 ±98.9	-50.9 ±52.2	-7.9 ±46.59	67.22 --	0.02 --	0.063 --
R_{67}	47.0 ±125.3	21.9 ±64.0	25.1 ±60.19	-13.57 --	-0.08 --	-0.023 --
R_{17}	427.0 ±213.6	327.1 ±109.9	84.1 ±103.32	35.91 --	-0.05 --	-0.122 --

[a]Unless otherwise specified, units are moles $\times 10^7$.

Table 9. Oxic and Anoxic Quantities: Changes in Quantities in the Hypolimnion which were Calculated to have Occurred where Oxic and Anoxic Conditions Prevailed in the Water; Changes are those due to Chemical and Biological Effects Only (units, moles × 10^7).

Reaction	Total CO_2		NH_3		NO_2		NO_3		Filtered alkalinity	
	Oxic	Anoxic	Oxic	Anoxic	Oxic	Anoxic	Oxic	Anoxic	Oxic	Anoxic
R_{12}	104.0	0	3.88	0	1.10	0	2.37	0	45.4	0
R_{24}	121.8	20.2	-1.35	3.88	-2.56	-0.023	4.11	-3.18	25.32	5.28
R_{45}	144.7	32.3	8.18	-5.34	0.18	0.022	0.94	-1.09	-15.47	6.37
R_{56}	-20.0	-23.0	0.87	3.48	2.63	0.124	0.448	-1.94	-3.19	-4.71
R_{67}	-11.6	58.6	-5.67	-1.42	0.74	-0.054	4.66	-3.64	21.62	3.48
R_{17}	338.9	88.1	5.92	0.60	2.09	0.069	12.53	-9.85	73.68	10.42

Reaction	PO_4		SiO_2		SO_4		Fe		Mn	
	Oxic	Anoxic	Oxic	Anoxic	Oxic	Anoxic	Oxic	Anoxic	Oxic	Anoxic
R_{12}	0.097	0	6.93	--	-12.5	--	0.57	--	3.59	--
R_{24}	0.170	0.735	3.20	3.45	-15.0	--	3.82	--	6.80	--
R_{45}	0.145	-0.210	6.97	1.24	35.6	--	1.89	--	1.13	--
R_{56}	0.260	-0.129	-4.79	2.02	13.0	--	3.64	--	0.21	--
R_{67}	-0.041	0.780	2.37	-0.256	-58.1	--	-2.10	--	2.23	--
R_{17}	0.631	1.176	14.68	6.45	-37.0	-26.4	7.84	3.80	13.96	3.79

Table 10. Oxic and Anoxic Rates:[a] Change in the Quantities of Certain Materials Present in the Hypolimnion due to Chemical and Biological Effects

Reaction	O_2	Total CO_2		NH_3		NO_2		SiO_2		Filtered alkalinity	
	Oxic	Oxic	Anoxic	Oxic	Anoxic	Oxic	Anoxic	Oxic	Anoxic	Oxic	Anoxic
R_{12}	-9.55	20.23	0	0.75	0	0.21	0	1.35	0	8.83	0
R_{24}	-11.91	11.24	25.0	-0.13	4.81	-0.24	-0.03	0.30	4.28	2.34	6.54
R_{45}	-21.81	31.73	62.6	1.79	-10.35	0.04	0.04	1.53	2.40	-3.39	12.34
R_{56}	-8.95	-5.57	-18.2	0.24	2.75	0.73	0.10	-1.33	1.60	-0.89	-3.73
R_{67}	-10.03	-2.48	26.4	-1.28	-0.64	0.17	-0.02	0.54	-0.12	4.88	1.57
R_{17}	-12.23	11.86	18.3	0.21	0.13	0.07	0.014	0.51	1.34	2.58	2.17

Reaction	PO_4		SO_4		Fe		Mn		Nitrifying bacteria $\times 10^5/m^2$	NO_3	
	Oxic	Anoxic	Oxic	Anoxic	Oxic	Anoxic	Oxic	Anoxic		Oxic	Anoxic
R_{12}	0.019	--	-2.4	--	0.11	--	0.70	--	3.64	0.46	--
R_{24}	0.016	0.911	-1.3	--	0.33	--	0.58	--	2.94	0.38	-3.94
R_{45}	0.032	-0.407	7.0	--	0.37	--	0.22	--	0.185	0.18	-2.11
R_{56}	0.073	-0.102	0.27	--	0.75	--	0.04	--	0.870	0.13	-1.53
R_{67}	-0.010	0.351	-8.7	--	-0.32	--	0.79	--	3.77	1.05	-1.64
R_{17}	0.022	0.245	-1.30	-5.5	0.27	0.79	0.49	0.79	2.11	0.44	-2.05

[a]Rates obtained by dividing the observed change in quantity by the area and time during which the change occurred. Unless otherwise specified, units are millimoles per square meter per day.

approximately 1 cm. The sedimented algae had obvious effects: they diminished the oxygen demand while they were alive (Figure 8), and the algal rains had a strong stimulating effect on the nitrifying bacteria populations and hence on the nitrate production. Values from Menon et al. (9) included in Table 10 show varying bacteria populations. There is a weak correlation between the nitrate production rate and the concentration of nitrifying bacteria on the bottom (very few nitrifying bacteria were found in the water column). This may have been because a very heavy sampling program is needed for quantitative bacteriology, owing to the high rates of population growth and decay. A program of the required size was not possible. Nevertheless, by using the average nitrate rate and the average nitrifying bacteria population found during the study, the estimate is made that the daily nitrate production rate was 1.9×10^{-9} moles of nitrate per bacterium. This value can be extended by including nitrite, giving the combined daily production rate of 2.1×10^{-9} moles ($NO_3 + NO_2$) per bacterium.

The sedimented algae provided an active site for extensive _Desulfovibrio_ and _Thiobacillus_ bacterial action (9). During the period from July 30 to August 12 (R_{12} and R_{24}), it appears that the _Desulfovibrio_ bacteria were fairly actively reducing sulfate. This is evident from the sulfate values in Table 9 and Figure 9. Examination of cores that were taken during that period revealed that the interface organic mat had an underlying black layer which thickened upward during the period. Presumably this black layer consisted of iron and manganese sulfides resulting from the reaction of hydrogen sulfide produced from sulfate reduction, with the Fe^{2+} and Mn^{2+} ions diffusing out of the sediments.

The reduction of sulfate under these conditions may be represented by the following equation (10, p. 432):

$$SO_4^{2-} + 2CH_2O + 2H^+ \rightarrow H_2S + 2CO_2 + 2H_2O \qquad (1)$$

where CH_2O represents organic matter (i.e., carbohydrate).

Some of the hydrogen sulfide thus produced would then react to form the observed black sulfides. This process is likely, since alkalinity increases (removal of protons) were observed with sulfate reduction, together with black sulfides occurring near the mud-organic matter interface. This relationship has been observed previously (11).

During the period between August 13 and 20 (R_{45} and R_{56}), the metallic sulfides were in contact with the overlying oxygenated water, as many black patches were observed to be at the surface of the organic mat. It is

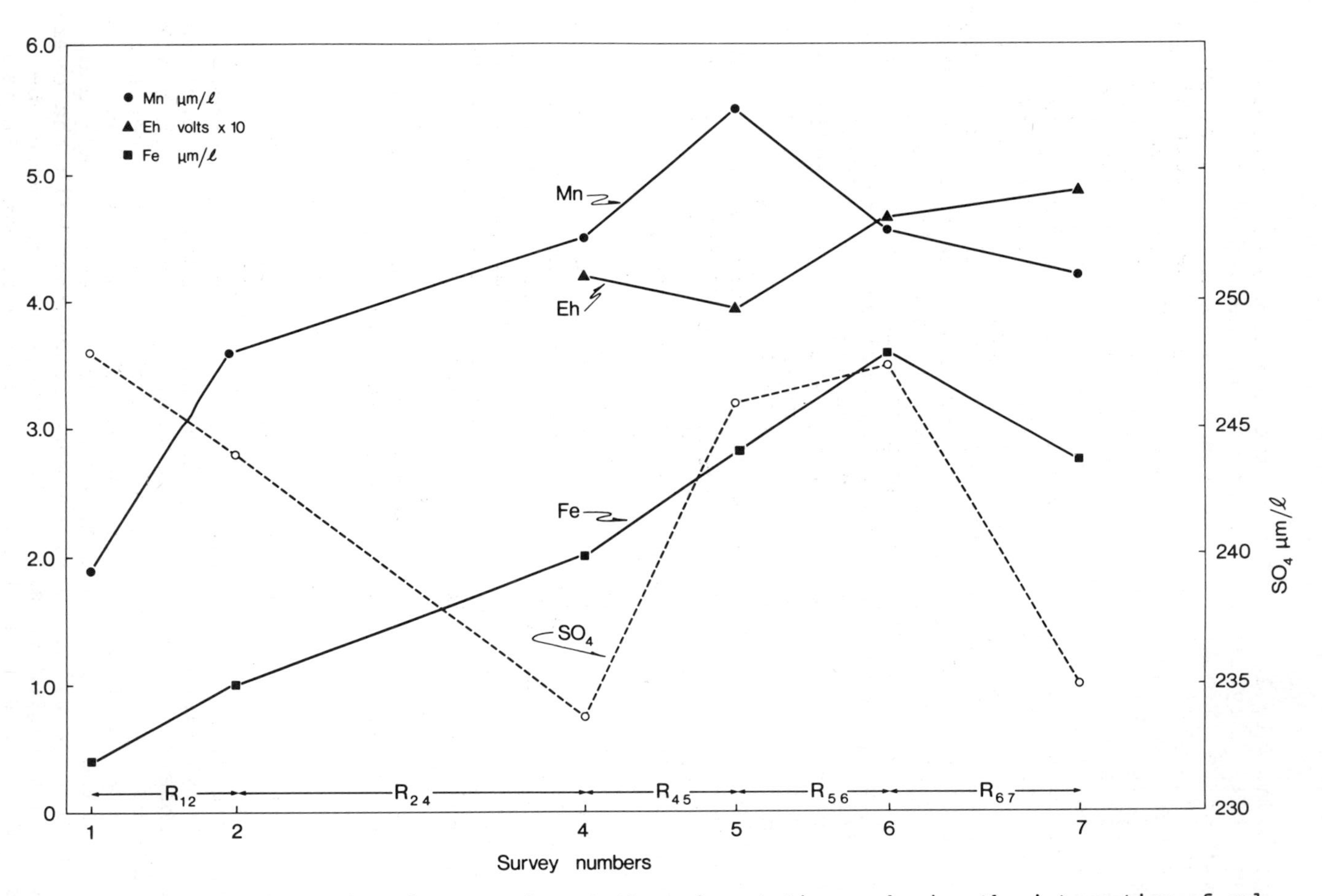

Fig. 9. Concentrations of various species at the major stations, showing the interaction of sulfate with iron and manganese.

suggested that during this period, especially during R_{45}, the following reaction occurred extensively:

$$2FeS + \frac{9}{2} O_2 + 4OH^- + H_2O \rightarrow 2Fe(OH)_3 + 2SO_4^{2-} \qquad (2)$$

This type of process would tend to increase the quantity of iron and manganese in the water and to reestablish the brown oxidized layer, produce sulfates and deplete oxygen at a high rate while reducing the alkalinity. Indeed, all these effects were observed (Figure 9, Tables 5 and 9). Prior to August 13 the black layers were generally buried fairly deep in the organic mat or were nonexistent, and Figure 9 shows that the sulfate concentrations decreased and the iron and manganese concentrations increased during this period. It is probable that sulfate reduction was proceeding, along with moderate release of iron and manganese from the sediments; there was probably very little dissolution of FeS. It seems that the appearance of the black layers close to the sediment surface is the controlling mechanism for the reaction of FeS with oxygen.

It is probable that a large part of the oxygen depletion occurred by way of the sulfate cycle. The oxidation of hydrogen sulfide can be simply represented by the reaction

$$H_2S + 2O_2 \rightarrow 2H^+ + SO_4^{2-} \qquad (3)$$

If the process represented by equation 1 occurred in the organic mat and the process represented by equation 3 occurred in the overlying water, it would be impossible to distinguish the results of this reaction sequence from those obtained from the oxic decay of organic materials. In other words, sulfate can act as a catalyst in the deoxygenation of bottom waters.

It is also likely that some elemental sulfur was formed from the interaction of sulfate and hydrogen sulfide. This possibility is described by Stumm and Morgan (10, p. 32). Clouds of white material, which were identified as bacteria (9), were observed on occasion in the water where the black patches were seen.

Oxygen Depletion Pattern

The oxygen areal depletion pattern is presented in Figure 10a-e. The overall oxygen loss was basin-wide during each survey. However, the rate of oxygen depletion is clearly documented by the foregoing figures as

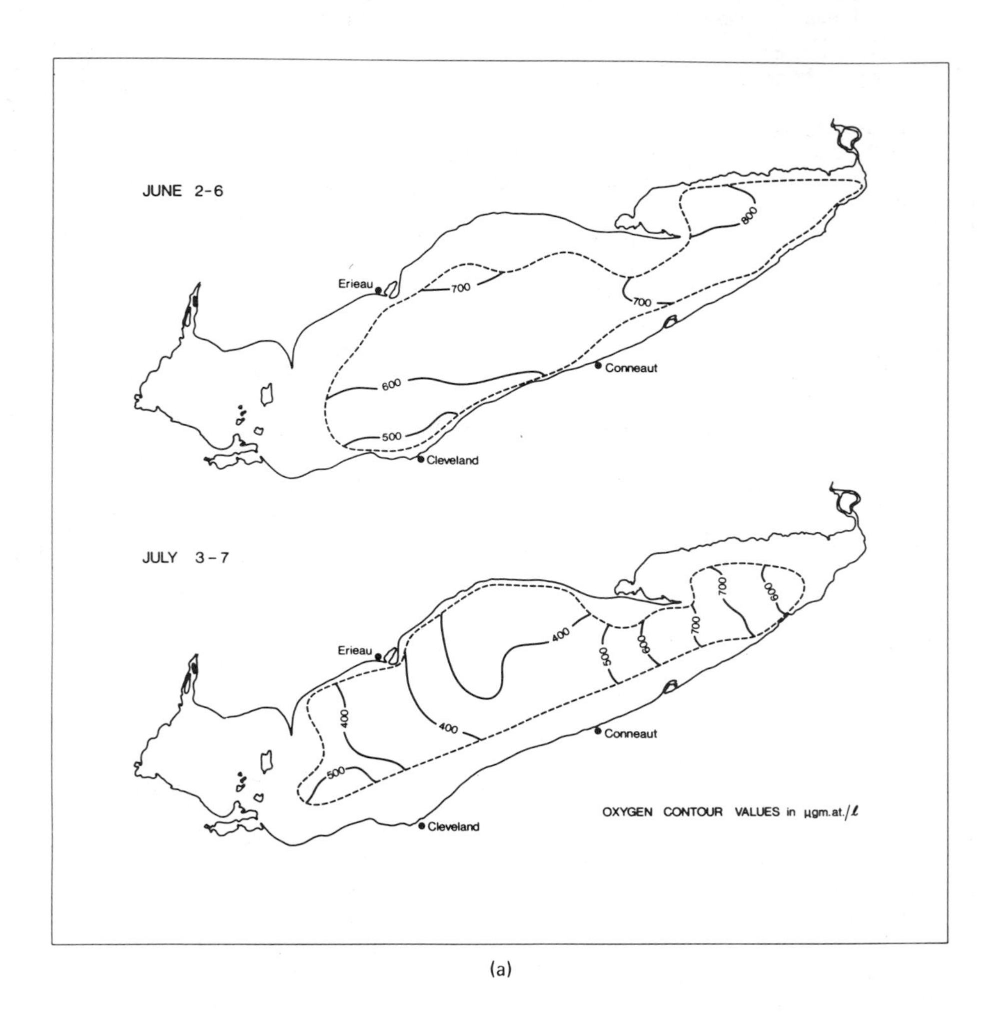

Fig. 10a. Hypolimnion oxygen concentration patterns: June 2 to 6 and July 3 to 7, 1970 (CCIW monitor cruises).

being highest in the western end and along the south shoreline, whereas the north shoreline, midlake, and the extreme eastern end of the basin had a lower rate of depletion. This observed pattern is confirmed by the findings from the sediment oxygen uptake studies (8), which indicated that the sediment oxygen uptake rates were higher in the western end and along the south shore than

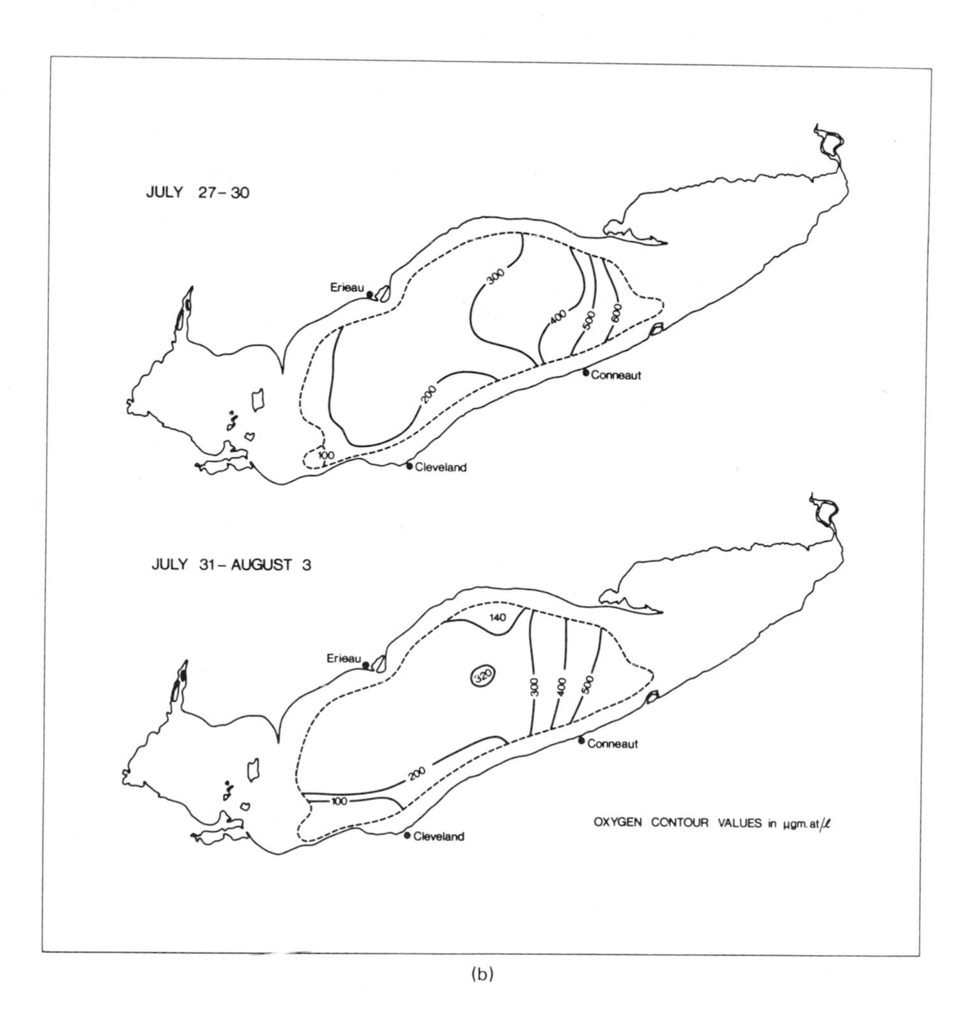

(b)

Fig. 10b. Hypolimnion oxygen cencentration patterns: July 27 to 30 (survey 1), July 31 to August 3, 1970 (survey 2).

midlake and the north shore. In addition, this pattern agrees with the biological findings (7) that indicated profuse productivity in the same areas of high rates of oxygen depletion.

The oxygen depletion pattern just cited resulted in the western stations becoming anoxic first, followed by the south shore stations, as demonstrated in surveys 4,

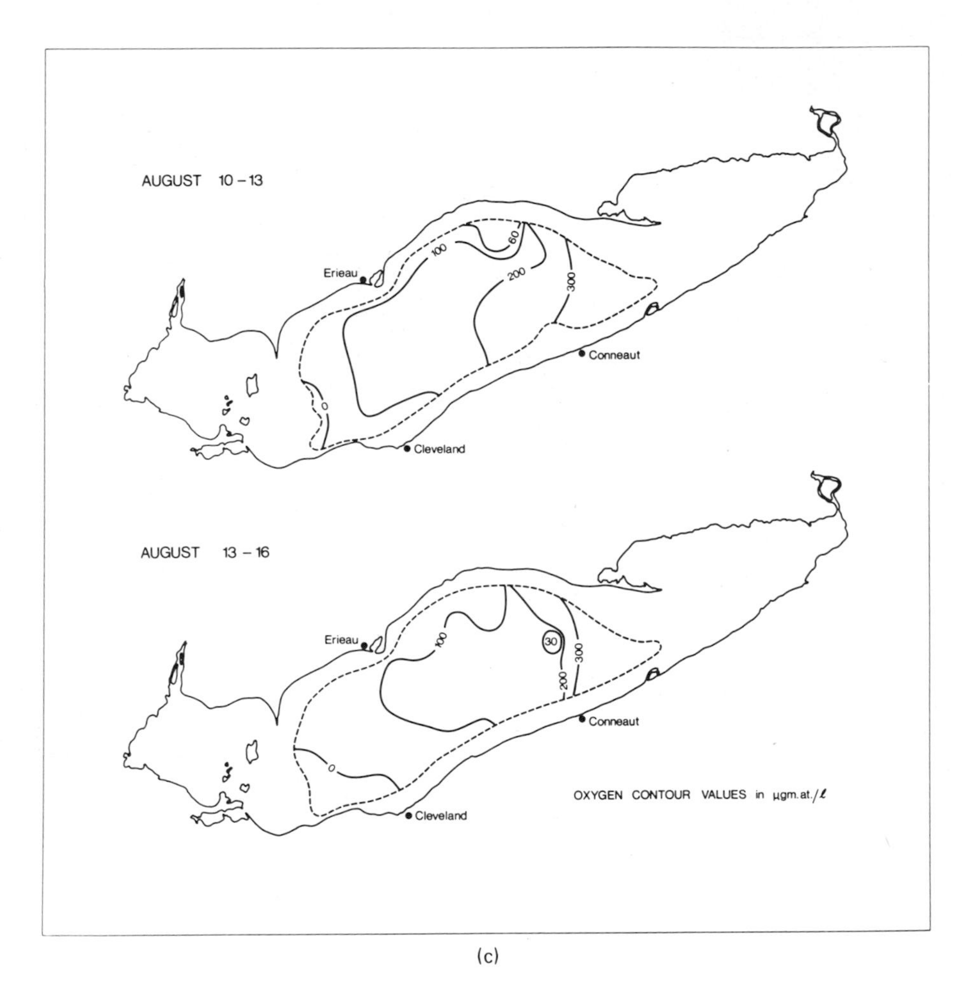

Fig. 10c. Hypolimnion oxygen concentration patterns: August 10 to 13 (survey 4), August 13 to 16, 1970 (survey 5).

5, and 6 (Figure 10c and d). Survey 7 indicated anoxic stations in the north shore area; midlake stations became depleted progressively from west to east. This depletion pattern was probably due to the heavy load of nutrients coming into the western end of the lake by way of the Detroit River (12). Finally, as shown in the September CCIW Monitor Cruise Survey (Figure 10e), the total

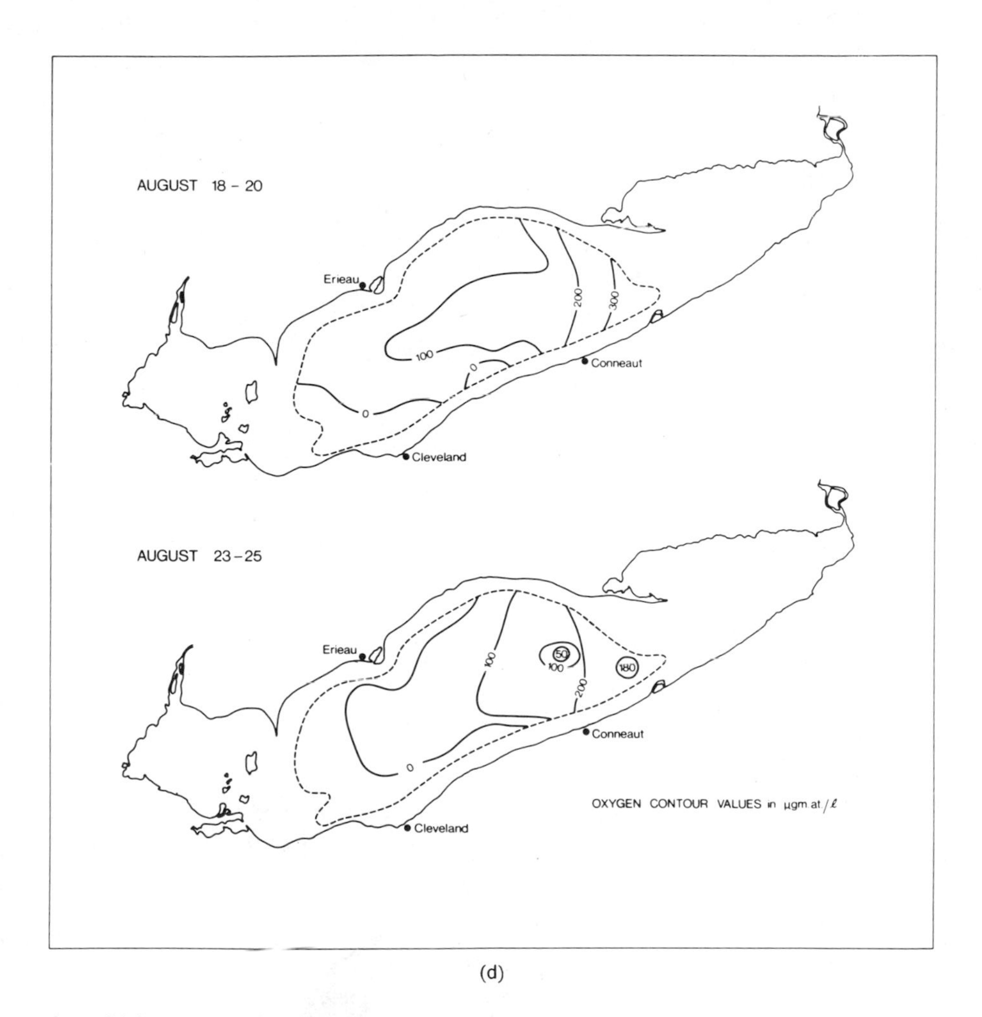

Fig. 10d. Hypolimnion oxygen concentration patterns: August 18 to 20 (survey 6), August 23 to 25, 1970 (survey 7).

hypolimnion (6600 km^2; 2500 miles^2) was anoxic, although it was diminished in area and volume by this late period in the year.

Two areas were observed to act independently from the pattern discussed previously. Station E, although a deep-water station and located on the eastern end of the basin, showed a low dissolved oxygen concentration (30 µg-at./

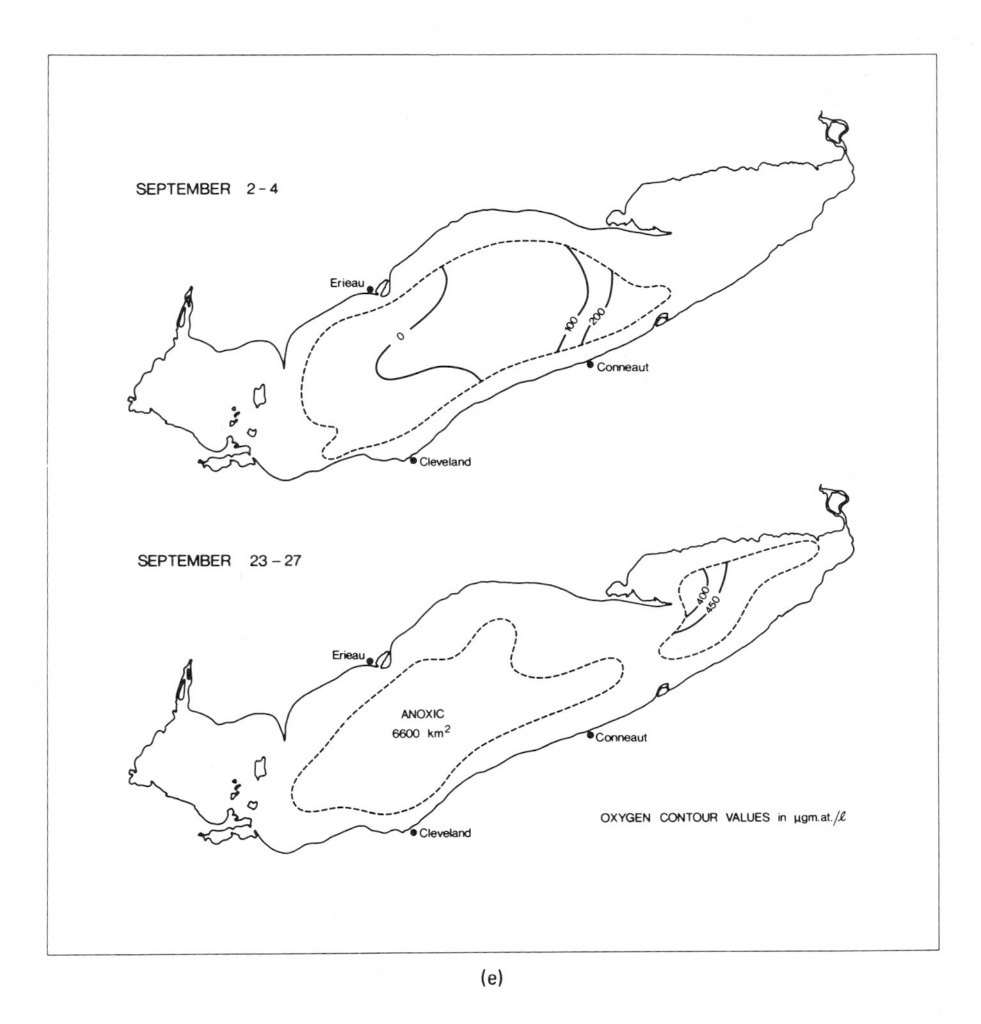

(e)

Fig. 10e. Hypolimnion oxygen concentration patterns: September 2 to 4, September 23 to 27, 1970 (CCIW monitor cruises).

liter) as early as survey 5 (Figure 10c). This situation was most probably caused by the valleylike topography of the lake bottom at this station, which prevented normal lake circulation of water in and out of the area and created a stagnant pool.

The extreme eastern stations appeared to be affected by a clockwise rotating current pattern that prevented development of anoxic water in the area of stations A, B, and C. This situation was probably influenced by a return flow pattern from the north shore and is demonstrated by the intrusion of a shoal along the lake bottom from Conneaut, Ohio, toward station G. This suggested current pattern agrees with currents at station G as described by Blanton and Winklhofer (3). However, their projected current patterns throughout the rest of the hypolimnion do not appear to have had a noticeable effect on the dissolved oxygen depletion pattern.

Oxygen Budget

Equation 1 shows that one mole of carbonate alkalinity will be gained for each mole of sulfate reduced. This relationship is also suggested by Berner et al. (11). Thus, using the values from Tables 8 and 9, the following budget for alkalinity can be calculated:

Total observed increased alkalinity	$= 84.1 \times 10^7$ moles
Alkalinity increase due to SO_4 reduction	$= -63.4$
Alkalinity increase due to NH_3 production	$= -3.3$
Therefore, alkalinity increase due to dissolved carbonate species	$= 17.4 \times 10^7$ moles

It should be borne in mind that when organic compounds degrade, they produce an appreciable quantity of water from the oxygen taken up; 30% of the oxygen used to oxidize the aliphatic chain of fatty acid will form water and cannot be detected in budget studies. Hutchinson (4) suggests that a realistic value for the respiratory quotient is 0.85. The respiratory quotient is defined as the ratio of the number of molecules of carbon dioxide produced to the number of molecules of O_2 consumed.

Equation 1 is not entirely correct because the formula for a carbohydrate chain is used. The quantity of CO_2 released per SO_4^{2-} ion reduced would be better expressed as $0.85 \times 2 = 1.70$ molecules of CO_2. Thus the total carbon dioxide budget can be estimated as follows:

Total CO_2 produced	= 427.0 × 10^7 moles
CO_2 from precipitated calcite or dissolved carbonates	= -17.4
CO_2 from sulphate reduction	= -107.1
Therefore, CO_2 from oxygen depletion	= 302.5 × 10^7 moles
Therefore, total O_2 depleted by decay processes	= 356.0 × 10^7 moles

It is now possible to calculate a detailed oxygen budget:

Total oxygen depleted from hypolimnion	= 409.00 × 10^7 moles
O_2 used in conversion of 2.68 × 10^7 moles NO_2 to NO_3	= -1.34 × 10^7
O_2 used in conversion of 2.16 × 10^7 moles NH_3 to NO_2	= -3.78 × 10^7
O_2 used in conversion of 0.455 × 10^7 moles organic P to PO_4 (assuming 50% of organic phosphorus is in oxidized form in the algae)	= -0.45 × 10^7
Balance of O_2 depleted	= 403.43 × 10^7
O_2 used in CO_2 production	= 356.0 × 10^7
Therefore, O_2 depleted by inorganic processes	= 47.43 × 10^7 moles
	= 12.2% of total O_2 depletion

From a chemical budget point of view, the conversion of hydrogen sulfide to sulfate cannot be considered to be one of the inorganic oxygen consuming processes because this would result in a net increase of sulfates in the water during the study. The actual decrease in sulfates during this period, indicated in Table 9, has already been taken into account in the foregoing budgets.

It is likely that the 46.0×10^7 moles of oxygen used in inorganic oxidation were consumed in the oxidation of ferrous species. An example of the many possible oxidation processes is

$$2Fe^{+2} + \frac{1}{2} O_2 + 2H^+ \rightarrow 2Fe^{+3} + H_2O \qquad (4)$$

Thus 0.25 moles of O_2 is used for each mole of iron that is oxidized. If all the iron that was oxidized remained in the water, this would give the iron an estimated concentration of 45 μmoles/liter. The observed final average concentration of iron at the oxic stations was 2.72 μmoles/liter, suggesting that in the order of 40 μmoles/liter had been oxidized during the study but that most had settled out of the water. The conditions at the oxic stations, having an Eh between 350 and 500 mV and a pH of approximately 7.4, are likely to keep the iron in the oxidized form and manganese in the reduced form (10, p. 533). Although only iron has been mentioned in this discussion, in reality the reacting species would have been a mixture of iron and manganese ions; however, this situation is not discussed further here.

Both Hutchinson and Mortimer (4, p. 644) have used hypolimnetic oxygen areal depletion rates as a measure of the trophic state of lakes, and their suggested criteria are as follows:

Area	Oxygen depletion rates, (mmoles $O_2/(m^2)(day)$)	
	Hutchinson	Mortimer
Oligotrophic lakes	<-5.3	<-7.7
Mesotrophic lakes	-5.3 to -10.3	-7.7 to -17.2
Eutrophic lakes	>-10.3	>-17.2

The average oxygen depletion rate measured during this study was -12.23×10^{-3} moles $O_2/(m^2)(day)$. If this value is compared with those listed previously, the central basin of Lake Erie can be considered to be eutrophic. There has been a marked increase in the oxygen depletion rates in recent years (13).

Since approximately 88% of the hypolimnetic oxygen was consumed in the decay of organic materials, it is probable that the massive algal bloom in the central basin during the last week of July 1970 was the major cause of the anoxic conditions that subsequently developed in the hypolimnion. The magnitude of the initial pulse of this bloom was most likely limited by the supply of soluble reactive phosphorus, because the survey completed on July 30 showed that the phosphorus had been reduced to the barely detectable level in the surface waters of 80% of the basin (Figure 11). It is probable that the extent of oxygen depletion would have been reduced if the supply of soluble reactive phosphorus had been much less during the last week of July.

Oxic Regeneration

From the estimate of 17.4×10^7 moles of CO_2 being of inorganic origin, it is possible to calculate that 96% of the CO_2 increase observed in the hypolimnion is of organic origin. Thus from the values in Table 9 the following oxic regeneration budget can be drawn up:

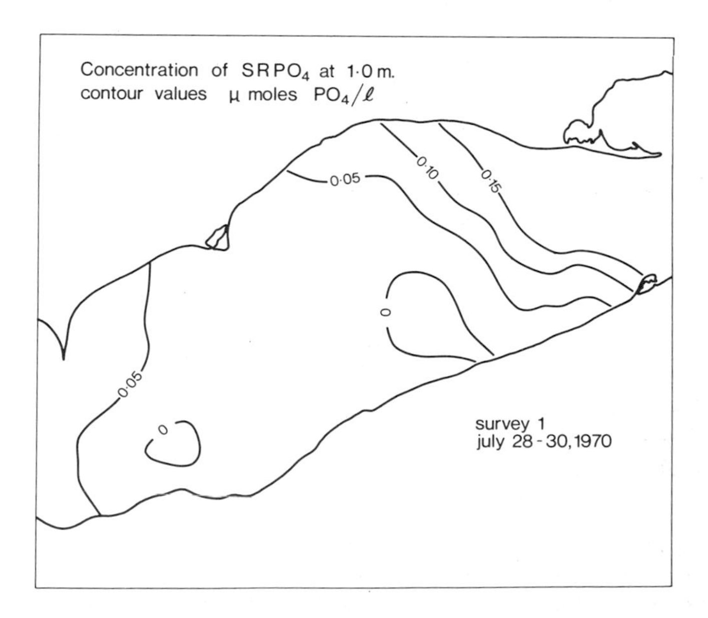

Fig. 11. Concentrations of soluble reactive phosphorus at 1.0 m depth in Central Lake Erie.

$$CO_2 \text{ (organic origin)} = 325.3 \times 10^7 \text{ moles}$$

$$NO_3 + NO_2 + NH_3 = 20.5 \times 10^7 \text{ moles}$$

$$PO_4 = 0.632 \times 10^7 \text{ moles}$$

which gives the following oxic regeneration rates (Table 11):

$$CO_2 = 11.4 \times 10^{-3} \text{ moles/(m}^2\text{)(day)}$$

$$\Sigma N = 0.72 \times 10^{-3} \text{ moles/(m}^2\text{)(day)}$$

$$PO_4 = 0.022 \times 10^{-3} \text{ moles/(m}^2\text{)(day)}$$

and the regeneration ratio is

$$(C:N:P)_r = 1:0.063:0.0019 \tag{5}$$

Table 11. Oxic Regeneration: The Quantities, Rates, and Proportions of Components that Returned to the Soluble Form in Oxygenated Hypolimnion Water

Component[a]	Quantity moles $\times 10^7$	Rate, moles $\times 10^{-3}$/(m^2)(day)	Ratio
PO_4	0.63	0.022	1.0
C	338.90	11.85	538.6
N	20.54	0.72	32.6
Si	14.68	0.51	23.2
Mn	13.96	0.49	22.2
Fe	7.84	0.27	12.5
C (org.)	325.34	11.38	517.3
C (inorg.)	13.56	0.47	21.4

[a]Org.= probable organic origin; inorg. = probable inorganic origin.

The ratio of carbon to nitrogen to phosphorus in the particulate organic material in the water sampled just above the mesolimnion was found to be

$$(C:N:P)_p = 122:18:1$$

$$= 1.0:0.148:0.0081 \tag{6}$$

This is presumably fairly close to the elemental composition of the organic material sedimenting into the hypolimnion.

Dividing equation 6 into equation 5 gives

$$\frac{(C:N:P)_r}{(C:N:P)_p} = 1:0.43:0.235$$

It would appear then, that for every atom of carbon which is mineralized only 0.43 or 43% of a nitrogen atom is mineralized and 0.235 or 23.5% of a phosphorus atom is mineralized. Thus it is possible to state that during the study, approximately 45% of the nitrogen and 25% of the phosphorus in the mineralized material returned to the water under oxic conditions: the balance of 55% of the nitrogen and 75% of the phosphorus must have remained complexed in some manner with the material on the bottom. Alternatively, the balance could have been part of the loss of nitrogen due to denitrification.

A possible explanation for the low percentage regeneration of phosphorus is that most of the orthophosphate from the organic decay is produced on the lake bottom in very close proximity to the precipitated ferric hydroxides. Thus there is probably a high concentration of orthophosphate at the interface in water containing much suspended ferric hydroxide; this would readily lead to the formation of insoluble ferric hydroxy-phosphate complexes. These complexes would most likely dissolve if conditions subsequently became anoxic, but they would remain insoluble if oxic conditions were maintained. The formation of some soluble organic carbon, nitrogen, and phosphorus compounds was probably a step in the mineralization process. The values in Table 6 show that there was a fairly significant increase in the quantity of soluble organic phosphorus. It is assumed that there were corresponding increases in the soluble organic carbon and nitrogen and that these organic components do not seriously affect the ratio argument outlined earlier.

A comparison of the oxic regeneration rates of phosphate, manganese, iron, and silica reveals that the molecules are regenerated or made soluble in the following ratio (Table 11):

$$PO_4:Mn:Fe:SiO_2 = 1:22:12:23$$

This proportionation is quite different from the anoxic regeneration ratios (Table 12), which are: $PO_4:Mn:Fe:SiO_2 = 1:3.2:3.2:5.5$.

Table 12. Anoxic Regeneration: The Quantities, Rates, and Proportions of Components that Returned to the Soluble Form in Oxygenated Hypolimnion Water

Component[a]	Quantity, moles × 10^7	Rate, moles × $10^{-3}/(m^2)(day)$	Ratio
PO_4	1.18	0.245	1.0
C	88.1	18.30	74.7
N	-9.2	-1.91	-7.8
Si	6.45	1.34	5.5
Mn	3.79	0.79	3.2
Fe	3.80	0.79	3.2
P (org.)	0.69	0.14	0.6
P (inorg.)	0.49	0.10	0.4
C (org.)	84.5	17.6	71.8
C (inorg.)	3.6	0.8	3.3

[a]Org. = probable organic origin; inorg. = probable inorganic origin.

Anoxic Regeneration

From the anoxic budgets for carbon, nitrogen, and phosphate and the extent of anoxic conditions observed during the study (4.82×10^{10} m^2 day), the following regeneration rates were calculated:

$$C = 18.3 \times 10^{-3} \text{ moles}/(m^2)(\text{day})$$

$$\Sigma N = -1.92 \times 10^{-3} \text{ moles}/(m^2)(\text{day})$$

$$PO_4 = 0.24 \times 10^{-3} \text{ moles}/(m^2)(\text{day})$$

The high rate of carbon dioxide production might be due in part to the dissolution of carbonates and bicarbonates. The negative value for nitrogen could be explained by suggesting that when anaerobic reduction of nitrate occurs, significant quantities of nitrogen gas are released to the water. If there had been no loss of nitrogen nutrients under anoxic conditions, the total quantity of nitrogen nutrients generated during the study would have been at least 20.54×10^7 moles; however, the quantity was only 11.4×10^7 moles. Thus the anoxic conditions caused a decrease of about 44% in the quantity of nitrogen nutrients generated. This is presumably part of the natural process of denitrification. Part of the anoxic phosphate regeneration was probably due to organic degradation.

If we assume that 96% of the carbon dioxide is of organic origin, as in the case of oxic regeneration, (i.e., 84.5×10^7 moles), and using the ratios given in equation 6, we see that the phosphorus associated with the degraded organic carbon would have been 0.69×10^7 moles of phosphorus. Since 1.176×10^7 moles was actually regenerated, the ratio of the phosphate that was regenerated to the phosphorus that was contained in the decayed organic material is 1.7:1. This means that almost one atom of phosphorus would have been extracted from the sediment for every atom produced by decay, assuming 100% return of the phosphorus in the decayed organic material to the water. If, however, there was less than 100% return of the phosphorus in the mineralized algae to the water, then the amount of the phosphorus from the sediment that returned to the water would be proportionately greater.

The anoxic regeneration rates for the various components are given in Table 12. The values in Table 6 show that there was a significant increase in soluble organic phosphorus as conditions in the hypolimnion water changed from oxic (0.2 μmoles P/liter) to anoxic (0.9 μmoles P/liter). Thus the total anoxic phosphorus regeneration was higher than the anoxic inorganic phosphate regeneration. Inorganic phosphate (soluble reactive phosphorus) forms the basis of the following discussion, and this means that all phosphorus estimates based on inorganic phosphate regeneration are conservative.

Vollenweider (14) has made an estimate of the anoxic regeneration rate for phosphate in a lake having two anoxic periods averaging 155 days each; the value obtained was 0.31 millimoles/(m^2)(day), which approximates the value of 0.24 millimoles/(m^2)(day) for Lake Erie. Thus the observed anoxic phosphate regeneration rate was 11 times greater than the oxic rate.

Kemp et al. (15) have given values for the atomic ratio of organic carbon to nitrogen to phosphorus in the top 3 mm of material from a core taken within 1 mile of station P in the central basin (see item 1 below). If we assume that the C:N:P ratios for the sedimenting algal material (item 2) are representative, then it is possible to calculate what the regeneration ratio of these elements should be (item 3). It is of interest to note that the ratio calculated on this basis falls between the values given previously for oxic and anoxic regeneration, which suggests that the material remaining on the bottom has undergone decay under both conditions.

1. Sediment (Kemp et al.) C:N:P = 68:8.6:1
2. Algae C:N:P = 122:18:1
3. Regeneration (calculated) C:N:P = 176:27.4:1
4. Regeneration (oxic) C:N:P = 517:32.6:1
5. Regeneration (anoxic) C:N:P = 72:-7.8:1

Figure 12 graphically represents the phosphate-oxygen relationships found on the first and last survey and gives some idea of the differences that can be caused by oxic and anoxic regeneration. The graph is complex because an attempt has been made to change from an oxygen scale to an Eh scale when the oxygen concentration reached a zero value. The judgement that an Eh of 450 mV was equivalent to zero oxygen content is purely arbitrary. On occasion when a sample had both an oxygen value above zero and a low Eh, the value was plotted according to the oxygen value.

It would appear that the phosphates started to increase in concentration in the water when it contained about 20 μmoles O_2/liter (0.64 mg O_2/liter). Presumably at this oxygen content in the water 1.0 m from the bottom, the sediment-water interface was already anoxic. This suggestion is largely in agreement with the bacterial findings, since the Desulfovibrio bacterial counts increased in the water column once the oxygen concentration had decreased to the range of 45 to 60 μmoles O_2/liter (9). Figure 12 demonstrates quite definitely that if the oxygen content of the hypolimnion water is kept above 30 μmoles O_2/liter (1.0 mg O_2/liter), anoxic phosphate regeneration can be averted.

Fate of Regenerated Quantities

There is little doubt that the nutrients released under oxic conditions will remain soluble when the hypolimnion water mixes with the surface water during the overturn. During this study a low concentration of iron was always found in the surface water, but the manganese was below the detection level in most instances. Perhaps this is because the colloidal oxidized form of iron is much less dense than the oxidized manganese form, which settles out. It is possible that manganese was a limiting nutrient in the surface waters at times during this study.

Phosphate concentrations of 1.3 μmolar and iron concentrations of 6 μmolar at a pH of 7.4 were about average for the anoxic hypolimnion water found in this study. If this anoxic water were mixed with equal quantities of

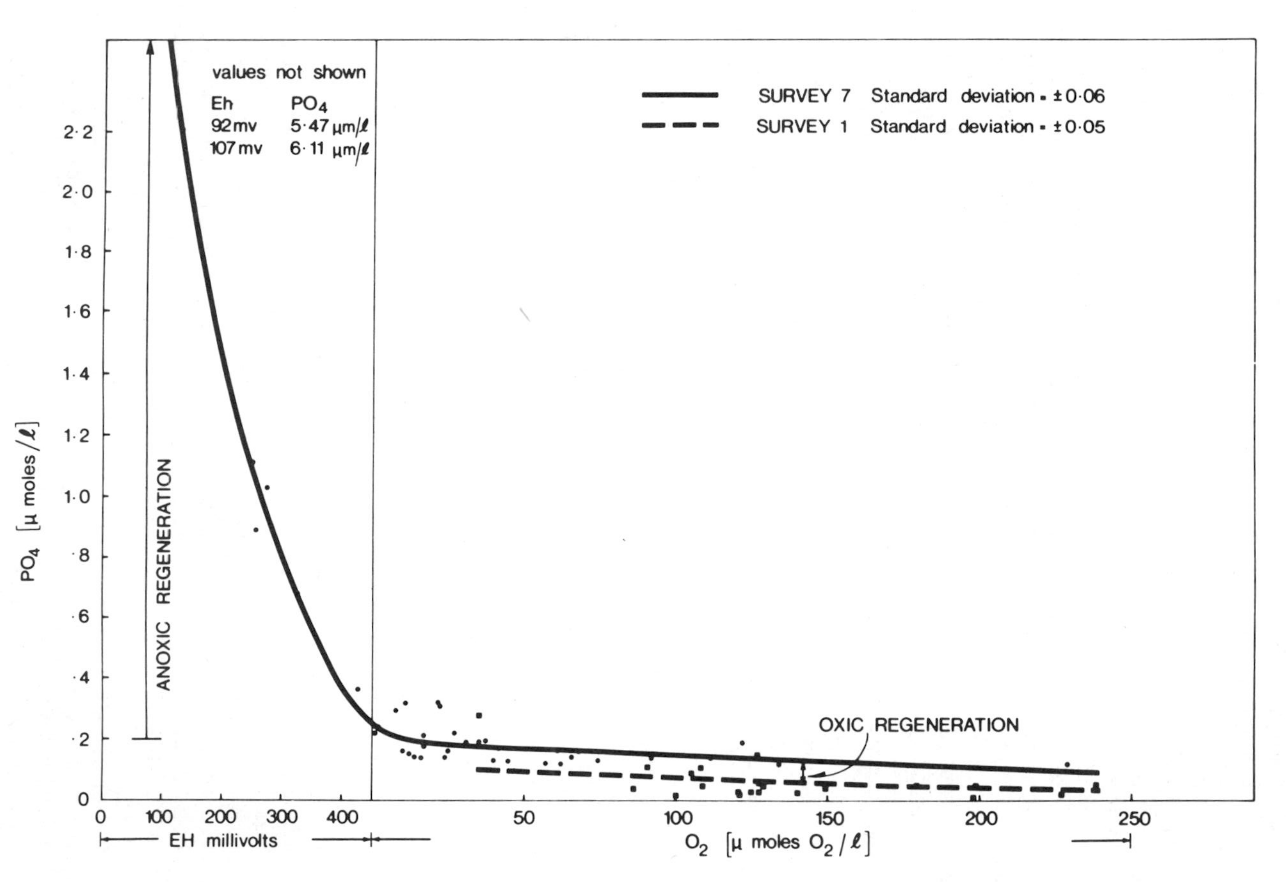

Fig. 12. Plot of oxygen-phosphate and Eh-phosphate relationships, surveys 1 and 7.

oxygenated surface water, the pH would rise to approximately 8, and it would appear from the equilibria described by Stumm and Morgan (10, p. 533) that the iron would precipitate out as the hydroxide and not as the phosphate. The ferric hydroxide could then adsorb, and exchange with some of the phosphate molecules, but it is difficult to estimate the extent of the absorption.

In this regard data are available from the CCIW monitor cruises. The results from two cruises showing the quantities and volume-weighted averages of soluble reactive phosphate, particulate phosphorus, and soluble organic phosphorus for September 1970 and October 1970 are listed in Table 13. The pertinent information in this case is that there was a small, completely anoxic hypolimnion remaining in the basin in September and that, during the time interval between the two cruises, the hypolimnion was eroded away. Therefore, the lake was completely unstratified and oxygenated in October.

The results in Table 13 indicate that the soluble reactive phosphorus decreased by approximately 10% during the overturn, whereas the decrease would have been approximately 53% if all the anoxic soluble reactive phosphorus had converted to the particulate form. In fact, the proportion of soluble reactive phosphorus to particulate phosphorus increased rather than decreased with the overturn. These results indicate that a significant part of the soluble phosphorus (both organic and inorganic) regenerated under anoxic conditions reenters the life cycle of Lake Erie.

Data from the September and October CCIW monitor cruises, together with data from this study, have made it possible to estimate the area and duration of anoxic conditions of 25.2×10^{10} m^2 days, during the summer of 1970 (Figure 13). Using the value obtained previously for the anoxic regeneration rate for phosphate, it is possible to estimate the quantity of phosphate that was regenerated under anoxic conditions during the summer:

$$= 25.2 \times 10^{10} \text{ m}^2 \text{ day} \times 0.245 \times 10^{-3} \text{ moles/(m}^2\text{)(day)}$$

$$= 6.17 \times 10^{7} \text{ moles of P}$$

$$= 1914 \text{ metric tons of phosphate}$$

The total content of the basin just before the onset of anoxic conditions was 15.5×10^7 moles of soluble reactive phosphorus (which is taken here as being equivalent to phosphate); thus the anoxic conditions raised the total quantity of available phosphate by 40% between

Table 13. Phosphorus Quantities in the Central Basin of Lake Erie, September and October 1970

Month	Volume, km^3	Concentrations, μmoles/liter			Quantities, moles × 10^6		
		Soluble Reactive PO_4	Soluble Organic P	Particulate P	Soluble Reactive PO_4	Particulate P	Soluble Organic P
September 1970	Hypolimnion - 9.4	2.00	0.394	0.81	18.8	7.6	3.7
	Epilimnion - 275.7	0.06	0.104	0.30	16.6	82.6	28.6
	Total - 285.1				35.4	90.2	32.3
October 1970	Total - 285.1	0.112	0.077	0.256	31.9	73.0	22.0

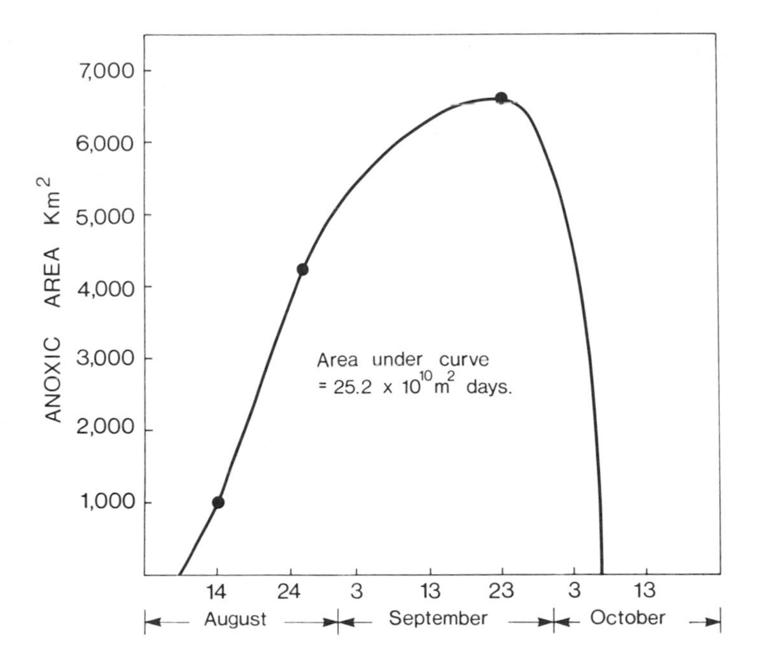

Fig. 13. Extent of anoxic conditions during the summer, 1970, Lake Erie central basin.

August and October. This was a period of peak algal growth as shown by Figure 14. The values for chlorophyll were corrected for phaeophytin content and represent the viable algae. These data (obtained on various monitor cruises) have been kindly made available by Dr. W. A. Glooschenko (CCIW). The volumes of the hypolimnion presented here are calculated on the basis of the monitor cruise data, and they differ somewhat from the volumes calculated using Project Hypo data. These differences are largely due to different station patterns and the use of different types of bathythermographs.

It can be seen that there is a correlation between hypolimnion volume decrease and chlorophyll increase. The upward entrainment of the anoxic hypolimnion water during September 1970 would have provided a supply of water enriched by all, or nearly all, the nutrients necessary for algal growth, resulting in the observed large growth pulse. The results of Braidech et al. (7) confirm

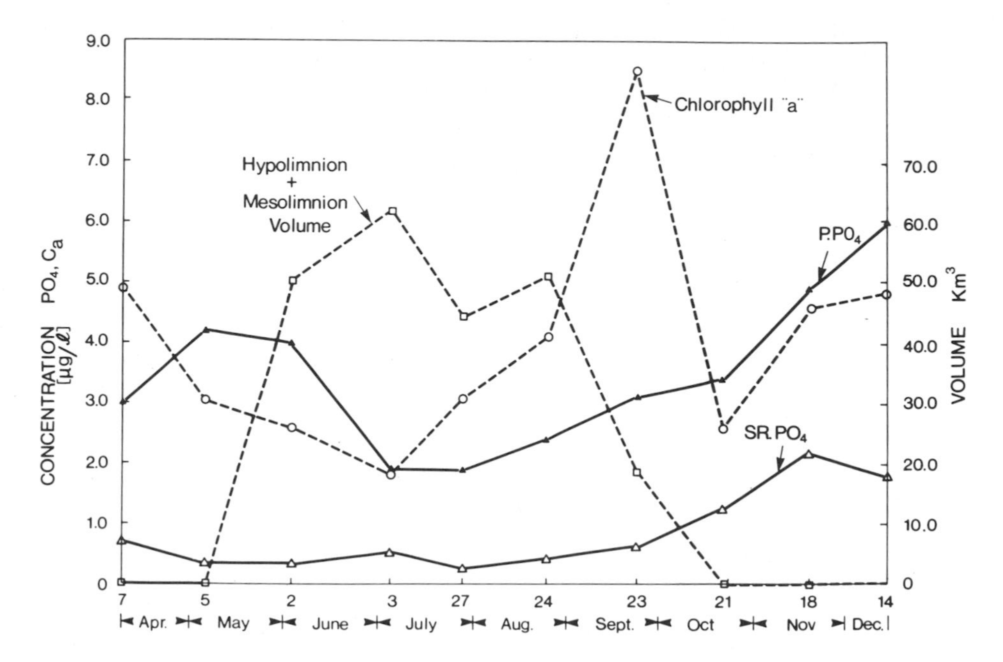

Fig. 14. Plots of soluble reactive phosphorus, particulate phosphorus, and chlorophyll "a" in the central basin epilimnion; hypolimnion plus mesolimnion volume also appears.

this, for they show that the highest algal concentrations encountered in the water during the project were found just above and in the mesolimnion, at stations where the hypolimnion water had recently become anoxic.

Internal Loading versus External Loading

It was demonstrated previously that the anoxic conditions added a large amount of phosphate to the lake at a critical period in the annual cycle. The question then arises, How does the internal loading of phosphate compare with the external loading of phosphorus?

The 1970 CCIW Lake Erie monitor cruises made it possible to estimate the loading of phosphorus into the central basin. The three monitor cruise stations on the western edge of the central basin show a fairly constant value of total phosphorus content in the water of

1.27 ± 0.1 μmoles P/liter. Knowing this, and finding that the annual flow of the Detroit and Maumee Rivers is 160 km^3/yr, we see that approximately 6300 metric tons of phosphorus enters the central basin each year from the western basin. In addition, the International Joint Commission report on Lake Erie (12) estimated the annual input of phosphorus to the basin from external sources to be 3950 metric tons/yr, giving a total external input to the central basin of 10,300 metric tons/yr or 860 metric tons/month.

The period of internal loading due to anoxic conditions lasted a little more than two months (from August 9 to ca. October 10, 1970). During this time, the extent of regeneration activity was 89.2×10^{10} m^2 day; of this, 25.2×10^{10} m^2 day was anoxic and 64.0×10^{10} m^2 day was oxic. Using these values, it is possible to calculate the following quantities for the two-month period prior to the fall overturn:

Two-months loading	Metric tons of phosphorus
External phosphorus loading	1720
Internal phosphate loading oxic regeneration	437
Internal phosphate loading anoxic regeneration	1914
Minimum total internal loading	2351
Total loading	4071

The tabulated values show that the central basin is no longer always acting as a settling basin for phosphorus but instead is becoming a production basin for the element during the period of summer stratification.

Two publications (12,16) recommend the reduction of phosphorus inputs to Lake Erie as a means of improving the water quality of the lake. If these assessments of the situation are correct and the phosphorus inputs to the lake are reduced by 80%, which is a preliminary recommendation of the Report to the International Joint Commission (12), then it is likely that the sediment oxygen demand will be diminished such that anoxic conditions will not occur during the stratified period (17). It is then possible to make some projections as to the probable phosphorus loading on the central basin using the value for the oxic regeneration rate of phosphate.

Estimated two-months loading	Metric tons of phosphorus
External phosphorus loading	342
Internal phosphate loading oxic conditions prevailing	607
Total loading	949

This indicates that if the external phosphorus loading on the central basin is reduced by 1378 metric tons, there will probably be a reduction of the total loading on the basin of at least 3122 metric tons during the two-month late-summer period.

SUMMARY

The study can be summarized most simply by following and explaining the observed sequence of events. From the monitor cruise data it is apparent that there was a fairly large erosion of the hypolimnion during the month of July 1970 (43.2 km^3 on July 3 to 27.6 km^3 on July 27). This seemed to be the initial factor in starting the growth period that lasted from the end of July until October 1970. The first bloom was heavy and resulted in a large fallout of algae onto the lake floor during the last week of July. These sedimented algae were photosynthetic for awhile after they fell onto the lake bottom.

Although the oxygen depletion was ameliorated by the oxygen production of living algae, organic decay continued to cause a net oxygen depletion. The aerobic heterotrophic and sulfate-reducing bacterial populations increased steadily while the living algae on the bottom died and matted down. By approximately August 10 most of the sedimented algae seemed to have died. The loss of the oxygen from photosynthesis, together with the activity of the large bacterial populations, caused a high rate of oxygen depletion from August 10 to 17; anoxic conditions were first observed on August 12. The oxygen depletion rate was again diminished on or about August 17 by a further period of algal rains, which again produced oxygen by photosynthesis. However, the photosynthesis effects of the algae were not sufficient to prevent the spread of anoxis which, by August 25, extended across approximately 4200 km^2 of the 12,700 km^2 hypolimnion area.

The anoxic conditions caused large-scale nutrient regeneration by the dissolution of inorganic forms, and a massive bloom resulted when these nutrients were mixed with surface water during September.

CONCLUSIONS

It is now possible to make a number of conclusions relevant to the initial aims of the study.

1. The massive algal bloom in the central basin during the last week of July 1970, caused a layer of algae 2 to 3 cm thick to be laid down on the floor of the basin and was the major cause of anoxic conditions that subsequently developed in the hypolimnion.
2. The net oxygen demand was variable (being influenced by the photosynthetic oxygen produced by the sedimented algae), but remained negative throughout the study; the average oxygen demand was -12.3×10^{-3} moles $O_2/(m^2)$(day), which is close to the demand expected of a eutrophic lake. The observed change of oxygen concentration in the water was approximately -40 μmoles/(liter) (day) or -3.9 mg O_2/(liter)(month).
3. Approximately 88% of the oxygen uptake was due to bacterial degradation of algal sedimentation, and 12% of the oxygen was taken up in the oxidation of reduced metallic species.
4. Nutrients cause organic growth, with phosphorus often being the limiting nutrient, and the oxygen depletion was largely due to organic decay; thus it can be concluded that a reduction of nutrients, especially phosphorus, would lead to a corresponding decrease in the oxygen depletion rate.
5. Anoxic regeneration of phosphate commenced only when the oxygen concentration in the water fell below 20 μmoles O_2/liter (0.6 mg O_2/liter).
6. Since phosphorus-deficient growth conditions are often encountered in Lake Erie and the internal loading of phosphate under oxic conditions is low, it is possible to conclude that Lake Erie would soon return to an acceptable state if phosphorus inputs to the lake were decreased such that oxygenated conditions were maintained all year in the lake.
7. The oxygen contour maps show that complete oxygen depletion started at the western end of the basin the week

of August 4 to 10 and moved eastward, with the eastward movement being more rapid in the shallow water and along the southern shore. By August 25 one-third of the hypolimnion was anoxic, and on September 23, 1970, the complete hypolimnion was found to be anoxic. The extent of observed anoxic conditions during the summer of 1970 was 25.2×10^{10} m^2 day.

8. The anoxic regeneration rate of soluble reactive phosphorus was 245.0 μmoles/(m^2)(day) and was approximately 11 times greater than the oxic rate, which was 22.0 μmoles/(m^2)(day).

9. The anoxically generated phosphate is largely the result of the solvation of inorganic complexes that were insoluble under oxygenated conditions. In the central basin of Lake Erie a large part of the phosphorus regenerated under anoxic conditions apparently reenters the life cycle of the lake.

10. During the oxic degradation of organic materials, approximately only 45% of the nitrogen and 25% of the phosphorus contained in the organic material returned to the water in soluble form.

11. The anoxic conditions caused an internal loading of phosphate to the central basin equal to 111% of the external loading of phosphorus during the same period. This quantity, together with the oxic regeneration in the complete basin for the same period, caused the two-month internal loading to equal 137% of the external loading. The central basin is now changing from being a settlement basin for phosphorus to a production basin for the element during the summer stratification period.

12. Under anoxic conditions there was a considerable loss of inorganic nitrogen nutrients, which were presumably converted to nitrogen gas. This loss decreased the quantity of nitrogen nutrients regenerated by 44%.

13. Manganese and iron came out of the sediments in the atomic ratio of Mn:Fe = 2:1 under oxic conditions, but the ratio under anoxic conditions was Mn:Fe = 1.0:1.0.

14. In the mesolimnion, the chemocline gradient varies with the chemical species and is frequently nonlinear with depth, even when the thermocline gradient is linear.

ACKNOWLEDGMENTS

We would like to thank the members of the Technical Operations Section and Water Quality Division who were involved in Project Hypo, and also the Captain and crew of the Limnos, who worked very steadily during a tiring assignment. The work of the Environmental Protection

Agency personnel in providing graphs and figures and almost immediate analysis of shore samples is much appreciated. We would like to thank Dr. Mary Thompson for anticipating both the programming and the ship-time needs of the project as well as ensuring that the necessary number of personnel were available. We would like to thank Dr. Jim Kramer not only for the data processing that was done, but also for suggesting methods for the management of the data.

REFERENCES

1. American Public Health Association, "Standard Methods for the Examination of Water and Wastewater," 12th ed. New York, APHA, 1965.
2. V. K. Chawla and W. J. Traversy, "Methods of Analyses on Great Lakes Waters," Proc. 11th Conf. Great Lakes Res., 524-530 (1968).
3. J. O. Blanton and A. R. Winklhofer, "Physical Processes Affecting the Hypolimnion of the Central Basin of Lake Erie," in "Report on Project Hypo," Canada Centre for Inland Waters, Paper 6; U.S. Environmental Protection Agency Techn. Bull. TS-05-71-208-24, 1971.
4. G. E. Hutchinson, "A Treatise on Limnology," Vol. 1. New York, Wiley, 1957.
5. W. Gardner and G. F. Lee, "Oxygenation of Lake Sediments," Int. J. Air Water Pollut., 9, 553-564 (1965).
6. R. P. Hartley, "Bottom Currents in Lake Erie," Proc. 11th Conf. Great Lakes Res., Int. Assoc. for Great Lakes Res., 398-405 (1968).
7. T. Braidech, P. Gehring, and C. Kleveno, "Biological Studies of Oxygen Depletion and Nutrient Regeneration Processes in the Lake Erie Central Basin," Proc. 14th Conf. Great Lakes Res., 805-817 (1971).*
8. A. M. Lucas and N. A. Thomas, "Sediment Oxygen Demand in Lake Erie Central Basin, 1970," Proc. 14th Conf. Great Lakes Res., 781-787 (1971).*
9. A. S. Menon, C. V. Marion, and A. N. Miller, "Microbiological Studies of Oxygen Depletion and Nutrient Regeneration Processes in the Lake Erie Central Basin," Proc. 14th Conf. Great Lakes Res., 768-780 (1971).*
10. W. Stumm and J. J. Morgan, "Aquatic Chemistry. An Introduction Emphasizing Chemical Equilibria in Natural Waters. New York, Wiley, 1970, 553 pages.
11. R. A. Berner, M. R. Scott, C. Thomlinson, "Carbonate

Alkalinity in the Pore Waters of Anoxic Marine Sediments," Limnol. Oceanogr., 15, 544-549 (1970).
12. International Lake Erie Water Pollution Board and International Lake Ontario-St. Lawrence River Water Pollution Board, "Report to the International Joint Commission on the Pollution of Lake Erie, Lake Ontario, and the International Section of the St. Lawrence River," Vol. 1, 1969, 150 pages.
13. H. H. Dobson and M. Gilbertson, "Oxygen Depletion in the Hypolimnion of the Central Basin of Lake Erie 1929 to 1970," Proc. 14th Conf. Great Lakes Res., 743-748 (1971).*
14. R. A. Vollenweider, "Scientific Fundamentals of the Eutrophication of Lakes and Flowing Waters, with Particular Reference to Nitrogen and Phosphorus as Factors in Eutrophication," Organization for Economic Co-operation and Development, DAS/CSI/68.27, 1968.
15. A. L. W. Kemp, C. B. J. Gray, and A. Mudrochova, "Changes in C, N, P, and S in the Last 140 years in Three Cores from Lakes Ontario, Erie, and Huron," Chapter VIII, this volume.
16. A. T. Prince and J. P. Bruce, "Development of Nutrient Control Policies in Canada," Chapter XV, this volume.
17. H. H. Dobson, M. Gilbertson, and T. R. Lee, "Phosphorus and Hypolimnial Dissolved Oxygen in the Central Basin of Lake Erie," in "Report on Project Hypo," Canada Centre for Inland Waters Paper 6; U.S. Environmental Protection Agency Tech. Bull. TS-05-71-208-24, 1971.

*This paper has been reprinted in "Report on Project Hypo," Canada Centre for Inland Waters Paper 6; U.S. Environmental Protection Agency Tech. Bull. TS-05-71-208-24, 1971.

Chapter VIII

CHANGES IN C, N, P, AND S IN THE LAST 140 YEARS IN THREE CORES FROM LAKES ONTARIO, ERIE, AND HURON

A. L. W. Kemp, C. B. J. Gray, and Alena Mudrochova

Canada Centre for Inland Waters,
Burlington, Ontario, Canada

The input of nutrients to the Great Lakes has increased since the settlement of the Great Lakes region. Changes in the environment and biota of the Great Lakes have been recently summarized by Beeton (1). Total dissolved solids have increased 43% in Lakes Ontario and Erie in the past century, as compared with only 9% in Lake Huron (1-4). The differences are due to the larger populations and resulting greater urban and industrial inputs into the former lakes. Sulfate concentration has doubled--from 15 ppm in 1850 to 30 ppm in 1967 for Lake Ontario, from 13 ppm in 1890 to 26 ppm in 1967 for Lake Erie, and from 7 ppm in 1850 to 14 ppm in 1967 for Lake Huron (1). Total dissolved solids and sulfate levels were constant up to about 1910 in each lake and have increased rapidly since that time. Because of a lack of early data, there is no information on the increase of organic carbon to the Great Lakes, and very little is known about the increasing load of phosphorus or nitrogen to the lakes. Data from the few open lake studies of the western basin of Lake Erie suggest that total phosphorus appears to have increased fourfold and that total nitrogen tripled between 1942 and 1967 (5).

Most of the input to the Great Lakes is eventually

deposited on the lake bottoms. Fresh sediment, mainly of organic origin, settles in the deeper parts of the lake basins and is then quickly decomposed. Diver observations, coupled with chemical analyses, have shown that a fresh fall of photosynthetic algae, 2 to 3 cm thick, decomposed rapidly during the summer months in the central basin of Lake Erie, liberating nutrients to the hypolimnion waters (6). This layer of algae and succeeding rains of algal matter were reduced to a thickness of a few millimeters after only nine months (7). The remaining surface layer of partly decomposed sediment becomes slowly buried under further fresh sediment, and the potentially available nutrients become less available to the overlying waters with increasing depth of burial. An examination of sediment cores should, therefore, reveal the history of the increasing inputs to the Great Lakes system in recent times.

Such an approach was advocated in 1941 (8), and there have been a number of studies since then, but these have focused mainly on organic carbon and phosphorus distributions in sediment cores from small lakes. In most cases, no clear evidence of cultural eutrophication was obtained from the organic carbon and phosphorus measurements. Recently changes in chemistry and sedimentation rates in two cores from Lago di Moterosi, Italy, were correlated with the building of the Via Cassia in 171 B.C. and the subsequent history of the region (9). However, organic carbon and nitrogen distribution data showed no evidence for cultural eutrophication in sediment cores from a number of small Wisconsin lakes (10), probably because of the mixing of the soft organic rich sediments deposited in recent times. An increase in organic carbon has been observed toward the sediment-water interface in the surface sediments of Lakes Ontario and Erie (11), Michigan, Superior, and Huron (12), and in southern Lake Michigan (13). Although the subsampling intervals were not closely spaced in any case, the change in organic carbon in the Great Lakes appears to represent up to a threefold increase.

This study reports the changes in total organic carbon (OC), carbonate carbon (CC), nitrogen (N), phosphorus (P), sulfur (S), pH, Eh, water content, and sediment texture in the top 50 cm of sediment in a single core from each of Lakes Ontario, Erie, and Huron. The study has several objectives:

1. To investigate changes in C, N, P, and S near the sediment-water interface in Great Lakes fine-grained muds.
2. To measure the increase in C, N, P, and S loading

to the sediments in the last 140 years by relating the measurements to a time scale based on estimated sedimentation rates.

3. To compare cores from regions close to industrial and urban sources of input (Ontario and Erie) with a core far removed from industrial and urban sources of input (Huron).

METHODS

Cores were taken from close to the deepest point in Lake Ontario, from the deepest point in the central basin of Lake Erie, and from the deepest point in South Bay, Lake Huron (Table 1). The cores from Lakes Ontario and Erie are fairly typical of the individual lake basins and were chosen from areas of high sedimentation rates and fine grain size. South Bay is a deep, narrow-mouthed body of water, receiving an input of Lake Huron waters under southwest wind conditions (14). The sediments are similar geochemically and mineralogically to those of the main basin of Lake Huron. The sedimentation rate is less and the sediment particle size greater than the Ontario and Erie core locations. The South Bay location was chosen because it is believed to represent a Great Lakes location more or less unaffected by cultural eutrophication. Half the bay is surrounded by an Indian reservation; the other half contains only a few cottages and farms.

Sampling was carried out with a modified Benthos triple corer, using 100-cm steel tubes with an internal diameter of 6.4 cm (15). One core was stored at 4°C for later subsampling and analysis of sediment particle size and pollen. Values of Eh and pH were measured on the second core, within 2 hr of retrieval, at intervals down the core; the sediment in this core was then stored at 4°C for later trace element analysis. The third core was subsampled at intervals within 2 hr of retrieval and was freeze dried immediately. All cores were subsampled at 3-mm intervals to 1 cm, then at 1-cm intervals to 10 cm, at 5-cm intervals to 20 cm, and at 10-cm intervals to 50 cm. The results are presented later, and the values are plotted as the midpoint of each section subsampled. The methods for extruding and subsampling the cores, removing the outside smear, and measuring Eh and pH are described in detail by the authors elsewhere (15). Each wet sediment subsample was weighed after extrusion and immediately freeze dried for later chemical analysis and measurement of water content. The dried sediment was sieved through a 35 mesh screen to remove any shells or large pieces of detritus; it was then ground with a ball mill grinder, to pass 280 mesh.

The water content based on the quantity of water removed by freeze drying is expressed as the percentage of water in the fresh wet sediment. The volume of sediment in each subsample was measured and utilized in the data for the dry weight of sediment per unit volume of core. Total carbon was determined by dry combustion at 1300°C in a Leco induction furnace, and organic carbon determined by dry combustion after carbonates had been removed with sulfurous acid (11). Total nitrogen was determined in duplicate by the modified Dumas method on a Coleman model 29A nitrogen analyzer, after the method of Keeney and Bremner (16). Total phosphorus was analyzed as orthophosphate after sodium carbonate fusion by the molybdic acid, ascorbic acid, and trivalent antimony method of Strickland and Parsons (17). After combustion in a Leco induction furnace, total sulfur was measured as sulfur dioxide by iodometric titration. Chemical data are expressed as percentage dry weight of sediment.

The sediment particle size was measured by a combination of pipette and settling tube analysis using the FAST (fast analysis of sediment texture) procedure (18). Total quartz and feldspar were determined gravimetrically as the residue after fusion with potassium pyrosulfate (19). The clay mineralogy was ascertained by X-ray diffraction (20). The *Ambrosia* pollen was measured quantitatively in each subsample at 2-cm intervals to a point usually 10 cm below the *Ambrosia* horizon. The procedures used were similar to those described by Waddington (21). Recent sedimentation rates were calculated, assuming a constant rate of sedimentation since the *Ambrosia* horizon, from the dry weight of sediment per square meter column above the horizon divided by the estimated age (B.P.) of the horizon.

RESULTS AND DISCUSSION

The recent sediments of Lakes Ontario, Erie, and Huron have been described in general (22); more detailed accounts are available for the sediments of Lake Ontario (20), Lake Erie (23), and South Bay, Lake Huron (24). The sample locations and general sediment characteristics are given in Table 1, and the results appear in Figures 1, 2, and 3.

Table 1. Sample Locations and General Characteristics of the Sediments

Lake	Date of sampling	Location		Lake water depth, m	Approximate postglacial mud thickness, m	Estimated[a] present-day sedimentation rate, $g/(m^2)(yr)$	Average sediment particle size distribution in top 10 cm, %		
		Latitude	Longitude				Sand	Silt	Clay
Ontario	May 1970	43°30.5'	77°54.3'	225	7.4	320	0	26	74
Erie	August 1970	42°0.2'	81°36.2'	24	18.0	540	0	25	75
Huron	July 1970	45°37.5'	81°52.7'	56	10.2	148	0	45	55

[a]Estimates from this study.

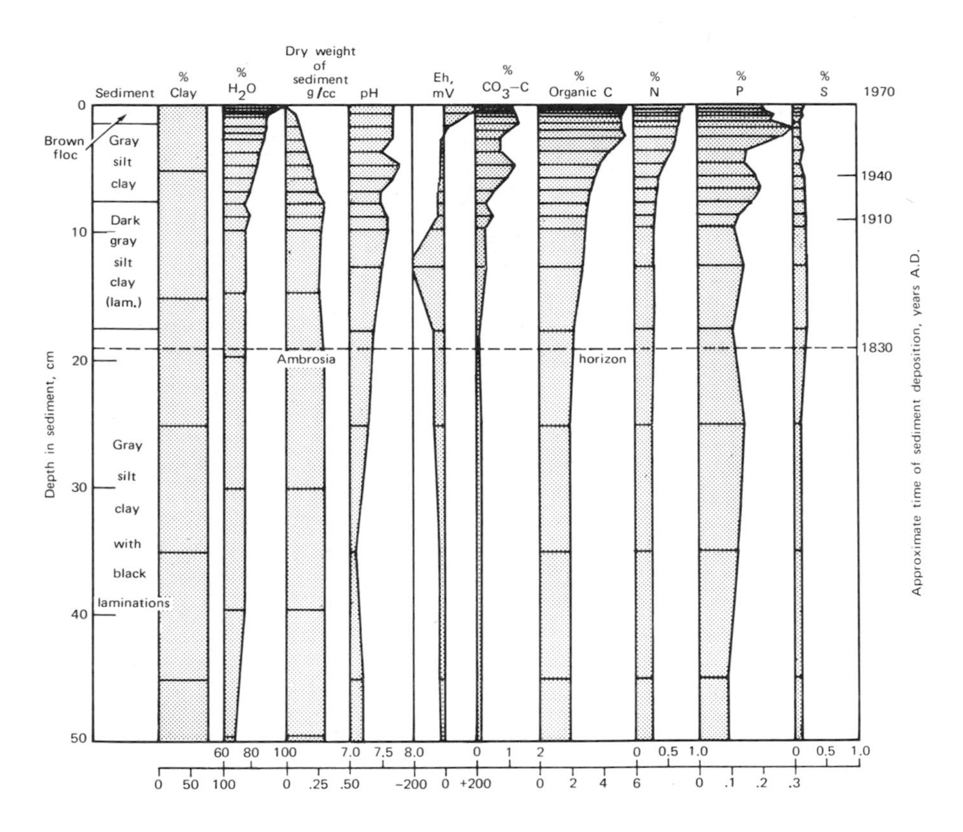

Fig. 1. Sample locations and general sediment characteristics of Lake Ontario.

Sediment Characteristics

Great Lakes muds are typically characterized by an orange-brown floc in the topmost few millimeters, overlying a soft gray ooze that becomes blackish in zones of very low redox potential. The sediment becomes firmer with depth of burial, and the water content usually decreases from more than 90% at the mud surface to about 60% at 50-cm depth. The dry weight of the sediment per cubic centimeter of wet mud usually increases from about 0.06 g at the surface to 0.30 g at 50 cm (Figures 1, 2,

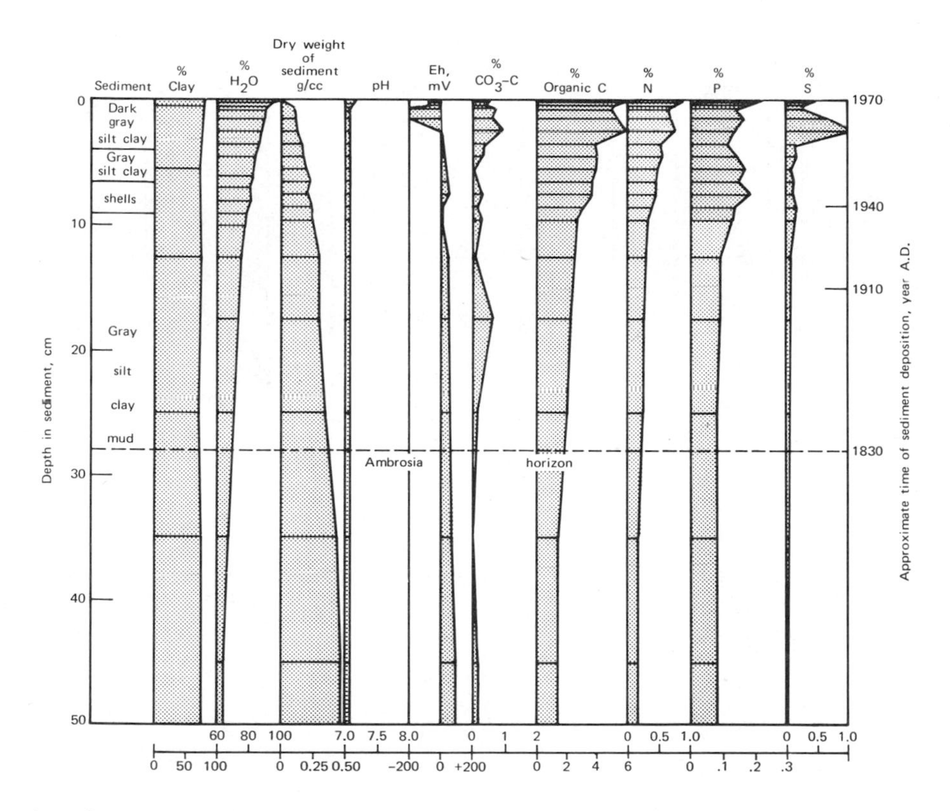

Fig. 2. Sample locations and general sediment characteristics of Lake Erie.

and 3). These characteristics have been observed at a large number of locations in each lake. Under anaerobic conditions, such as are often found in the central basin of Lake Erie during the summer months, the orange floc disappears and is replaced by a dark gray coloration and a negative redox potential at the sediment surface (Figure 2). Local blackish reduced lenses and laminations are frequently observed in the Great Lakes mud column and have been encountered as deep as 8 m in the sediment (25).

A limited bottom faunal population is encountered in the deep lake muds, comprised mostly of amphipods,

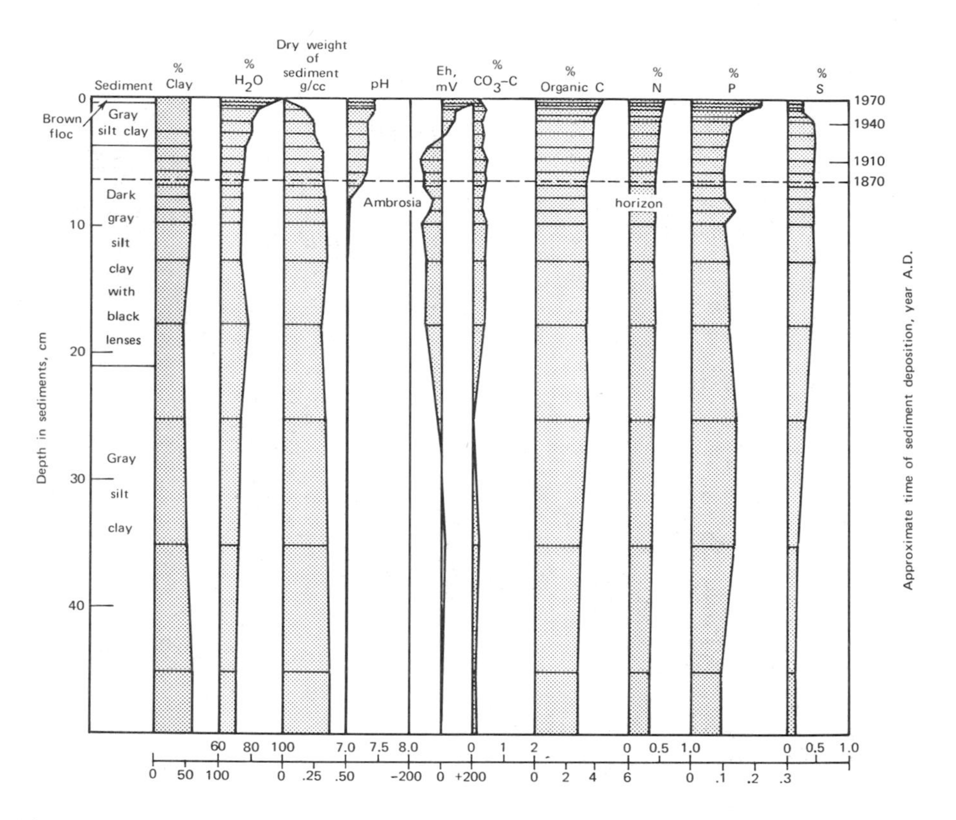

Fig. 3. Sample locations and general sediment characteristics of Lake Huron.

oligochaetes, and chironomids (26). The greatest numbers are found in the nearshore and intermediate depth zones. Tubificid tubes were present at the surface of each of the cores described, and a number of amphipods were swimming in the immediate bottom waters, but no chironomids were observed. Bottom organisms are rarely observed below 2 cm in the deep Great Lakes mud cores, probably because oxygen is lacking in the reduced muds. Microbial populations were greatest at the sediment-water interface and were greatly reduced at a depth of 5 cm in a number of mud cores from the Kingston basin of Lake

Ontario (27). Similar results were obtained by the authors in mud samples from the deepest point in Lake Ontario.

Several differences are apparent from the sediment description and the sediment particle size distribution of the three cores (Figures 1, 2, and 3). The Ontario and Erie cores have a uniformly high clay size content (< 0.004 mm), between 71 and 78% throughout the sediment column; the remainder of the sediment is silt size (0.004 to 0.062 mm). The high clay content and the deep-water environment indicate fairly even deposition of sediment under low-energy conditions (Figures 1 and 2). However, a layer of shells mixed with the clay between 7 and 9 cm in the Erie core suggests a separate event within the normal sedimentation processes. This is discussed later.

The Huron core has a generally coarser texture, with a varying clay content--54% at the surface, decreasing to 43% at 18 cm, and then increasing to 60% at 45 cm (Figure 3). The remainder of the sediment was again of silt size. The data suggest a fluctuating pattern of sedimentation at this location with slight variations either in energy conditions or in the input of detrital material. All samples had a zero sand size fraction (0.062 to 2.0 mm). The color of the sediment indicated the Eh throughout the cores; orange-brown predominated at the sediment-water interface, gray was seen where the Eh was close to zero, and black appeared where the Eh was very negative. Each of the cores showed black lenses or black laminations irregularly down the core column.

The water content paralleled sediment textural differences closely in the top 10 cm of each core, being greater with the finer grained sediments of Ontario and Erie. The cores showed no appreciable differences below 10 cm, and the water content decreased as compaction increased. The top centimeter of each core consisted of a watery "floc" that had to be pipetted from the core tube. The mud below this zone, although soft, was firm enough to extrude. The mud appeared to become firmer at about 85% water content, which was at 4 cm in the Ontario and Erie cores and at 1 cm in the Huron core. The muds became very firm at 75% water content and retained their shape indefinitely when extruded. The dry weight of sediment in each core increased as the water content decreased, ranging from 0.02 g/cc at the surface to 0.46 g/cc at 45 cm in the Erie core (Figures 1, 2, and 3).

The mineralogical data for the top 3 cm of sediment in each core are presented in Table 2. Organic matter was calculated by multiplying the carbonate carbon by 1.7, and calcium carbonate was determined by multiplying the

Table 2. Mineralogical Characteristics of the Top 3 cm of the Sediment Cores, Calculated as Percentage Dry Weight of Sediment

Lake	Organic matter	Carbonates, calculated as calcium carbonate	Quartz and feldspars	Total mineral clay	Clay minerals		
					Illite	Chlorite	Kaolinite
Ontario	8.5	9.3	26.7	55.5	34.2	11.0	10.3
Erie	9.2	5.8	26.6	58.4	43.6	5.3	9.4
Huron	6.1	2.6	43.4	47.9	30.7	8.6	8.6

carbonate carbon by 8.3. Total mineral clays were calculated by subtracting the organic matter, carbonates, and quartz results from 100%. In the fine-grained sediments of the Great Lakes, Thomas (28) demonstrated a high degree of correlation between the fraction with less than 2 μ grain size and the mineral clay content, which he interpreted as being a function of sorting and deposition from suspended particulate material. Mineral clay as expressed here is thus believed to be a more accurate measure of the true clay content of a sediment than values derived from alternate methods based on size fractionation procedures.

Feldspar contents, measured at a number of Lake Ontario locations, were found to be about 5% at the Lake Ontario core location (29). The results show that all the cores have a similar mineral composition, but the Huron core has a higher quartz content and a corresponding higher silt content (Table 1). Illite was the predominating clay mineral, and there were lesser quantities of chlorite and kaolinite. Traces of vermiculite were also present in the clay mineral fraction at each location, but no montmorillonite was observed. The clay minerals were determined in each subsample of the Lake Ontario core down to 45 cm and were found to be uniform.

Sedimentation Rates

The *Ambrosia* or ragweed pollen horizon reflects the increase in *Ambrosia* pollen due to the clearing of the land by the early settlers. In each case, the *Ambrosia* horizon increased sharply, more than tenfold, from a low baseline at the depths indicated in Figures 1, 2, and 3. The increase, with minor fluctuations, continued uniformly to the sediment-water interface. The increase in *Ambrosia* is dated at around 140 years ago for Lakes Ontario and Erie (30). The horizon is estimated to be about 100 years old for Manitoulin Island, based on the local population increase between 1870 and 1880. The horizon was sharp in each case; for example, the *Ambrosia* count in Lake Ontario rose from a few hundred at 20 cm to 10,000 at 19 cm. The sudden increase in each case, which is strongly indicative of the lack of mixing of the sediment at the three locations and has also been observed at a number of other sample locations in the three Great Lakes (31). The absence of mixing is discussed later.

Modern sedimentation rates were calculated for the three cores, assuming that sedimentation has been constant since the *Ambrosia* horizon. Average present-day sedimentation rates of 320, 540, and 148 $g/(m^2)(yr)$ were found for the Ontario, Erie, and Huron cores, respectively. These values do not coincide with the total postglacial mud thicknesses at each location (Table 1). Estimates of average postglacial sedimentation rates were made, assuming that sedimentation has been constant since the retreat of the last glaciers. Calculations were made from the postglacial mud thickness and water content of the sediment from piston cores in each of the lake subbasins. Sedimentation rates of 440, 1000, and 655 $g/(m^2)(yr)$ were calculated for the Ontario, Erie, and Huron core locations, respectively. The differences among these calculations and those from the *Ambrosia* horizon indicate that sedimentation in postglacial times has been uneven.

Carbon dates were obtained at depths of 101, 131, and 161 cm on a Benthos core taken at the Ontario location at the same time that the triple core was retrieved. Carbon dates of 3800, 4200, and 5500 years B.P., respectively, were found at the same three subsample locations. The results indicate that the rate of sedimentation has increased at the Ontario site in the last 140 years, and it is likely that a similar situation exists in the other Great Lakes. The low modern sedimentation rate in South Bay is due probably to the small drainage area of the region, which has a very small river input of sediment.

In order to interpret the chemical results, estimates were made of the time of deposition of the sediment for each subsample that was analyzed. These estimates assume a constant rate of sedimentation and no mixing of the sediment; it is also assumed that the bulk of the inert inorganic sediment material remains in situ once buried. The time scale estimates appear on the right margins of Figures 1, 2, and 3. As previously mentioned, the assumption of a constant rate of sedimentation is probably incorrect; however, the time scale does give an indication of the age of the sediment. Studies are presently underway in our laboratories to locate more recent time horizons in the Great Lakes sediments. The zone of shells (sphaerridae) in the Erie core suggest that these may have been deposited at a time of an exceptional storm. This zone has been observed in a number of cores from the central basin of Lake Erie (32) and should prove to be a useful marker horizon. Our estimated date for this zone is for some time between 1940 and 1950. The reliability of the estimated recent sedimentation dates is discussed later.

Hydrogen Ion Concentration

The pH decreases down each core from surface highs of 7.7, 7.1, and 7.4 in the Ontario, Erie, and Huron cores, respectively, to a common value of about 7.0 at the base of each core (Figures 1, 2, and 3). Concurrent bottom water pH measurements of 7.8, 7.2, and 7.5 were recorded at each of the sampling locations. The surface sediment pH values are close to the average values found in the surface sediments of the three lakes (22); they also appear to be related to the average water depths of the lakes under discussion. The pH parallels the carbonate carbon content of the Ontario core, but it does not show such a close relationship in the Erie and Huron cores. The difference in pH of the surface sediments and their relation to carbonates is discussed in the section on carbonate carbon.

Redox Potential

Values of Eh show similar trends in the Ontario and Huron cores (Figures 1 and 3). The Eh is most positive at the sediment-water interface, decreasing to zero at about 2 cm and then becoming negative with the lowest values at about 10 cm. Toward the bottom of the cores, the Eh becomes close to zero. In both lakes, the hypolimnion waters are usually saturated with oxygen; the lowest saturation value recorded for Lake Ontario was 70%, in the Kingston basin (3). The orange-brown color of the surface sediments in each of these cores is believed to be attributable to the presence of hydrated iron oxides. The most negative potentials are accompanied by a black coloration of the muds. On exposure to the air, the black muds become rapidly oxidized to a brown color, and the black coloration is believed to be due to the presence of iron sulfides. The sulfur concentration, which closely parallels the redox potential in each of the cores, is discussed in the sulfur section.

The trends in Eh observed in the Ontario and Erie cores have been noted in many cores taken from the muds of the three Great Lakes. The redox potential is very sensitive to positioning of the electrode. A movement of 1 mm often yields a value 20 mV different from the original position; therefore, the absolute values recorded must not be taken literally. However, the trends observed from very positive to very negative are the important features. In order to define the general chemical and biochemical processes taking place in the Great Lakes sediments, a knowledge of the redox potential is mandatory. For example, denitrifying bacteria and sulfate-reducing bacteria operate between definite Eh limits. On a micro scale, the Eh is quite variable, and this subject requires close attention in future studies.

The Erie core was taken in August while the central basin hypolimnion waters were depleted in oxygen (6). However, cores taken by the authors in the central basin at other times of the year showed Eh trends similar to those of the Ontario and Huron cores. The negative potentials at the sediment-water interface are attributed to the hypolimnion oxygen depletion and are accompanied by black coloration and a content of over 1% total sulfur (Figure 2). A residual negative peak in Eh is observed at about 10 cm in the core. As fresh algae rained on the sediment-water interface, anoxic conditions rapidly prevailed, and blacking of the fresh algae commenced from the bottom of algal layer upward (6). It would appear that changes in sediment Eh are rapid under these circumstances, and care must be taken in interpreting Eh data. Mortimer (33) recorded similar rapid changes in the surface sediments of Esthwaite Water under oxygen depletion. The reducing conditions in the Great Lakes sediments are probably due to the activity of bacteria that oxidize organic matter (34). Anaerobic bacterial counts were highest in the reduced zone of the recent sediments of the Kingston basin of Lake Ontario (27).

The redox potential is also important in terms of regeneration of chemical nutrients. Mortimer (33) found a rapid increase in ammonia and phosphate concentrations after the development of anoxic conditions. Redox potential accounted directly for the mobilization of manganese, nickel, cobalt, chromium, vanadium, and uranium in sediments from the Pacific Ocean (35). Iron, manganese, phosphorus, and ammonia concentrations increased in the hypolimnion waters of the central basin of Lake Erie after the anoxic conditions had commenced (6). The variation of phosphorus and sulfur in these three Great Lakes cores are also attributable to the redox potential and are discussed later.

Carbonate Carbon

Carbonate carbon (CC) is low at the base of the Ontario and Erie cores and shows a general increase toward the surface, but no significant change occurs in the Huron core (Figures 1, 2, and 3). The CC concentration in each of the subsamples displays a strong correlation with pH and calcium concentration in the Ontario core and moderate correlation in the other cores (Table 3). X-Ray-diffraction data indicate that most of the carbonate is in the form of calcite, and it is believed that most of it is precipitated biochemically.

Table 3. The Correlation between Calcium and pH, Carbonate Carbon and Calcium, and Sulfur and Eh in the Three Great Lake Cores

	Correlation Coefficients		
Lake	Ca vs. pH	Ca vs. CC	S vs. Eh
Ontario	0.824	0.980	0.596
Erie	0.069	0.336	0.496
Huron	0.509	0.633	0.763

Two general factors are believed to control the pH and CC values observed in the cores: (1) lowering of sediment pH due to carbon dioxide produced by the decomposition of sediment organic matter and (2) the volume of the hypolimnion. The low pH values at the base of each core are probably due to the upward migration of carbon dioxide from the surrounding and underlying sediment. A decrease in pH has been observed by the authors in numerous cores from each of the three lakes, usually to a uniform value of about 6.7 at one meter. The higher pH values near the sediment-water interface reflect the influence of the hypolimnion waters superimposed on the normal lower pH of the underlying sediments. The pH of the hypolimnion waters of each lake basin is controlled by the volume of the hypolimnion and the quantity of organic matter decaying at the sediment-water interface.

The quantity of organic matter is similar at the sediment-water interface in the Ontario and Erie cores (Figures 1 and 2), yet the interface pH is 7.7 in Ontario and 7.1 in Erie. The pH differences are believed to be due to the hypolimnion volume at each location. Carbon dioxide produced by the decay of organic matter at the sediment-water interface exerts little influence on Lake Ontario, which has a maximum depth of 240 m. On the other hand, in central Lake Erie (maximum depth 18 m), the large quantities of carbon dioxide released at the sediment-water interface control the pH of the hypolimnion waters and are responsible for the lower pH values found at the surface of the Erie core. South Bay with an intermediate depth has an intermediate pH value (Figure 3).

The changes in CC versus the approximate time of deposition are plotted in Figure 4. It can be seen that CC has generally increased in Ontario and Erie and has remained more or less constant in the Huron core. The increases coincide with the increasing input of chemicals to the lakes in about 1910 reported by Beeton (1). The increase in CC for Ontario and Erie also coincides with the increase in organic carbon (see below). The correlation between pH, Ca, and CC in the Ontario core (Table 3) suggests that carbon dioxide, formed by the decomposition of the increasing load of organic matter in the lake, is being preserved as calcite at the higher pH conditions at the sediment-water interface. The lesser amounts of CC in the Erie core Figure 4) are probably due to the lower hypolimnion pH and the solubilization of precipitated calcite. The jaggedness of the curves could be the result of an additional input of detrital calcite to the sediments. The detrital calcite could be of biogenic origin (e.g., broken up shells from organisms), or it

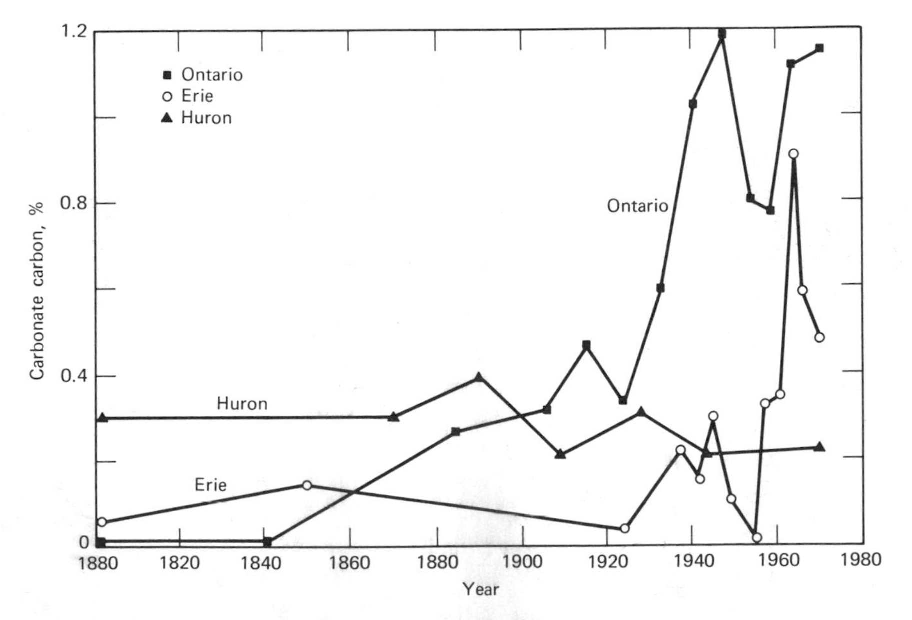

Fig. 4. Changes in carbonate carbon versus approximate time of deposition (Lakes Ontario, Erie, and Huron).

could have come from erosion of nearshore or shoreline glacial tills. Undoubtedly, the reason for the uniformity of CC in the Huron core from 1870 to the present day is that there has been little change in South Bay during the last century. The rise in CC has been observed near the surface in a number of cores from Lakes Erie and Ontario and is a useful horizon for the increase in nutrient loading to these lakes. The nature of the carbonate material is being investigated in our laboratories.

Organic Carbon

The organic carbon (OC) content of the Ontario and Erie cores increases regularly from a low of about 1.6% at the base of the cores to a high of over 5% at the surface (Figures 1 and 2). In the Huron core, the increase is much less from a low of 2.8% at the base of the core to a high of 3.9% at the surface (Figure 3). In the Ontario and Erie cores, the OC content is uniformly high to a depth of 2.5 cm, whereupon it decreases smoothly. The high surface values are attributed to mixing of the sediment in the top 2.5 cm by bottom organisms and by bottom-feeding fish. Tubificid tubes were numerous at both locations, and examination indicated that they did not extend more than a few millimeters below the sediment-water interface. This mixed zone would compact to a few millimeters when buried to a depth of 10 cm.

The change in OC content from the surface to the bottom of the core is due to two main processes: an increase of organic loading to the sediments in recent years, and degradation of the organic matter by microorganisms.

An examination of Figure 5 indicates the extent of the two processes. The change in OC content of the Huron core is linear with time, and this is interpreted as slow degradation of the organic matter after burial and mixing at the surface. The change from 3.9 to 3.3% in 100 years in equivalent to a 15% loss in 100 years, owing to normal diagenetic processes. Similarly, the OC content of the Ontario and Erie cores increases linearly to about 1900 and then increases rapidly to 1970. The linear portion, which is attributed to normal degradation processes, represents a 30% loss for Ontario and a 33% loss per 100 years for Erie. The increase from about 1900 represents the increasing load of organic matter to the sediments since that time. The smoothness of the curves suggest a steady increase in organic loading. These results match the curves of Beeton (1) for the increase of dissolved solids in Lakes Ontario, Erie, and Huron. The

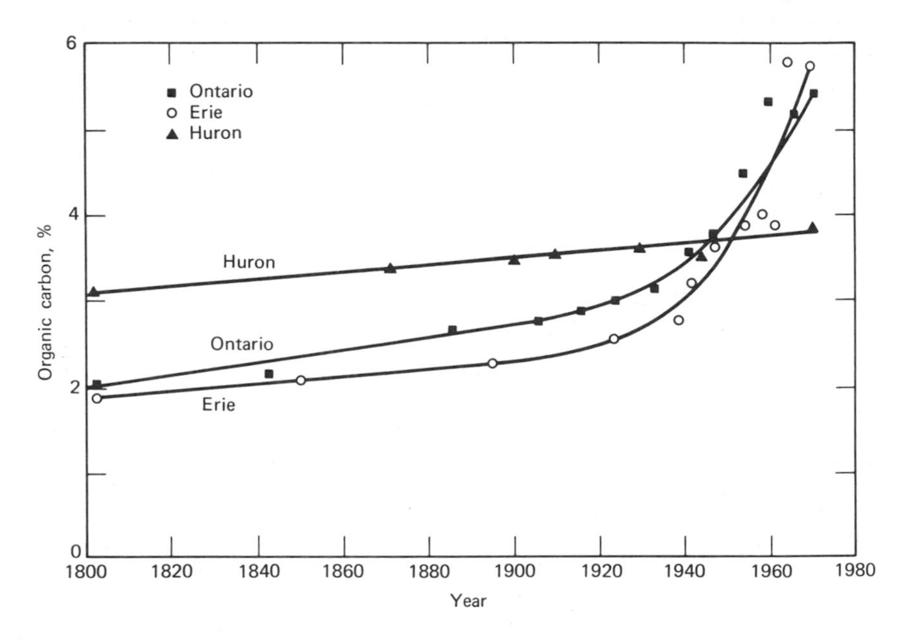

Fig. 5. Changes in organic carbon versus approximate time of deposition (Lakes Ontario, Erie, and Huron).

results suggest that increasing inputs to Lakes Ontario and Erie, and resulting stimulated primary production, are preserved in the sediment column as an increase in OC. The linear increase in the Huron core indicates no increase in primary production in South Bay in the last century.

Harlow (36) estimated that 90% of the organic matter in Lake Erie is autochthonous, mainly diatoms, and that only 10% is allochthonous. It is believed that a similar situation exists in Ontario and in Huron. Microscopic examination of the cores showed plankton fecal pellets, diatoms, blue-green algal remains, bacterial colonies, and brown humuslike material. Gray and Kemp (37) estimated that 99% of the primary chlorophylls were degraded before burial in the sediments of Lakes Ontario and Erie. A soil fractionation analysis of the top centimeter of sediment from selected basin mud samples from Ontario and Erie showed that 30% of the surface organic matter was in the form of fulvic and humic acids and that 60% of the organic matter was in the form of humins (25). These results and the studies of Burns and Ross (6) indicate that most of the organic matter is degraded at the sediment-water interface and in the water column after the death of the plankton. Thus, the slow diagenesis of organic matter suggested by these core results is not surprising, since the material is humified before burial and is, therefore, fairly stable. The previously-mentioned increases in OC recorded by other investigators in the cores of Lakes Michigan, Superior, and Huron can presumably be accounted for in a like manner.

Total Nitrogen

Total nitrogen (N) in the cores from the three lakes parallels that of OC (Figures 1, 2, 3, and 6). Ninety-five percent of the nitrogen in the surface sediments is organic N, and the reasons for the differences within

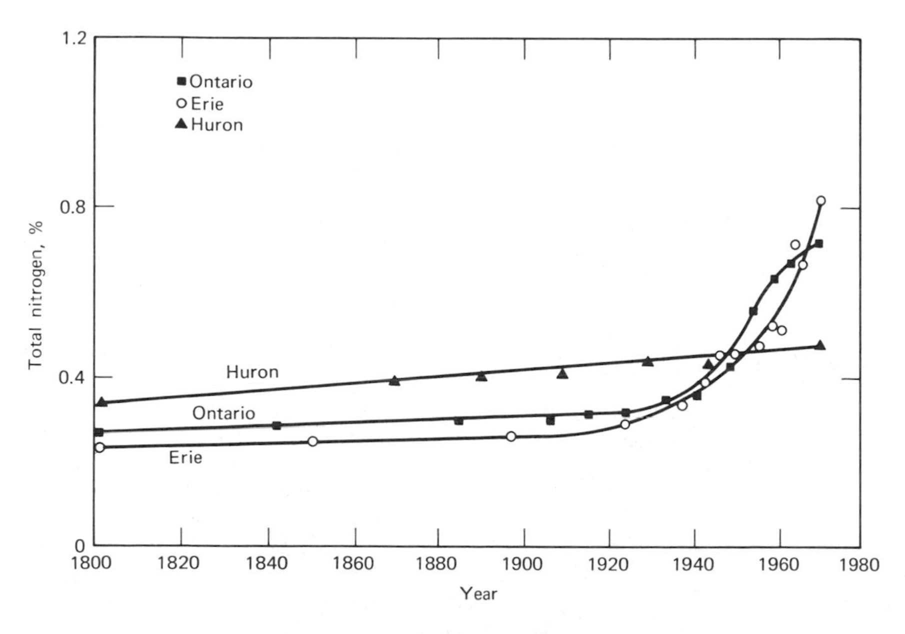

Fig. 6. Changes in total nitrogen versus approximate time of deposition (Lakes Ontario, Erie, and Huron).

each core are similar to those already discussed with regard to OC. The OC:N ratios in the three cores increase from the surface values of 7.5, 6.8, and 8.6 in Lakes Ontario, Erie, and Huron, respectively, to a general value of about 9 at a depth of 15 cm. They then decrease again to the bottom of the cores. The sediment surface OC:N ratios parallel those of the recent sedimentation rates. The higher value in the Huron core is probably due to the slow sedimentation rate and preferential decomposition of protein, with subsequent release of nitrogen before burial.

Nitrate, nitrite, exchangeable ammonia, and fixed ammonia were measured in the Ontario core. Nitrate decreased from 28 ppm at the surface to a few ppm at 2 cm and remained uniformly low throughout the core. Exchangeable ammonia increased uniformly from 16 ppm at the surface to 56 ppm at 50 cm. Fixed ammonia similarly increased from 248 ppm at the surface to 424 ppm at 50 cm. It is believed that the decrease in the OC:N ratio below 15 cm is due to the decay of organic matter which releases carbon dioxide; this freely migrates upward or precipitates out as calcite. The fixation of the released ammonia by the clays may also cause the decrease. A similar situation probably exists in the Erie and Huron cores.

Total Phosphorus

Total phosphorus (P) increases from a uniform baseline below the _Ambrosia_ horizon to the mud surface in each of the cores (Figures 1, 2, 3, and 7). However, the increases are not smooth as encountered with OC and N, except for the Huron core. The linearity of the CC, OC, and N curves for the Huron core (Figures 4, 5, and 6) suggest that the input of phosphorus to South Bay is unlikely to have changed significantly in the last century. It is probable that the increasing P concentration at the Huron sediment-water interface is due to upward migration of soluble phosphate ions, released from the lower zones of negative Eh, which is then trapped at the sediment surface. This is likely to be a natural process and not due to any increasing phosphorus loading to South Bay.

The increases in CC, OC, and N (Figures 4, 5, and 6) in the Ontario and Erie cores suggest that P should have increased similarly. It seems likely that the results reflect two continuing processes: (1) upward migration of soluble phosphate ions from the reduced zones of mud

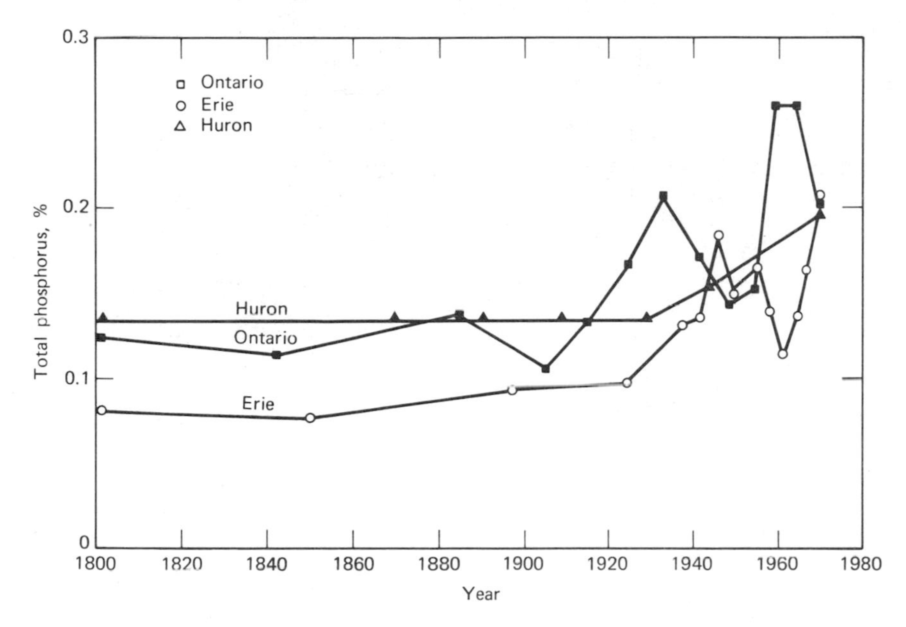

Fig. 7. Changes in total phosphorus versus approximate time of deposition (Lakes Ontario, Erie, and Huron).

as in the Huron core and (2) an increasing load of P to the two lakes. An examination of the change in P in the Ontario core (Figure 7) shows that P increases in a zig-zag fashion from 1200 ppm at 19 cm to 3050 ppm at 2 cm and then decreases to 2000 ppm at the sediment-water interface. This represents a 2.5-fold increase to the 2-cm level from the *Ambrosia* horizon. The increase is similar to that of OC, except that the P increase is uneven (Figures 5 and 7). The unevenness of the P curve (Figure 7) is interpreted as loss of phosphate by upward migration in the low points of the curve and is not due to a fluctuating input of P to the sediments since 1900.

An examination of the P highs and lows in the Ontario core indicate that up to 25% of the element may have migrated upward and may have been regenerated to the lake under favorable conditions. A similar interpretation is given for the Erie core (Figure 7). Calculations show that the OC:N:P ratio for the surface 3 mm of the Erie core is 68:8.6:1. The OC and N are much less than the average for Lake Erie phytoplankton where the OC:N:P ratio is 122:18:1 (6). This difference probably reflects a greater retention of P in the sediments than OC and N, and an input of P to the sediments other than from the phytoplankton remains. The jaggedness of the P increase in the Ontario and Erie cores indicates that the increasing P inputs to the two lakes is not truly reflected in the sediment cores. The phosphorus cycle in the Great Lakes sediments is currently under a more detailed examination at the Canada Centre for Inland Waters.

Total Sulfur

Although the sulfate concentration has doubled in the three lakes since 1830, total sulfur (S) shows no observable increase in the cores (Figures 1, 2, 3, and 8). Instead, S is closely correlated with Eh (Table 3) in

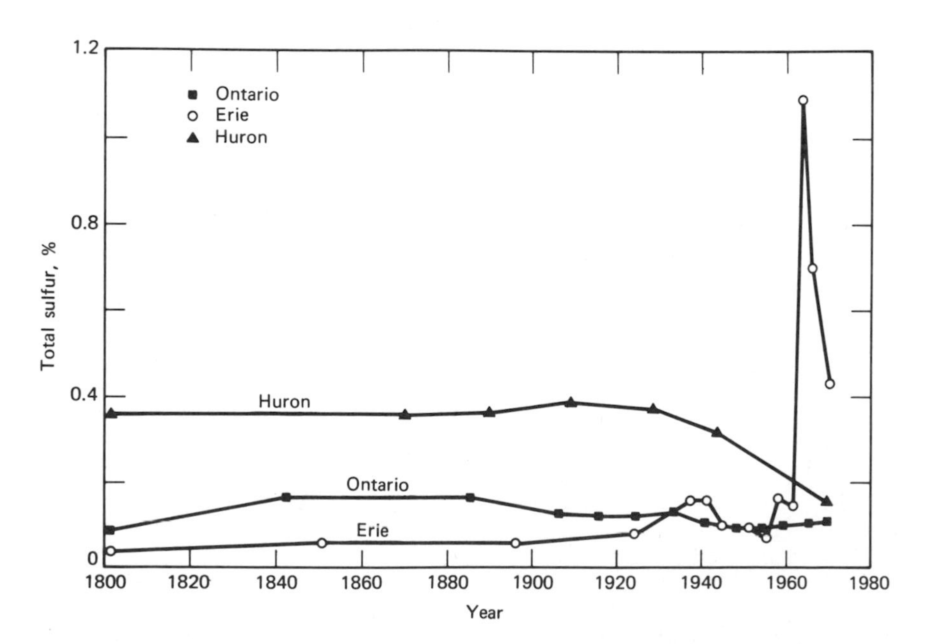

Fig. 8. Changes in total sulfur versus approximate time of deposition (Lakes Ontario, Erie, and Huron).

each of the cores. Although the sulfur cycle has not been studied in the cores, an explanation can be given for the observed results. Sulfate is readily reduced by desulfovibrio when the redox potential drops below +100 mV (34). Hydrogen sulfide is precipitated out as iron sulfide at negative potentials. An increase in redox potential frees the sulfide, which is bacterially oxidized to free sulfur and sulfate, and these migrate upward. Thus S can be expected to have values above a normally low baseline only in zones of negative Eh, as was observed in the three cores. The bacterial S cycle at the sediment-water interface was studied in the central basin of Lake Erie at the time of the collection of our core. The microorganisms necessary for the complete recycling of sulfate were abundant under the appropriate Eh and pH conditions (38). The increasing inputs of sulfate to the Great Lakes, together with the increasing load of organic S, are not reflected in the three cores examined (Figure 8).

Preservation of the Sediment Record

The recent sediments of the three lakes consist of quartz, feldspars, and clay minerals, which are relatively inert to chemical changes, and organic matter, carbonates, and micro quantities of minerals, such as phosphates, which are readily subject to changes. These changes can be biochemical, chemical, or physical. The observed changes in C, N, P, and S levels have already been discussed and have been attributed to increased loading in Lakes Ontario and Erie, with chemical and physical changes superimposed on the increasing load. It is also possible that the higher surface C, N, and P values could be due to sediment mixing processes that enrich the surface sediments. Are the estimated times of deposition of the sediment shown in the diagrams reasonable, or has the sediment been disturbed since settling on the lake bottom?

Sediment mixing processes have recently been reviewed by Lee (39). The processes that might enrich the surface sediments are: turbulent mixing, movement of gases or solutes, and activities of bottom organisms.

Turbulent mixing of the sediments as a surface-enriching process is unlikely except in cases of nitrogen and phosphorus, which could be sorbed at the surface after mixing. The core locations were deliberately chosen in the center of large subbasins of fine-grained sediments, which are present because the environment on a long time scale is a low-energy one. This process is unlikely in the Great Lakes, since the sediments are generally low in organic matter and considerable energy is required to set the dense sediment in motion. Diver observations at our core location in Lake Erie indicated strong horizontal bottom currents in August. However, the fresh, fluffy algal material appeared to be strongly bound to the bottom and was unaffected by the bottom currents (7).

Upward migration of gases is likely to push the sediment aside in the course of upward movement, but sediment particles probably will not be transported upward by this means. However, gaseous movement opens up a route for the rapid transfer of nutrients such as ammonia or phosphate to the sediment surface. No gaseous activity was observed in the three cores on retrieval. Cores from small lakes, rich in organic matter, often yield gases from the underlying sediment immediately on retrieval. It is probable that gaseous movement plays only a small role in the Great Lakes muds. Most of the gases formed are likely to be at the sediment-water interface, where decay of fresh organic matter is rapid. Movement of soluble OC is unlikely in the Great Lakes sediments as the material is rapidly humified or sorbed on the clays on deposition. The OC and N results suggest that the organic matter remains in situ and that some of the ammonia released by organic decay is fixed by the clays. Phosphorus and sulfur are likely to be mobile in the buried sediments; their relationships have already been discussed in terms of upward migration.

The small bottom faunal populations reported for midbasin sediments in the Great Lakes, the highly reduced muds below 2 cm, and the observed absence of organisms below 2 cm at the three locations suggest that bottom fauna are unlikely to be responsible for the high C, N, and P concentrations in the surface Great Lakes muds. However, the effects of bottom fauna cannot be discounted, and they are certainly present to a much greater depth in the sediment and in greater numbers in the shallower water and the coarser grained sediments of the Great Lakes. The OC values in the top 2 cm of the Ontario and Erie cores suggest that this is the major mixing zone at these locations. The biological enrichment of surface sediment in the Great Lakes is an area that needs much further study.

The results presented suggest that the increasing loading of nutrients to the lakes is preserved in the sediment record and that mixing processes other than chemical migration through the interstitial waters are negligible in the offshore muds. The parallel between

OC and N changes in the last 140 years (Figures 5 and 6) and the increase in dissolved solids to the lakes reported by Beeton (1) indicate that the sediment particles have remained in situ after burial. The argument is supported by the sharpness of the _Ambrosia_ horizon in these cores and a number of other cores from the Great Lakes that were examined in our laboratories. Recent studies at CCIW show a sharp mercury horizon at 6 cm at the Ontario location, lending further credence to the idea that little mixing occurs in the sediments (40). It is concluded that the muds in the center of subbasins of fine-grained sediment accumulation reflect the impact of cultural eutrophication on the region and are not disturbed again to any significant extent after burial.

CONCLUSIONS

The results of this study are summarized in Table 4, which relates the maximum OC, CC, N, P, and S values in the surface sediments with the concentrations at the _Ambrosia_ horizon. The values of OC, N, and P have increased about threefold in the Erie core and just under threefold in the Ontario core, suggesting a threefold increase of loading to the sediments of the two lakes. The slightly greater increase in the Erie core reflects the greater loading of the Erie sediments, as the cores are mineralogically and texturally similar. The South Bay, Lake Huron, core shows no appreciable change in OC, CC, and N, and this area appears to be unaffected by the increasing loading of Lake Huron. It is believed that the increase in P in the Huron core is due to the upward migration of phosphate from the reduced sediments. The increases in OC and N in the Ontario and Erie cores parallel the increase in dissolved solids to the two lakes. The increase commenced in about 1900 and seems to be accelerating. Changes in sediment OC and N appear to reflect the increased loading of the Great Lakes, as well as measurements of water quality.

The matching of our data with the water quality data of Beeton (1) indicates that our estimated sedimentation rates for the three cores are valid. Changes in CC and P are less regular and require interpretation; however, both parameters register the increasing loading. The increasing load of S to the lakes is not reflected in the sediment cores, its presence in the sediment column being controlled by Eh.

The changes in all the parameters measured suggest that it will be necessary for future investigators to

Table 4. The Ratio of Maximum Recent OC, CC, N, P, and S Concentrations to those Deposited 140 Years Ago in Lakes Ontario and Erie and 100 Years Ago in Lake Huron

Lake	OC	CC	N	P	S
Ontario	2.7	13.5	2.8	2.5	0.6
Erie	3.2	8.0	3.8	2.7	16.8
Huron	1.2	0.7	1.3	1.6	1.1

subsample Great Lakes sediment cores at close intervals in order to interpret the various geochemical cycles. Finally, it is concluded that cultural eutrophication of Lakes Ontario and Erie has resulted, so far, in a threefold increase in sediment organic matter, nitrogen, and phosphorus above the natural sediment levels. South Bay, Lake Huron, which is expected to be relatively unaffected by cultural eutrophication, shows no appreciable changes in sediment concentrations of carbon, nitrogen, phosphorus, and sulfur.

ACKNOWLEDGMENT

The authors are indebted to Dr. J. M. McAndrews for assistance and advice with the pollen data, to Dr. C. F. M. Lewis for discussions of postglacial mud thicknesses and rates of sedimentation, and to Dr. N. M. Burns for discussions on the chemical results. We would like to thank Mrs. N. Harper for the carbon and phosphorus analyses, Miss Janet Swegles for the pollen analyses, and Miss Louise Homer for the mineralogical analyses. We would also like to thank members of the Canada Centre for Inland Waters staff for assistance in this study. Finally we would like to thank Drs. R. L. Thomas and P. G. Sly, who read the manuscript and offered constructive criticism and helped in many other ways during the course of the investigation.

REFERENCES

1. A. M. Beeton, "Changes in the Environment and the Biota of the Great Lakes." in "Eutrophication: Causes, Consequences, Correctives," National Academy of Sciences, Washington, D. C., 1969, pp. 150-187.

2. J. R. Kramer, "Theoretical Model for the Chemical Composition of Fresh Water with Application to the Great Lakes," Proc. 7th Conf. Great Lakes Res., Int. Assoc. for Great Lakes Res., 1964, pp. 147-160.
3. H. Dobson, "Principal Ions and Dissolved Oxygen in Lake Ontario," Proc. 10th Conf. Great Lakes Res., Int. Assoc. for Great Lakes Res., 1967, pp. 337-356.
4. R. R. Weiler and V. K. Chawla, "Dissolved Mineral Quality of Great Lakes Waters," Proc. 12th Conf. Great Lakes Res., Int. Assoc. for Great Lakes Res., 1969, pp. 801-818.
5. "Report to the International Joint Commission on the Pollution of Lake Erie, Lake Ontario and the International Section of the St. Lawrence River." Queen's Printer, Ottawa, Canada, 1969, p. 151.
6. N. M. Burns and C. Ross, "Oxygen-Nutrient Relationships in Central Lake Erie," in "Nutrients in Natural Waters," H. E. Allen and J. R. Kramer, Eds. New York, Wiley, 1972.
7. N. M. Burns, Canada Centre for Inland Waters, Burlington, Ont., Canada, personal communication.
8. B. M. Jenkin, C. H. Mortimer, and W. Pennington, "The Study of Lake Deposits," Nature, 147, 496 (1941).
9. U. M. Cowgill and G. E. Hutchinson, "Chemistry and Mineralogy of the Sediments and Their Source Materials," Trans. Am. Phil. Soc., 60, 37-101 (1970).
10. J. G. Konrad, D. R. Keeney, G. Chesters, and K. L. Chen, "Nitrogen and Carbon Distribution in Sediment Cores of Selected Wisconsin Lakes," J. Water Pollut. Control Fed., 42, 2094-2101 (1970).
11. A. L. W. Kemp and C. F. M. Lewis, "A Preliminary Investigation of Chlorophyll Degradation Products in the Sediments of Lakes Erie and Ontario," Proc. 11th Conf. Great Lakes Res., Int. Assoc. for Great Lakes Res., 1968, pp. 206-229.
12. E. Callender, "Geochemical Characteristics of Lakes Michigan and Superior Sediments," Proc. 12th Conf. Great Lakes Res., Int. Assoc. for Great Lakes Res., 1969, pp. 124-160.
13. N. F. Shimp, J. A. Schleicher, R. R. Ruch, D. B. Heck, and H. V. Leland, "Trace Element and Organic Carbon Accumulation in the Most Recent Sediments of Southern Lake Michigan," Illinois State Geological Survey, Environ. Geol. Notes, 41, 25 (1971).
14. R. A. Bryson and C. R. Stearns, "A Mechanism for the Mixing of the Waters of Lake Huron and South Bay, Manitoulin Island," Limnol. Oceanogr., 4, 246-251 (1959).
15. A. L. W. Kemp, H. A. Savile, C. B. J. Gray, and

A. Mudrochova, "A Simple Corer and a Method for Sampling the Mud-Water Interface," Limnol. Oceanogr., 16, 689-694 (1971).
16. D. R. Keeney and J. M. Bremner, "Use of the Coleman Model 29A Analyzer for Total Nitrogen Analysis of Soils," Soil Sci., 104, 358-363 (1967).
17. J. D. M. Strickland and T. R. Parsons, "A Manual of Sea Water Analysis," Fish. Res. Bd. (Canada) Bull. 125, Queen's Printer, Ottawa, 1965, p. 203.
18. N. A. Rukavina and G. A. Duncan, "F.A.S.T.--Fast Analysis of Sediment Texture," Proc. 13th Conf. Great Lakes Res., Int. Assoc. for Great Lakes Res., 1970, pp. 274-281.
19. L. J. Trostell and D. J. Wynne, "Determination of Quartz (Free Silica) in Refractory Clays," J. Am. Ceram. Soc., 23, 18-22 (1940).
20. R. L. Thomas, A. L. W. Kemp, and C. F. M. Lewis, "The Distribution, Composition and Characteristics of the Surficial Sediments of Lake Ontario," J. Sediment. Petrol., in press.
21. J. C. B. Waddington, "A Stratigraphic Record of the Pollen Influx to a Lake of the Big Woods of Minnesota," Geol. Soc. Am. Spec. Pap. 123, 263-282 (1969).
22. A. L. W. Kemp, "Organic Carbon and Nitrogen in the Surface Sediment of Lakes Ontario, Erie and Huron," J. Sediment. Petrol., 41, 537-548 (1971).
23. C. F. M. Lewis, "Sedimentation Studies of Unconsolidated Deposits in the Lake Erie Basin," Ph.D. thesis, University of Toronto, Canada, 1966, p. 134.
24. A. L. W. Kemp, unpublished data.
25. A. L. W. Kemp, "Organic Matter in the Sediments of Lakes Ontario and Erie," Proc. 12th Conf. Great Lakes Res., Int. Assoc. for Great Lakes Res., 1969, pp. 237-249.
26. R. O. Brinkhurst, A. L. Hamilton, and H. B. Herrington, "Components of Bottom Fauna of the St. Lawrence, Great Lakes," Great Lakes Institute, University of Toronto, Rep. PR33, 1968, p. 50.
27. J. M. Vanderpost and B. J. Dutka, "Bacteriological Study of the Kingston Basin Sediments," Proc. 14th Conf. Great Lakes Res., Int. Assoc. for Great Lakes Res., 1971, pp. 137-156.
28. R. L. Thomas, "A Note on the Relationship of Grain Size, Clay Content, Quartz and Organic Carbon in some Lake Erie and Lake Ontario Sediments," J. Sediment. Petrol., 39, 803-809 (1969).
29. R. L. Thomas, "The Qualitative Distribution of Feldspars in Surfical Bottom Sediments from Lake Ontario," Proc. 12th Conf. Great Lakes Res., Int. Assoc. for

Great Lakes Res., 1969, pp. 364-379.

30. J. H. McAndrews, personal communication. Royal Ontario Museum, Toronto, Canada.
31. J. H. McAndrews, A. L. W. Kemp, and C. F. M. Lewis, unpublished results. Canada Centre for Inland Waters, Burlington, Ont.
32. C. F. M. Lewis, personal communication. Geological Survey of Canada, Canada Centre for Inland Waters, Burlington, Ont.
33. C. H. Mortimer, "The Exchange of Dissolved Substances Between Mud and Water in Lakes, III and IV," J. Ecol., 30, 147-201 (1942).
34. C. E. ZoBell, "Studies on Redox Potential of Marine Sediments," Bull. Am. Assoc. Petrol. Geol., 30, 477-513 (1946).
35. E. Bonatti, D. E. Fisher, O. Joensuu, and H. S. Rydell, "Post-depositional Mobility of Some Transition Elements, Phosphorus, Uranium and Thorium in Deep Sea Sediments," Geochim. Cosmochim. Acta, 35, 189-201 (1971).
36. G. L. Harlow, "Task Which Lies Ahead in the Lake Erie Basin," Proc. 23rd Ind. Waste Conf., Purdue University, 1968, pp. 856-864.
37. C. B. J. Gray and A. L. W. Kemp, "A Quantitative Method for the Determination of Chlorin Pigments in Great Lakes Sediment," Proc. 13th Conf. Great Lakes Res., Int. Assoc. for Great Lakes Res., 1970, pp. 242-249.
38. A. S. Menon, C. V. Marion, and A. M. Miller, "Microbiological Studies of Oxygen Depletion and Nutrient Regeneration Processes in Lake Erie Central Basin," Proc. 14th Conf. Great Lakes Res., Int. Assoc. Great Lakes Res., 1971, 768-780.
39. G. F. Lee, "Factors Affecting the Transfer of Materials Between Water and Sediments," Literature Review No. 1, University Wisconsin, Water Resources Center, Madison, 1970, 50 pages.
40. R. L. Thomas, "The Distribution of Mercury in the Sediments of Lake Ontario," presented at the 14th Conf. Great Lakes Res., Int. Assoc. for Great Lakes Res., Toronto, April, 1971.

Chapter IX

EFFECTS OF SEDIMENT DIAGENESIS AND REGENERATION OF PHOSPHORUS WITH SPECIAL REFERENCE TO LAKES ERIE AND ONTARIO

J. D. H. Williams and Tatiana Mayer

Canada Centre for Inland Waters
Burlington, Ontario, Canada

REVERSAL OF EUTROPHICATION

The deterioration in water quality in Lakes Erie and Ontario, which is most visibly expressed by the enormous increase in the algal and aquatic macrophyte populations (1) has been the subject of intensive discussion and documentation in recent years. The increasing eutrophication of these lakes can be reversed only by a program of nutrient control, with the specific objective of drastically reducing the input of phosphorus into these lakes (2,3). The International Joint Commission report (2) also discussed the need to assess the capacity of the sediments in the lakes to return phosphorus to the overlying waters for algal growth, a process termed "regeneration," in the event of a substantial reduction in input of phosphorus into Lakes Erie and Ontario.

To date, there have been few situations available for study in which eutrophication has been reversed by nutrient control. The return of Lake Washington, Seattle, to a more oligotrophic condition following diversion of nutrient sources has been described by Edmondson (4). Vollenweider (5) proposed a model to predict the rate at which new equilibrium concentrations of soluble substances were attained following changes in the rate of supply of these substances to the lake. If inputs of phosphorus and nitrogen were reduced or increased to a new constant level, the concentrations of total phosphorus and nitrogen in the lake water would asymptotically approach a new mean steady-state value.

A relatively simple situation involved no movement of substance across the interface separating the overlying lake water from the sediments (henceforth, "interface") and no exchange of substance between the lake and the atmosphere. In this case, the time required for the new mean steady-state values to be attained (called the "transition period" in this chapter) would be determined by considering two factors: (1) the replenishment time of water in the lake, calculated from its volume and rate of inflow of water, and (2) how completely inflowing waters mixed with the main body of lake water. In the case of phosphorus, the transition period would be shorter if there was a net sedimentation of phosphorus throughout the period. Conversely, it would be longer if a net regeneration of phosphorus from the sediments occurred. To date the only situations observed in which regeneration has exceeded sedimentation have been associated with seasonal changes from oxic to anoxic conditions in hypolimnetic waters. Such situations, which can last for several weeks or months, are ultimately reversed by reoxygenation of the hypolimnion (6-8).

Provided oxic conditions are maintained, reduction in phosphorus input should result in more rapid attainment of new mean steady-state phosphorus concentrations than would be predicted if the role of sedimentation is ignored. Vollenweider (5) found that, after sewage was diverted, the Zellersee appeared to improve at a much faster rate than was indicated by the rate of addition and subsequent mixing of the inflow water.

Possible Influence of Regeneration of Phosphorus from Sediments on Time Required to Reverse Eutrophication

The replenishment times for Lakes Erie and Ontario are 2.6 and 7.9 years, respectively; thus the time required for 90% removal of a conservative pollutant (i.e., one that is not sedimented), assuming a high degree of mixing, is 6 to 7 years for Erie and 21 to 22 years for Ontario (2). Even if mixing were incomplete, however, the phosphate concentration of the waters of these lakes would probably respond more rapidly than this because of the importance of sedimentation in removing phosphorus from solution. The International Joint Commission (2) estimated that 84% of the phosphorus entering Erie and 77% of that entering Ontario is sedimented. From this information it might be possible to estimate the rate at which Erie and Ontario would revert to a more oligotrophic condition following a decrease in phosphorus input. The accuracy of the estimate, however, will depend in part on whether regeneration of phosphorus from the sediments affects the net rate of sedimentation of the element during the transition period.

There is little correlation in general between the phosphorus content of lake sediments and the trophic state of the overlying waters. High concentrations of phosphorus, ranging up to 7000 μg/g and thousands of times greater on a volume basis than in the overlying water, were found in the sediments of even highly oligotrophic lakes. Clearly, however, regeneration of phosphorus was not significant for the growth of aquatic organisms in these lakes. This was partly because the phosphorus was combined in the sediments in various chemical modes, particularly with iron (9). The record of sediment cores taken from Erie and Ontario indicates an appreciable rate of accumulation of phosphorus in the sediments even before man's activities induced eutrophication (10).

The oligotrophic condition of these lakes in earlier times, as evidenced by records of biota populations (1), shows that regeneration of phosphorus from the sediments was then insufficient to affect algal growth markedly. Thus, if regeneration of phosphorus from the sediments is to be a significant process in the future, when phosphorus inputs have been reduced to levels recommended by the International Joint Commission (2), at least two conditions must be fulfilled. First, phosphorus must be accumulating in the sediments at a more rapid rate at present than in earlier, more oligotrophic times, thereby creating what may be termed an excess of phosphorus in the sediments. Evidence that this is so is presented elsewhere (10). Second, some of the excess phosphorus must return from the sediments to the overlying water following reduction in phosphorus input into the lakes. The extent to which this excess phosphorus may be regenerated under conditions of lower phosphorus input will depend on the degree to which the sediments act as "traps," retaining all sedimented phosphorus, or as "buffers," with the concentration of phosphorus in the sediments fluctuating in response to changes in the concentration of phosphorus in the overlying waters.

The ability of sediments to act as "buffers" when in

intimate contact with a restricted volume of water is well known. Carritt and Goodgal (11) and Williams et al. (12), using estuarine and lacustrine sediments, respectively, have shown that orthophosphate may be adsorbed onto or desorbed from sediments, depending on the relative amounts of phosphate in solution or sorbed onto the solid phase. Exchange of ^{32}P-phosphate between water and sediments has been demonstrated by Hayes and co-workers (13) and by Olsen (14). The total quantity of phosphate in sediments that is exchangeable with ^{32}P-phosphate in this manner is large and may be in the order of hundreds of micrograms per gram of dry sediment (15). The existence of a large pool of highly mobile phosphate bound to the solid phase of sediments and in equilibrium with a relatively much lower concentration of phosphorus in solution seems to explain why the phosphorus of sediments is readily available to algae growing in intimate contact (16) but is not readily available if sediment and algae are physically separated by a dialysis membrane (17). The relative importance of these exchange reactions is perhaps greater in laboratory conditions than in large lakes because exchange reactions in the larger system occur against a background of more or less continuous phosphorus sedimentation.

Net Regeneration of Phosphorus from Sediment Columns in Idealized and Actual Situations

Under conditions of absolutely uniform sedimentation, with no variation of any kind, there can be no net regeneration of phosphorus from the sediments. This statement holds irrespective of any concentration gradient in total phosphorus or forms of phosphorus associated with the solid phase of the sediment or with the interstitial water. The situation is illustrated by Figure 1, which indicates how the total amount of phosphorus in a sediment core increases with time for an idealized situation. Substitution of time instead of the more usual depth for the y-axis simplifies discussion by eliminating complications due to increasing degree of compaction of sediments with increasing degree of burial. If depth is substituted for time in Figure 1, the resulting phosphorus concentration profile resembles that of cores taken from a few little-disturbed locations in the Great Lakes (10).

Figure 1 represents a sediment column of fixed unit area such that one gram of material containing 1800 μg of phosphorus, represented by $A_1A_1'B_1'B_1$, accumulates at the top of the sediment column during one year. (The

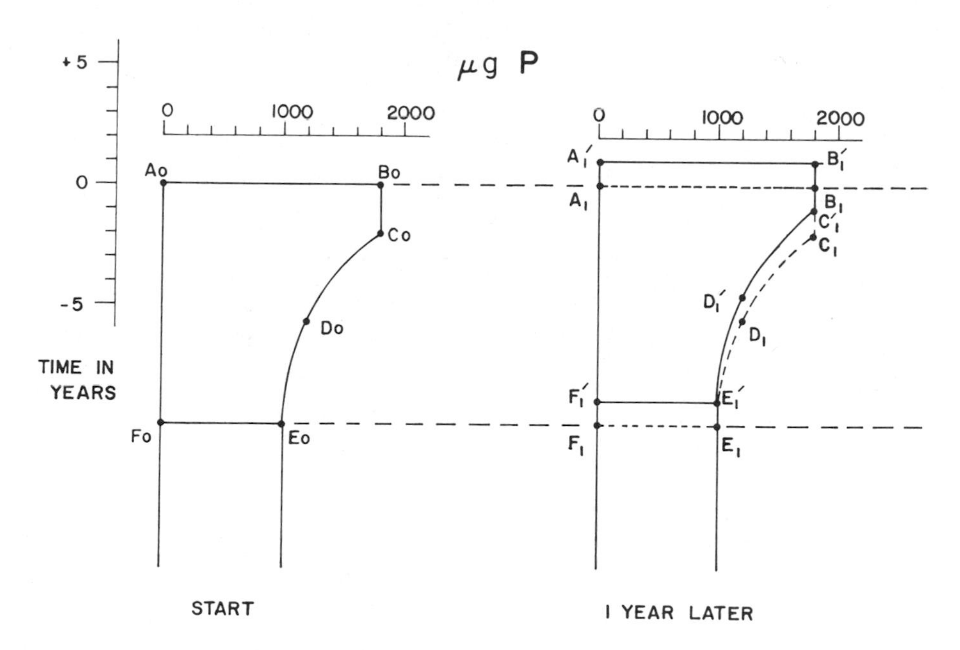

Fig. 1. Phosphorus addition to and movement within an idealized sediment column.

numbers are chosen to be representative of Great Lakes conditions.) Because a steady-state condition exists, the region in the column of variable phosphorus content, represented by $A_0B_0C_0D_0E_0F_0$, moves upward to $A_1'B_1'C_1'D_1'E_1'F_1'$. Meanwhile, the total net amount of phosphorus that is added to the sediment column during the year is represented by the area $F_1F_1'E_1'E_1$. Suppose this amount is 1000 μg, as in the piston core from Lake Ontario discussed below. In this case, of the 1800 μg added to the head of the sediment column, 1000 μg is derived from the overlying waters and 800 μg by migration of phosphorus upward from underlying sediment layers. Area $E_1E_1'D_1'C_1'C_1D_1$ represents the amount of phosphorus that migrates upward into the uppermost part of the sediment column during the year. This account is derived from the description by Emery and Rittenberg (18) of how interstitial water may move within a sediment core without escaping through the sediment-water interface. The discussion implies that compaction of sediments does not cause regeneration of phosphorus directly.

A large number of collected sediment cores resemble the idealized core of Figure 1 by showing a decline, often irregular, in total P content with depth in the top of the core (10,19,20). It is difficult to relate higher total P values at the top of the core to increased quantities of phosphorus that have been measured in the overlying waters in recent years. Diagenetic processes may account for variation in total P in a sediment column even in conditions of absolutely uniform sedimentation. This is discussed further below.

One possible cause of decline in total phosphorus with depth is release of phosphorus from a sediment layer as, with increasing burial, it passes from a region of positive Eh at the head of the core to a less positive Eh region. Rittenberg et al. (21) noted a sharp increase in soluble orthophosphate in the interstitial waters of a marine sediment core. The increase occurred at the point in the core where Eh became negative. They suggested that the increase was related to the influence of Eh on the ferric to ferrous ratio. This situation resembles the release of phosphorus from sediments into newly anoxic hypolimnia, which is also related to conversion of ferric to ferrous iron (6,8).

The total P profile in sediment cores is rarely as regular as in the idealized model. Livingstone (22) assumed that variations in the total phosphorus content of bands in a sediment core from Linsley Pond, Connecticut, were the result of a constant rate of phosphorus deposition with a variable rate of subsequent regeneration. On this basis he calculated that 45% of the sedimented phosphorus was regenerated. On the other hand, Wentz and Lee (20) interpreted the variation of "available P" with depth in a core from Lake Mendota, Wisconsin, as being the result of variations in the rate of deposition of phosphorus relative to organic matter, carbonates, and clastic materials. These variations were due to such factors as commencement of agriculture in the surrounding watershed and the raising of lake level.

It is clear that the total P concentration on a weight basis of any sediment layer must represent the balance between the total phosphorus content at the time the layer was at the interface and the loss of phosphorus from the layer subsequently, the resulting P value being expressed in terms of the net total of all forms of sedimented material. Thus the total P content of any part of the sediment column may be the result of the interaction of many environmental processes.

In the idealized situation of absolutely uniform sedimentation, regeneration of phosphorus from sediments can manifest itself as a process distinguishable from sedimentation only if the regenerated and sedimented forms of phosphorus are distinguishable chemically. Regeneration may affect algal growth if phosphorus is sedimented in association with solid materials (such as algal remains) and then returned to the overlying water in soluble form.

When conditions are not uniform, regeneration may become apparent in other ways. One such situation may occur when sediments are disturbed and resuspended. Seasonal variations may have such a great effect on the rate at which sedimented phosphorus returns to the overlying water that marked seasonal variations in the net sedimentation rate are caused. As already mentioned, regeneration may exceed sedimentation during the development of anoxic conditions in hypolimnia. Finally, regeneration may have some influence on the net sedimentation rate of phosphorus in the transition period following reduction in phosphorus input loadings.

Mechanisms for Transfer of Nutrients from Sediments

Transfer of phosphorus from sediment to overlying water can occur if the sediment is disturbed so that the solid phase and interstitial water are mixed with overlying lake water. This was apparently the mechanism in some observed cases of regeneration of nutrients from sediments. In extremely shallow Upper Klamath and Agency Lakes, southern Oregon, which have a mean depth of 2.4 m, regeneration of nutrients maintains eutrophic conditions (23). Short-term algal blooms were observed following the disturbance of sediments in experimental ponds (24). Kramer et al. (25) recorded increases in soluble orthophosphate and suspended mineral material in water samples collected from western Lake Erie following a period of rough weather. They suggested that the increase in phosphorus was due to release of phosphorus from disturbed and possibly resuspended sediments. Skoch and Britt (26) concluded that the top 5 to 7.5 cm of sediment at a locality in western Lake Erie, with water depth of 11 m, was well mixed because of wave action and currents.

Lee (27) has reviewed the processes that can cause disturbance and mixing of sediments, and the extent to which mixing redistributes the sedimented materials in the sediment column, making the historical record of the column become "out of focus." It is clear that regeneration of phosphorus by this mechanism may be more significant for shallow lakes, such as Erie, than for relatively deep lakes such as Ontario, because of greater effects on the sediments due to wind, wave, and current action.

A second mechanism for regeneration is diffusion of soluble phosphorus out of the sediments as a result of a difference in concentration of soluble phosphorus between the interstitial water of the sediment and the overlying water, followed by dispersion of phosphorus by bulk transport (via currents, eddies, etc.) or by diffusion. The soluble phosphorus regenerated may be formed from sedimented particulate phosphorus by diagenetic processes in the sediments. Rittenberg et al. (21) found concentrations of soluble reactive phosphate in the interstitial water of marine sediments that were up to 50 times as great as in the overlying seawater. The increase in concentration of soluble phosphate at the interface was very sharp.

Alternatively, a marked variation in regeneration rate, possibly resulting from seasonal variations, may have a detectable effect on net sedimentation rate over a given period. Skoch and Britt (26) found slight seasonal variation in the concentrations of phosphorus and iron in cores taken monthly over three years from a locality in western Lake Erie. It is possible that this variation reflects changes in the net rate of sedimentation of phosphorus and iron, and this in turn may be related to varying regeneration rates. Kanishige (28), who also found seasonal variation in the composition of sediments from Lake Mendota, Wisconsin, tentatively attributed this to translation of sediments.

The rate of removal of interstitial phosphorus from the sediment surface layers is determined by the porosity of the mud and by the circulation of water over the mud surface (29). Phosphorus that diffuses out of the sediment is available for algal growth only if it moves into the trophogenic zone by diffusion or bulk transport processes. Since diffusion processes are slow in comparison with bulk transport processes, dispersion of the regenerated phosphorus may be greatly slowed if mixing of overlying waters is poor, due, for example, to thermal stratification (27).

The role of the oxidized microzone, frequently found at the interface, in controlling regeneration of phosphorus has not been fully elucidated. It has been suggested that the microzone acts as a barrier to the exchange of phosphorus across the interface because of the presence of ferric iron. It is questionable, however, whether the barrier is effective when the microzone is unconsolidated and has a high water content.

A third mechanism by which phosphorus may travel from sediment to overlying water is incorporation into organisms. Examples are bottom-feeding fish such as carp and perch species. Again, phosphorus moves up from the sediments during the transformation of *Chironomus* species from larval to adult form. The phosphorus that moves in this form is not immediately available for plant growth, and thus the significance to lake nutrient budget and eutrophication studies is probably not great; however, this phosphorus may be an appreciable item in the total budget of the lake.

It may be permissible to distinguish between two stages of regeneration of phosphorus from sediments on the basis of kinetic factors. The first stage is the conversion of sedimented phosphorus into soluble phosphorus at the interface without incorporation of the sedimented material into the sediment column proper. This process is exemplified by the initial fairly rapid decomposition of algal remains. Diver observations of the Central Basin of Lake Erie (30) suggest that sedimented algal remains are completely decomposed within a few months. Burns and Ross (6) found that a quarter of the phosphorus liberated by decomposition of algal remains returned to the overlying water, the remainder presumably being retained by the sediments.

The second stage is the regeneration of soluble phosphorus liberated within the sediment column as a result of diagenetic processes. The rate at which soluble phosphorus is liberated at the interface by processes associated with the first stage should diminish rapidly in response to reduced phosphorus inputs. It is phosphorus regenerated during the second stage, possibly extending over many years, that may cause delay in attaining the anticipated trophic state after the phosphorus input into the lake has been reduced.

Forms of Phosphorus in Sediments

The capacity of sediments to act as "traps" or "buffers" for phosphorus may depend in large measure on the chemical forms in which sedimented phosphorus reaches the interface, as well as on the subsequent diagenesis of these forms in the sediments. Rittenberg et al. (21) considered that the phosphorus in sediments from marine basins off California was predominantly inorganic and present as detrital minerals such as apatite and phosphate adsorbed on "...hydrolyzates and oxidates (e.g., clays and ferric hydroxide)." A small amount of "organic P" (organic esters of phosphoric acid), originating from organic debris, was also thought to be present. Mackereth (31) and Wentz and Lee (20) postulated six modes of deposition of phosphorus in benthic sediments. These were detrital phosphate minerals derived from the water shed, phosphate coprecipitated with iron and manganese, sorbed phosphate, phosphate associated with carbonates, and phosphorus in combination with autochthonous or allochthonous organic matter.

The solid-phase forms of phosphorus listed by Stumm and Morgan (32) as being of possible significance in natural water systems include the mineral phases hydroxyapatite and carbonate fluorapatite [$Ca_{10}(PO_4)_6X_2$, where X = OH, F, 1/2 CO_3], variscite ($AlPO_4 \cdot 2H_2O$), strengite ($FePO_4 \cdot 2H_2O$) and wavellite [$Al_3(OH)_3(PO_4)_2$]. No mineralogical evidence has yet been presented demonstrating the existence of variscite, strengite, and wavellite in modern sediments. Bache (33) has contended that variscite and strengite dissolve incongruently at pH values greater than 3.1 and 1.4, respectively. Stumm and Morgan (32) also estimated that, at pH values exceeding about 6.5, hydroxyapatite supported a lower concentration of phosphate in solution than either variscite or strengite, provided the calcium concentration in solution exceeded about 40 mg/liter. Apatite, which exists in many varieties, is the most common phosphate mineral in sedimentary environments (34).

Vivianite [$Fe_3(PO_4)_2 \cdot 8H_2O$] has been reported several times in lacustrine sediments (31,35-37). Mackereth (31) suggested that vivianite formed only under strongly reducing conditions, associated particularly with the decomposition of organic materials. Rosenqvist (37) observed vivianite only in regions where Eh values clustered around -390 mV and pH values were around 7.4. He concluded that the occurrence of vivianite in sediments was erratic because of competing reactions resulting in the formation of ferrous sulfide, siderite ($FeCO_3$), and apatite.

Recent work on Wisconsin lake sediments (9,38-40) indicated that phosphorus was present predominantly as apatite, organic phosphorus, and orthophosphate ions covalently bonded to short-range-order (X-ray amorphous) materials related in composition to some form of hydrated iron oxide. All three forms of phosphorus appear to be present in major amounts in surficial Great Lakes sediments (see below). It is the opinion of the authors that any apatite found in fine-grained lacustrine sediments toward the centers of the depositional basins in

Lakes Erie and Ontario is predominantly of authigenic or diagenetic rather than detrital origin (see below).

In general, the waters of Erie and Ontario are undersaturated with respect to hydroxyapatite except in the vicinity of major phosphorus input sources (41-43). However, diagenetic processes in the sediments may raise the concentration of phosphorus in the interstitial water to the point at which precipitation of apatite commences. However, because of the very great difference in concentration on a volume basis between phosphorus in the interstitial water and phosphorus associated with the solid phase of sediments, the amount of phosphate required to induce precipitation of apatite might be a quite small part of the solid-phase phosphorus. For a number of Erie and Ontario sediments, the concentration of soluble Ca^{2+}, PO_4^{3-}, and OH^- ions in the expressed interstitial waters did in fact correspond very closely to the solubility product of hydroxyapatite (44). Precipitation of apatite in the sediments is thus a process acting in opposition to the regeneration of soluble phosphorus from sediments to the overlying water. Once apatite is formed, release of phosphate from this mineral (even into greatly undersaturated waters) is unlikely to be rapid, as the release will be determined by the specific rate of solution per unit surface area of the apatite crystallites, among other factors. Apatite, although in disequilibrium with most noncalcareous terrestrial soils, can nevertheless persist for hundreds or thousands of years in this type of environment until completely weathered (45).

Organic phosphorus is present in sediments, and mineralization of this material is one way by which soluble orthophosphate may be released into interstitial waters. Burns and Ross (6) found that the rapid mineralization of sedimented algal remains accounted for a regeneration rate of 0.01 to 0.1 mmoles P/(m^2)(day) in the hypolimnetic waters of the central basin of Lake Erie during a four-week period in the summer of 1970.

Equilibrium between sorbed forms of phosphate and phosphate in solution can occur over a wide range of concentrations of phosphate in solution--from zero to the point where precipitation of a discrete phosphate commences. By contrast, if the concentration of phosphate in solution falls below the value determined by the solubility product of a mineral phase such as apatite, equilibrium is no longer maintained and dissolution of the mineral phase occurs. The positions of equilibrium between sorbed phosphate and phosphate in solution are controlled by sorption and desorption reactions and can be described by isotherms resembling those of Langmuir or Freundlich, as has been demonstrated for estuarine sediments by Carritt and Goodgal (11) and for Great Lakes sediments by Gumerman (46).

Experiments using ^{32}P-phosphate (15,47) indicate that the half-lives of the adsorption and desorption reactions are of the order of minutes or hours. It would seem that the quantity of sorbed phosphate in the sediment could respond rather rapidly to changes in the concentration of phosphate in the surrounding interstitial water. In comparison, however, the response of the interstitial water to changes in the phosphate concentration of the overlying lake water may be relatively slow, being controlled by the rate of transfer of phosphate across the interface.

Forms and Amounts of Phosphorus in Great Lakes Sediment

There is little information concerning the nature or quantity of the phosphorus in Great Lakes sediments. Total P in the top 3 cm of Erie and Ontario sediments ranged from 2500 ppm dry weight of sediment in the finest silty clay muds to 600 ppm in the sands (48). Total P concentrations paralleled the sediment particle size distribution in the two lakes, and the greatest values were associated with the fine-grained sediments. Total P ranged from 4740 ppm in eutrophic Hamilton Harbor muds and 1200 ppm in Ontario silty clay muds to 630 ppm in Ontario sands when samples of the top 10 cm of sediment were taken (49).

Skoch and Britt (26) found that total P decreased from about 1000 ppm at the sediment-water interface to about 600 ppm 15 cm below, at the same sample location. Schleicher and Kuhn (50) investigated the variation of total P with depth for six cores taken on a traverse across Lake Michigan from Waukegan to Benton Harbor. In one core total P was constant and equal to approximately 1250 ppm from 100 to 300 cm below the interface.

The need for a rapid assessment of the status of phosphorus in Erie and Ontario sediments led to a study of the forms of phosphorus in the recent sediments of these lakes and to an investigation into the diagenetic changes involving phosphorus occurring in the sediments. This report presents (1) preliminary information on the forms of phosphorus in the topmost centimeter of sediments from a representative sample location in each of the main basins of mud accumulation in Erie and Ontario, and (2) a description of the forms of phosphorus in a long sediment core taken by piston gravity corer from the center of a basin of mud accumulation in central Lake Ontario.

MATERIALS AND METHODS

The bathymetry, surficial sediment distribution, and general characteristics of Lakes Erie and Ontario are presented in Figures 2 and 3 and Table 1. Lake Erie, the shallowest of the five Great Lakes, is clearly divided into three distinct basins of mud accumulation (Figure 2). The eastern and central basins are bounded by sand-veneered ridges of glacial deposits which trend northwest from Erie and Lorain. The western basin is separated from the remainder of the lake by a chain of residual bedrock islands trending north from Sandusky. The main basin of Lake Ontario is a simple, elongated trough trending west to east (Figure 3). Two low cross-lake rises divide the main basin into three separate subbasins of mud accumulation (Figure 2). The Rochester basin in the east is deepest and largest, with a maximum depth of 244 m. The Mississauga basin in the center and the Niagara basin to the west are successively shallower and less extensive. The shallower Kingston basin is separated from the main lake basin by an east-west trending chain of islands.

The mud in Lakes Erie and Ontario occurs as a continuous offshore deposit, becoming finer grained toward the center of each basin. The mud is characteristically a soft grey silty clay with a redox potential of about -50 to +50 mV. The mud surface is usually coated with a thin brown oxidized microzone (not exceeding 2 cm in thickness), which is in contact with the overlying water. Gravel, sand, and silt occur as a veneer over residual glacial clay deposits on the ridge crests between basins and in the nearshore zones around the margins of the major mud deposits. The redox potential is usually more positive in these zones, approaching that of the hypolimnion waters. The sediment characteristics and distributions in the two lakes are more fully described elsewhere (51-54).

A piston core was collected from the Mississauga basin of Lake Ontario (Figure 3). The top 5 m of the core consisted of soft grey postglacial silty clay muds overlying stiff glacial clays, 10 m thick. The age of the uppermost glacial clays is approximately 10,000 years B.P. The uppermost part of the sediment column was lost during sampling because of the violent entry of the piston corer.

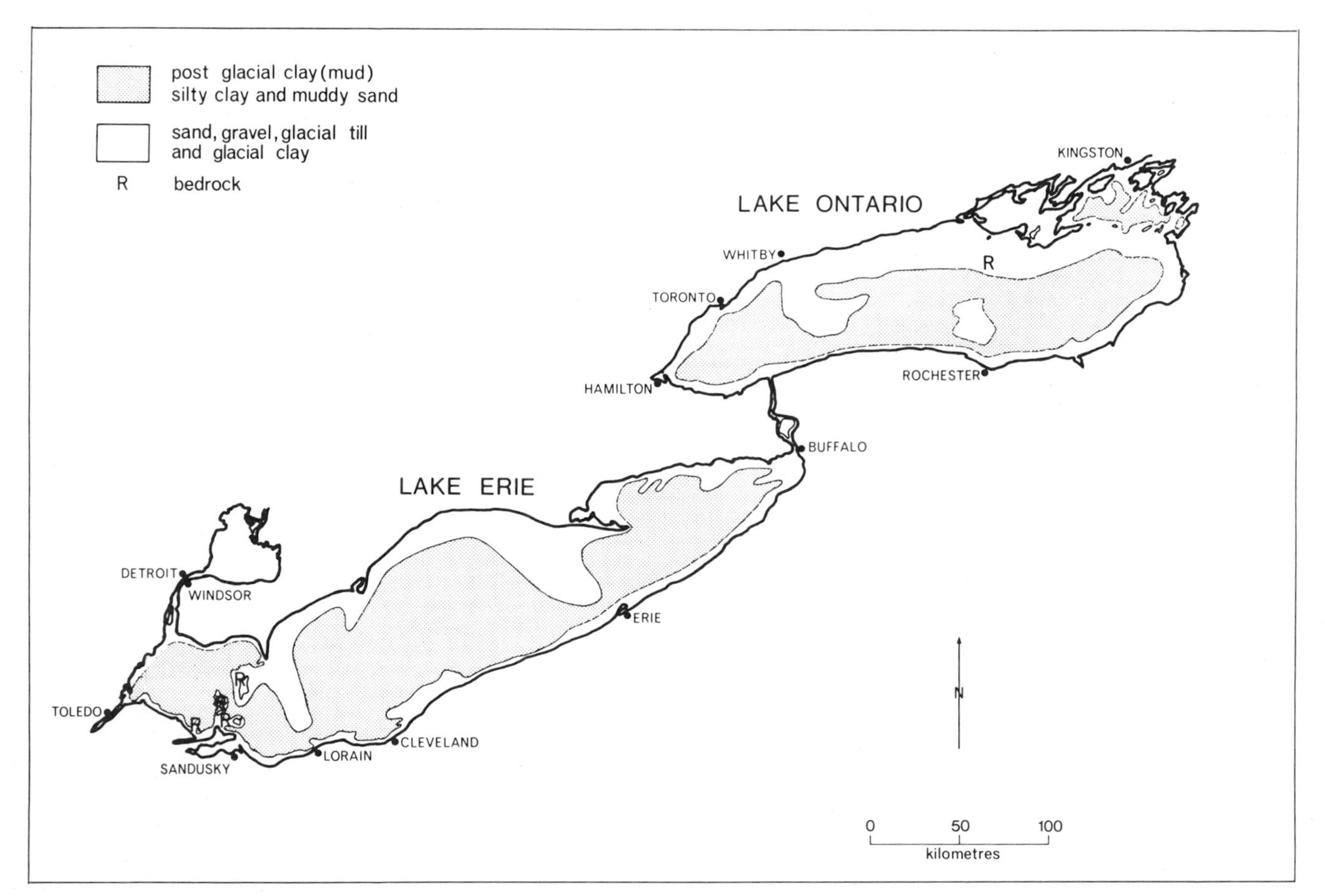

Fig. 2. Sediment distribution in Lakes Erie and Ontario.

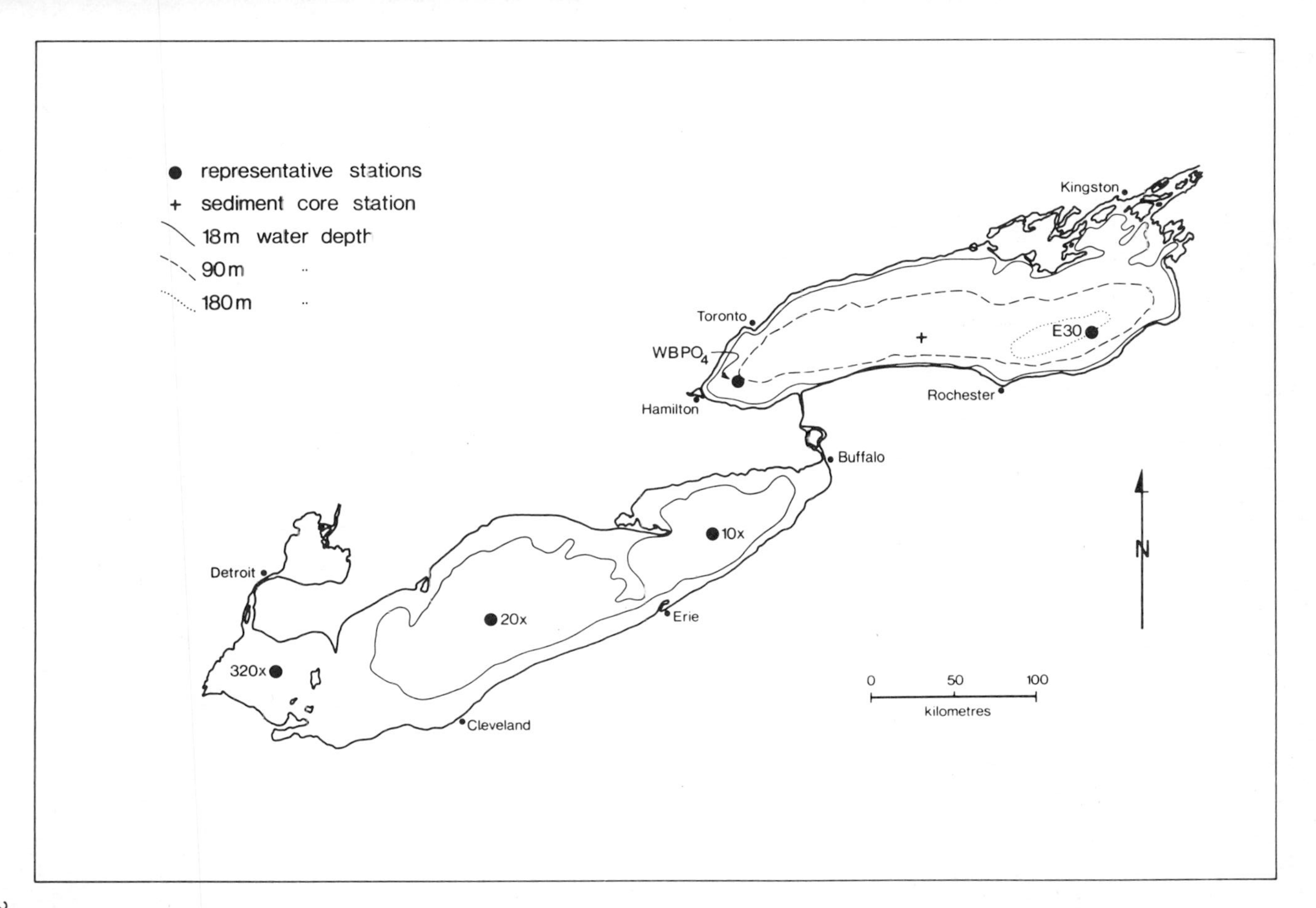

Fig. 3. Bathymetry and sediment sample locations in Lakes Erie and Ontario.

Table 1. General Characteristics of Lakes Erie and Ontario

Lake	Surface area, km^2	Maximum water depth, m	Mean water depth, m	Volume, km^3	Trophic state[a]
Erie	25,821	64	18	458	Eutrophic, tending to oligotrophic in the eastern basin
Ontario	19,009	244	84	1638	Oligotrophic and eutrophic characteristics

[a]Beeton (1).

The thickness of sediment lost is not precisely known and is denoted by x cm in this chapter. Comparison of organic C and CO_3-C values for the piston core with values for a Benthos core taken at the same location indicate that x probably did not exceed 20. At five representative stations, two in Ontario and three in Erie (Figure 3), 0 to 1 cm material was recovered to study the forms of phosphorus present. All core measurements were taken from the apparent sediment-water interface of the piston core as retrieved, and hence sediment depth is used. The contact between glacial clays and more recent sediments was between x + 478 and x + 497 cm below the original interface. The 0 to 3 cm sample, although regarded as part of core 17-5, was recovered at the same location using a Shipek bucket sampler.

From each sample, subsamples were taken and freeze-dried immediately. Details of the subsampling procedures have been given elsewhere (52,53). The freeze-dried samples were ground to <250 mesh prior to analysis.

Organic and carbonate carbon were analyzed by dry combustion in a Leco induction furnace (52). Total P was analyzed by sodium carbonate fusion of the sediment. The fusion cake was dissolved in 9N sulfuric acid and neutralized with sodium hydroxide (55). Orthophosphate was determined by the molybdic acid-ascorbic acid-trivalent antimony method of Strickland and Parsons (56). Organic P was determined by the method of Mehta et al. (57).

In the study of phosphorus in terrestrial soils, procedures have been developed whereby the material is extracted with a sequence of different reagents and inferences are made about the amounts of different forms of phosphorus from the amounts of phosphorus present in each reagent at the end of the extraction. Most recent sequential extraction studies of phosphorus in soils have been based on the procedure of Chang and Jackson (58). This approach has been applied to lake sediments (39,40, 59), and one attractive feature of the method is its potential for distinguishing between small amounts of phosphorus in different modes of chemical combination. This is particularly so for modes of combination in which phosphate is sorbed onto solid phase surfaces or is associated with short-range-order materials.

The fractionation procedure used in this study, which is a development of that of Williams et al. (40), is outlined in Figure 4. Full details of the procedure will be presented elsewhere (60). The scheme depends on the following postulates, based on previous work (9,38,39,58,59, 61,62):

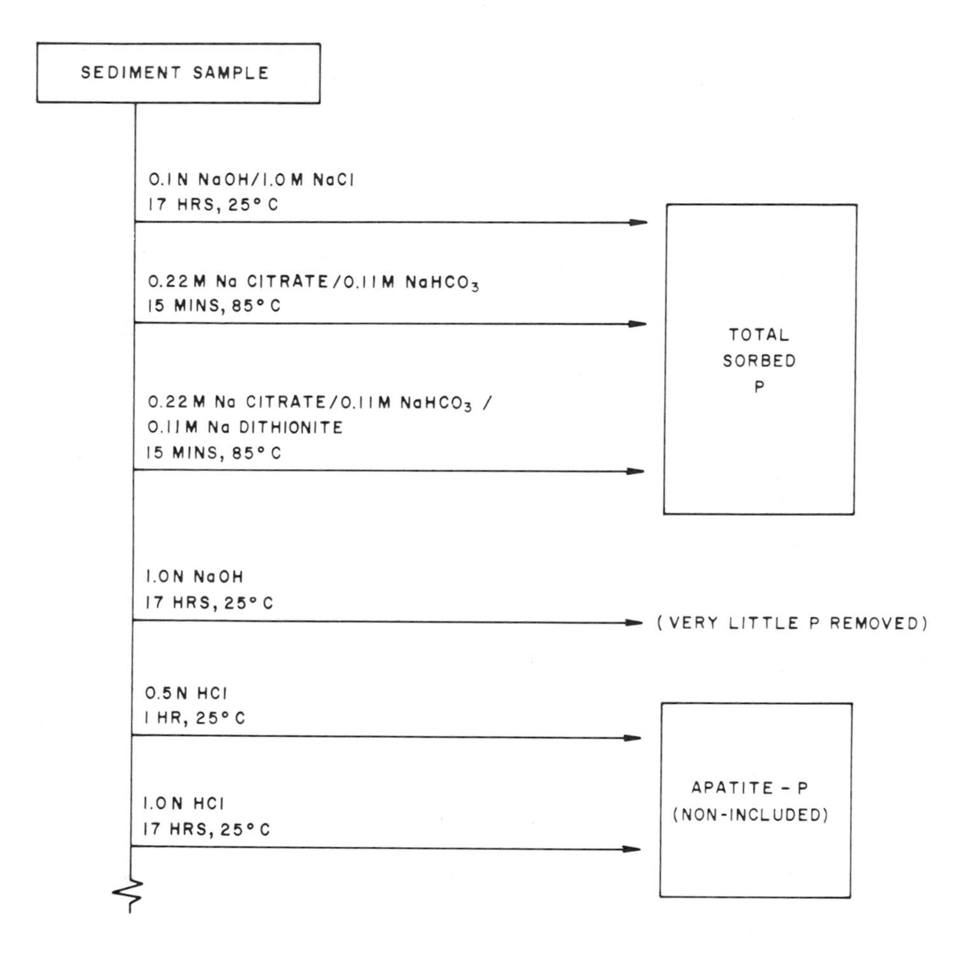

Fig. 4. Outline of fractionation scheme for inorganic P in sediments.

1. The predominant forms of acid- and/or alkali-soluble inorganic phosphorus in the sediments are apatite minerals and orthophosphate ions sorbed onto the surfaces of solid phases of high iron and/or aluminum content, including, in particular, phases closely related chemically to hydrated iron oxides. Forms of phosphorus assumed to be absent include other calcium phosphates such as monetite ($CaHPO_4$) and brushite ($CaHPO_4 \cdot 2H_2O$) (63) and ferric and aluminum phosphates of well-defined miner-

alogical composition such as variscite, strengite, and wavellite. In comparison with sorbed phosphate, other acid- and/or alkali-soluble forms of phosphorus, such as orthophosphate ions sorbed onto calcite surfaces and ions occluded within the matrices of crystalline iron- and aluminum-hydrated oxides and within calcite, are assumed to be of minor importance.

2. The combination of four extractions with sodium hydroxide and sodium citrate removes all forms of dilute acid-soluble orthophosphate from the sediments except for phosphate present in apatite. The phosphorus in these extracts is thus thought to be predominantly sorbed phosphate, particularly the phosphorus removed by the first sodium hydroxide extraction. The four extractions should also remove most, if not all, of any phosphate present as occluded phosphate within iron and aluminum oxides or calcite. It should also suffice to extract all phosphate from discrete iron and aluminum phosphates such as strengite and variscite.

3. The amount of phosphate extracted from the apatite of the sediments by the sodium hydroxide and citrate reagents is a small or negligible proportion of the total apatite-phosphate fraction.

4. The two hydrochloric acid extractions remove all apatite-phosphate except for material protected from acid extraction by being included within resistant minerals such as feldspars and ilmenite.

Values of apatite-P, total sorbed P (thought to be predominantly iron bound) and the amount of sorbed phosphate removed by the first sodium hydroxide extraction of the fractionation procedure ("NaOH-extractable sorbed P") are discussed below. The apatite-phosphate and sorbed phosphate forms appeared to account for more than 90% of the total inorganic phosphorus in each sediment sample.

RESULTS AND DISCUSSION

Piston Core

Total P was fairly constant throughout the core, ranging from 950 to 1250 μg/g, except for two samples (Figure 5). The surface sample had by far the highest value, close to 2000 μg/g. The three samples from below the glacial clay contact suggest that total P was virtually constant in this region. Organic P in the ($\underline{x}$ + 7) to ($\underline{x}$ + 13) cm sample was less than half that of

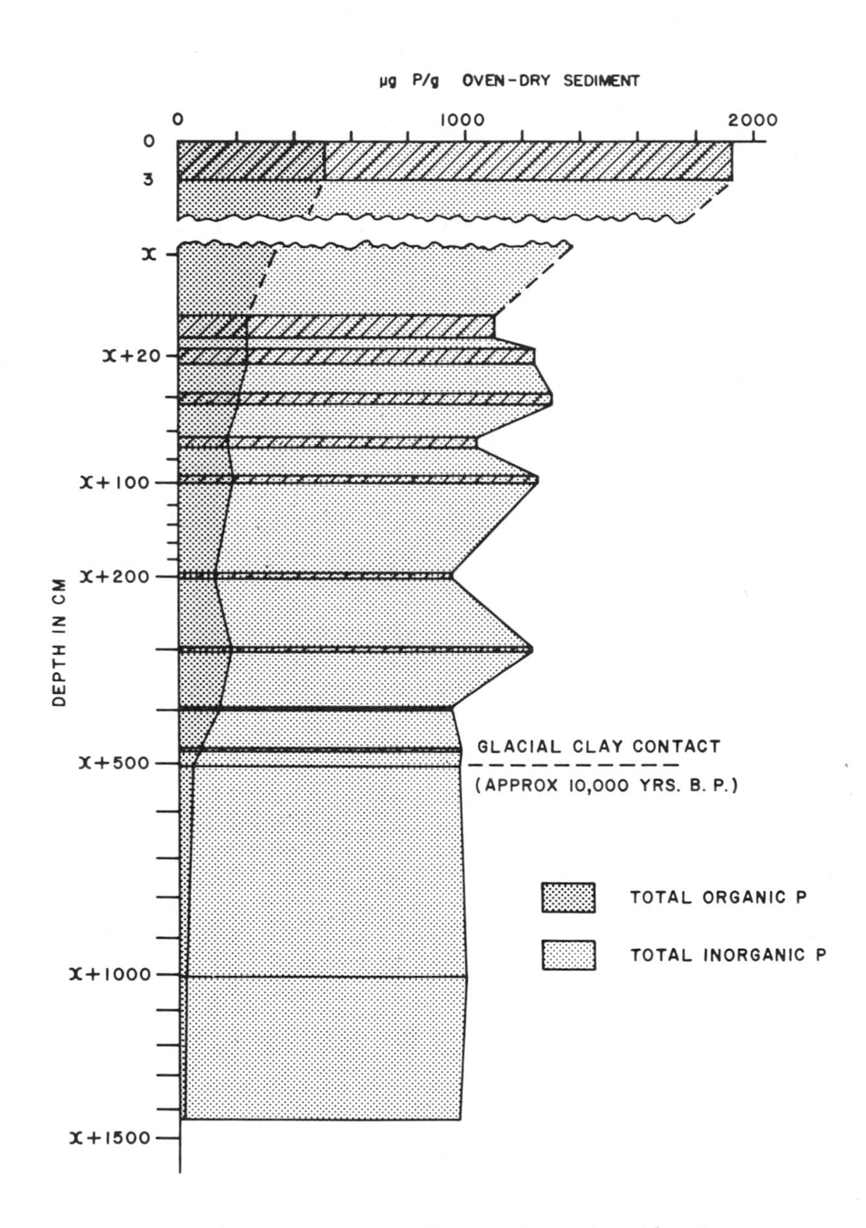

Fig. 5. Total P, organic P, and inorganic P in the piston core.

the 0 to 3 cm sample, and this form of phosphorus declined still further with depth, becoming less than 50 μg/g in the glacial clay samples (Figure 5). A high in total P and organic P at $\underline{x}$ + 300 cm is paralleled by a high in sediment organic C and is attributed to the Hypsithermal period. By contrast, apatite-P increased steadily with depth, from 302 μg/g (16% of total P) at the 0 to 3 cm level to 748 μg/g (> 76%) in the bottom sample (Figure 6). The regularity with which organic P decreases and apatite-P increases with depth is in contrast to the irregularities in the total P profile.

Similar trends were exhibited by NaOH-extractable and total sorbed P with depth in the piston core. Like organic P, these two sets of values tended to decrease with increasing depth. Unlike organic P, however, the sorbed P values showed marked irregularities that closely resembled the irregularities shown by the total P profile, particularly in the range $\underline{x}$ + 10 to $\underline{x}$ + 400 cm (Figure 7). The manner in which much of the irregularity in total P values can be matched by the sorbed P values is reminiscent of how sorbed phosphate accounted for much of the variation in total phosphorus in a group of Wisconsin lake sediments (39). The tendency for both organic and inorganic P to decline with depth at the top of the core is in contrast to the conclusion of Rittenberg et al. (21) that a decrease in organic phosphorus would be compensated by an increase in inorganic phosphorus with increasing depth in sediment cores.

Organic C, like organic P, decreased with depth, with the exception of a marked increase in the $\underline{x}$ + 300 cm sample (Figure 8). Throughout the core, the ratio of organic C to organic P by weight was in the order of 100, a value similar to that for terrestrial soils of high microbiological activity (64). The CO_3-C values were irregular, suggesting that the increase in apatite with depth was not due to an increase with depth of the percentage of sedimented materials that were originally derived from carbonate rocks. Instead, the increase seems to indicate diagenesis of other forms of phosphorus to apatite in the sediments.

Representative Samples

The total P values of the five representative 0 to 1 cm samples tend to increase from west to east (Table 2). However, the phosphorus loading of Erie is greater than for Ontario, and the major sources of phosphorus inputs for both lakes are concentrated at their western ends (2). Therefore, the properties of the representative samples

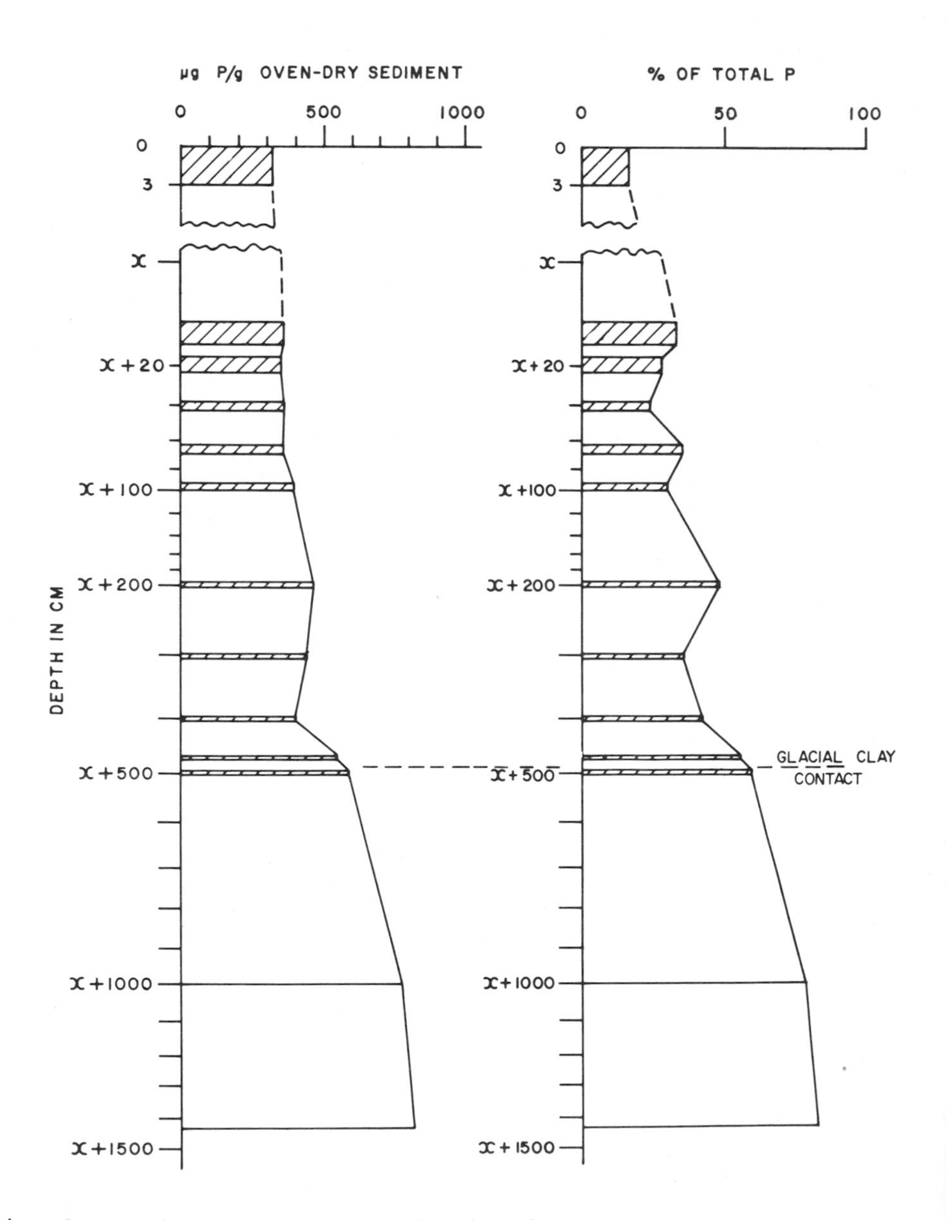

Fig. 6. Nonincluded apatite-P in the piston core.

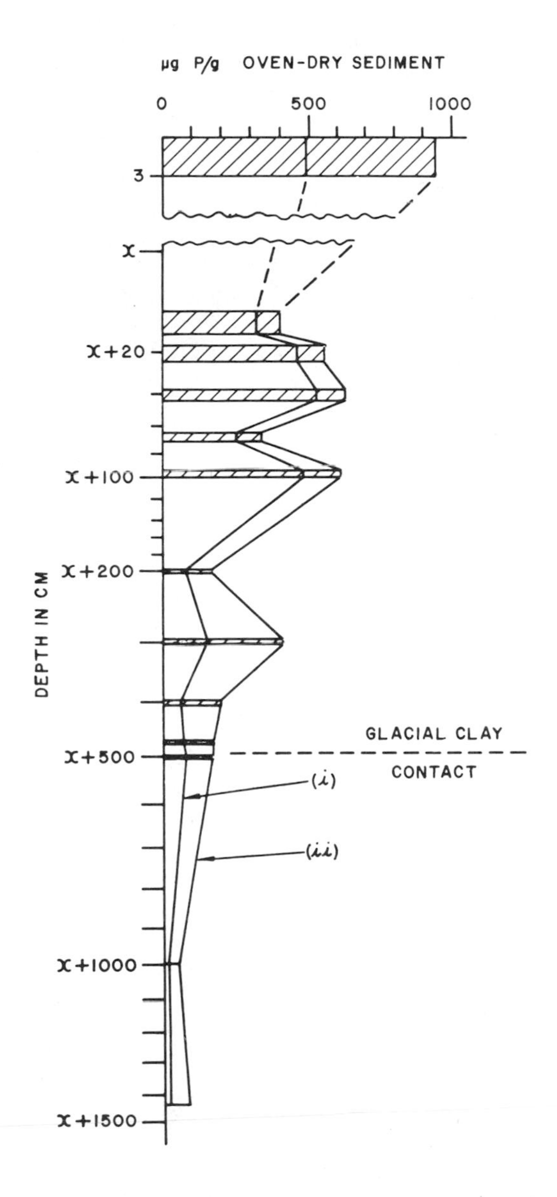

Fig. 7. Variation of NaOH-extractable sorbed P (i) and total sorbed P (ii) with depth in the piston core.

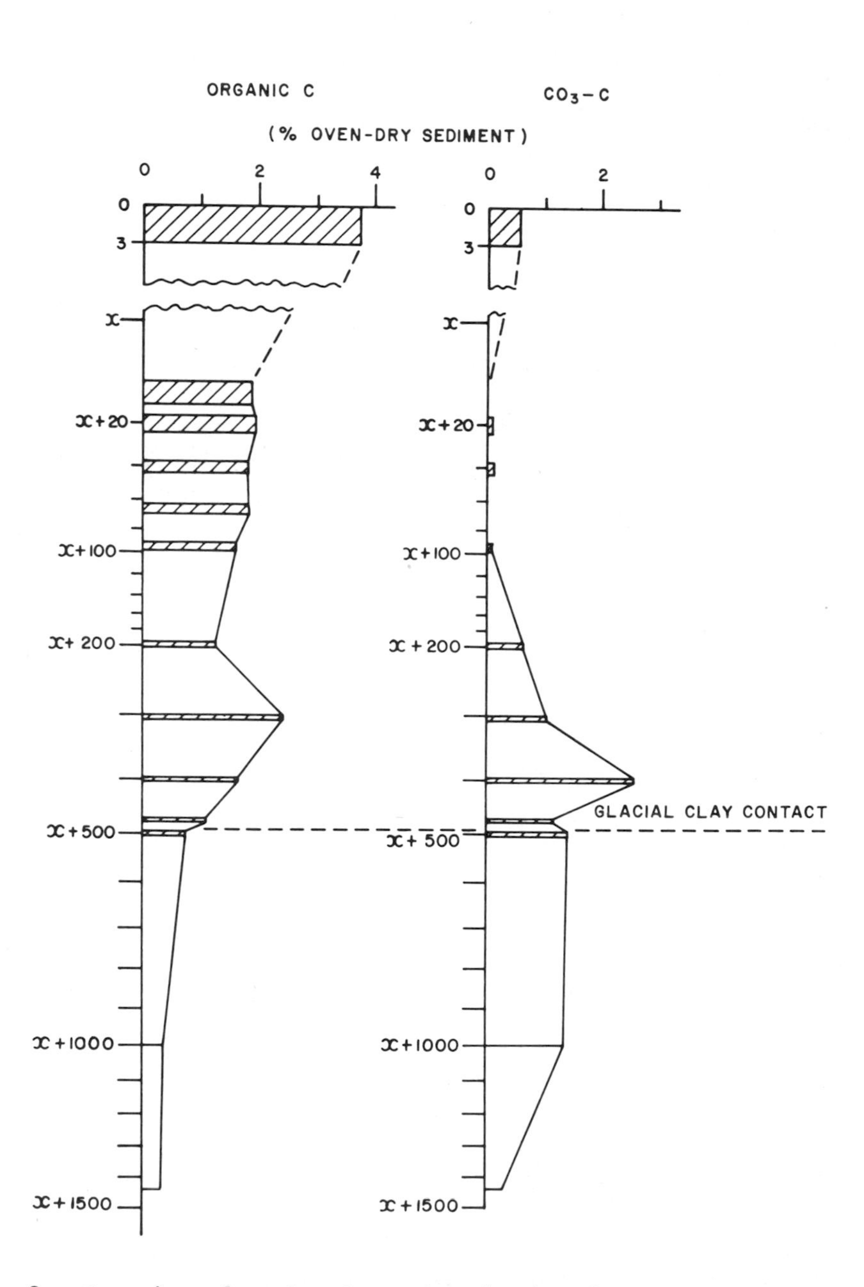

Fig. 8. Organic and carbonate carbon in the piston core.

Table 2. Phosphorus Values of Representative Surface Samples, μg/g

Sample	Lake	Basin	Total P	Organic P	Apatite-P	NaOH-Extractable sorbed P	Total sorbed P
320X	Erie	Western	1080	123	360	226	390
20X	Erie	Central	1010	208	386	137	272
10X	Erie	Eastern	1600	220	511	427	696
$WBPO_4$	Ontario	Niagara	1810	471	514	290	683
E30	Ontario	Rochester	1850	409	326	403	866

may have played a larger role in controlling their total phosphorus contents than the concentration of phosphorus in the overlying waters. The inorganic P fractionation values of the five samples indicate that there are appreciable amounts of sorbed phosphate and apatite-phosphate as well as organic phosphorus in the surface sediments of both lakes. In particular, the presence of large amounts of apatite in all five sediments suggests that apatite is a stable phase of the surface sediments throughout these lakes, irrespective of the Ca^{2+}, OH^-, and PO_4^{3-} concentrations in the overlying waters.

The pH values of the surface sediments of Erie average 7.3 to 7.5 and those of Ontario average 7.6 to 7.9 (48). Krumbein and Garrels (65) stated that the pH boundary for pure calcium phosphate formation from seawater is 7.5. Apatite appears to be stable in terrestrial soils whose pH values exceed 7.0, especially if free $CaCO_3$ is present. The present findings support those of Sutherland et al. (44), who found that the interstitial waters extracted from three sediment samples from the Great Lakes were almost exactly saturated with respect to hydroxyapatite. Factors that could account for the interstitial waters having a higher degree of saturation than the water above include a higher concentration of orthophosphate (due perhaps to phosphorus transformations in the sediments) and higher pH values, or Ca^{2+} concentrations.

GENERAL DISCUSSION

This chapter presents information on the forms of phosphorus in sediments from Lakes Erie and Ontario and suggests some of the processes involving transformation and movement of phosphorus which occur in the sediments. Apatite is a major form of phosphorus in the sediments, and the presence of this heavy mineral in the center of areas of sediment deposition suggests an origin that is authigenic or diagenetic rather than detrital. The stability of the mineral under the prevailing environmental conditions, particularly with respect to pH, is further indicated by the manner in which other forms of phosphorus in the sediments are ultimately converted into apatite.

The decline of organic P with depth in sediments, paralleled by a decline in organic C, indicates that processes resulting in mineralization of organic molecules continue at a slow rate long after the sediment has been buried. The decline of sorbed phosphate with depth may be due to more than one chemical process, resulting in conversion of one or more mineral components retaining the sorbed phosphate into other compounds. The processes of mineralization of organic phosphorus and release of sorbed phosphate both tend to increase the concentration of phosphate in the interstitial waters. On the other hand, precipitation as apatite tends to reduce the phosphate concentration. These processes, plus diffusion of phosphorus in the interstitial water, may greatly influence features of sediment cores such as the distribution, with depth, of total phosphorus associated with the solid phase and of phosphate that is in solution in the interstitial waters.

Although there would be a tendency for processes leading to a rise in the concentration of soluble phosphorus in solution to be counterbalanced by precipitation of apatite, the existence of at least three processes in sediments acting in opposite directions on the concentration of phosphorus in the interstitial waters may explain why the concentration of interstitial phosphate was variable with depth in one of three sediment cores taken from marine basins (21). Figures 9 and 10 illustrate the transformations of phosphorus in Great Lakes sediments and the resultant effect on the variation with depth of apatite, sorbed phosphate, and organic phosphorus in an idealized Great Lakes sediment core. In practice, irregular variations with depth of the phosphorus parameters, particularly sorbed phosphate, are probably common,

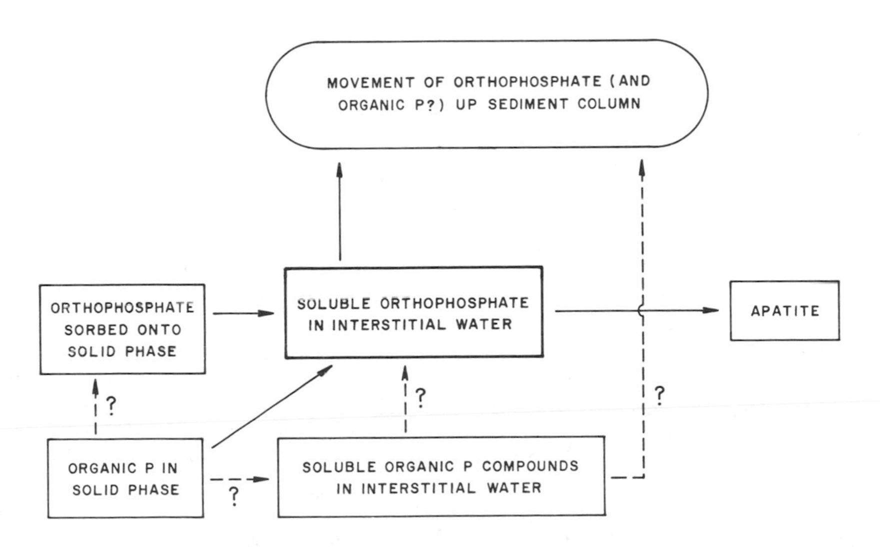

Fig. 9. Transformation of P within Great Lakes sediments.

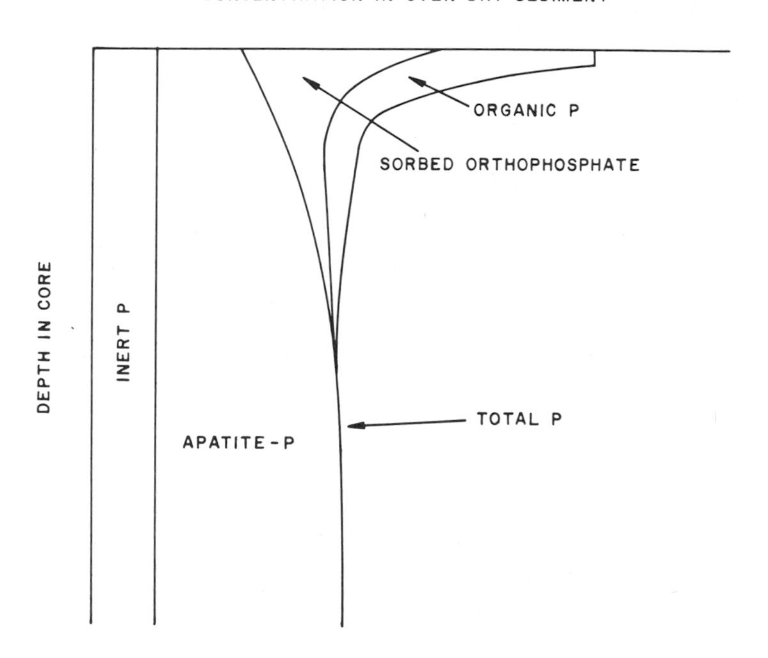

Fig. 10. Variation with depth of forms of P in an idealized Great Lakes core, assuming uniform conditions throughout the period of accumulation. "Inert P" includes very stable mineral species such as monazite.

reflecting changing conditions of deposition.

The evidence of the piston core suggests that in some sediment columns there may be a dividing line below which total phosphorus shows no particular trend with depth and above which total phosphorus increases with decreasing distance from the interface. The position of the dividing line has not been established for the core examined here. Soluble phosphorus released from the sorbed phosphate or organic phosphorus modes of combination in sediment layers lying above this dividing line may be converted into apatite in situ, or it may migrate up the sediment column. For sediment layers below this line, it is likely that little of the same released phosphorus migrates upward. Instead, the released phosphorus is almost entirely converted into apatite in situ.

It is possible that the rate of migration of phosphorus upward through the lower sediment layers is much less because the rate of release of soluble phosphorus from the sorbed and organic modes of combination is much slower than at the top of the column. Thus the steepness of the concentration gradient of soluble phosphate in the interstitial water decreases with depth. This further suggests that the fate of phosphorus released from sorbed or organic modes of combination (conversion into apatite, migration upward through the sediment column or even across the interface) is primarily determined by kinetic factors.

During recent years, the rate of input of phosphorus into Lakes Erie and Ontario has greatly increased. The effect of this on rate of sedimentation of phosphorus in these lakes is not known, although it is reasonable to assume that an increase has occurred. We now know that the summer months have been marked by a tremendous increase in the deposition of phosphorus on the sediment-water interface, in the form of algal remains, and that a high proportion of this algal-derived phosphorus is retained by the sediments during this period (6).

On the other hand, the evidence of Wisconsin lake sediments (9) suggests that the amount of sedimented phosphorus that remains in the sediments after the initial phase of decomposition has passed will be determined primarily by the amounts of iron and aluminum inorganic components capable of reacting with orthophosphate and organic phosphorus, which are sedimented simultaneously. This suggests that the levels of phosphorus in the surface sediments should by no means be proportional to phosphorus input, as indicated by the evidence that the ratios readily extractable-iron:sorbed-orthophosphate and extractable-aluminum:organic-phosphorus in sediments are relatively insensitive to the trophic state of the lake (9). Furthermore, much of any excess of sorbed and organic phosphorus in the sediments of Lakes Erie and Ontario over and above what would be present under oligotrophic conditions is probably being converted into apatite at the present time and hence would not be regenerated from the sediments should phosphorus inputs into these lakes be reduced. (This discussion presupposes a constant rate of sedimentation of the iron and aluminum components that retain orthophosphate and organic phosphorus in the sediments. It is also assumed that oxic conditions are maintained throughout.)

This discussion suggests that only a small fraction of the total phosphorus now being sedimented in Erie and Ontario is likely to be regenerated during the transition period following reduction in phosphorus loadings. Nevertheless, the known capacity of sediments to sorb phosphate from solution indicates that excess phosphorus will continue to build up in the sediments for as long as the present excessive loadings prevail. The longer the period of time that elapses before phosphorus input control is instituted for these two lakes, the greater will be the influence of regenerated P during the transition period. This may even result in more time being needed for the lakes to clean themselves up once phosphorus input control has been established.

Despite these qualifications, regeneration of phosphorus from sediments following reductions in phosphorus input is not likely to affect the ultimate outcome. It has sometimes been suggested that eutrophication of lakes occurs slowly but inevitably under natural conditions and that the process can be accelerated, but not reversed, by man. This may be true for very shallow lakes (23); but the hypothesis does not seem to be valid for Lakes Erie and Ontario, in the light of what is now known about how the sediments of these lakes retain phosphorus. In Erie and Ontario, regeneration of soluble phosphorus from sediments can only modify the rate at which the predominant process, sedimentation, occurs.

The more prevalent view is that, given a sufficient period of time and provided the rate of input of phosphorus is controlled sufficiently, even the most eutrophic lake will revert to an oligotrophic condition. If this is done, regeneration of a part of the excess phosphorus that accumulated in the sediments during the eutrophic conditions may extend the transition period and delay the attainment of oligotrophic conditions, but the ultimate trophic state of the lake should not be affected.

ACKNOWLEDGMENTS

I am indebted to A. L. W. Kemp for discussion and assistance at all stages of the project and for supplying the samples. The manuscript was read by N. M. Burns. The piston core was collected by R. L. Thomas, A. L. W. Kemp, and C. F. M. Lewis. Approval for publication was given by the Chief, Lakes Division, Department of the Environment and the Director, Canada Centre for Inland Waters.

REFERENCES

1. A. M. Beeton, "Changes in the Environment and Biota of the Great Lakes," in "Eutrophication: Causes,

Consequences, Correctives." National Academy of Sciences, Washington, D. C., 1969, pp. 150-187.
2. International Joint Commission, "Report to the IJC on the Pollution of Lake Erie, Lake Ontario and the International Section of the St. Lawrence River, Vol. 1, Summary." Queen's Printer, Ottawa, 1969.
3. A. T. Prince and J. P. Bruce, "Development of Nutrient Control Policies in Canada," Chapter XV, this volume.
4. W. T. Edmondson, "Phosphorus, Nitrogen, and Algae in Lake Washington after Diversion of Sewage," Science, 169, 690-691 (1970).
5. R. A. Vollenweider, "Moglichkeiten und Grenzen elementarer Modelle der Stoffbilanz von Seen," Arch. Hydrobiol., 66, 1-36 (1969).
6. N. M. Burns and C. Ross, "Oxygen-Nutrient Relationships within the Central Basin of Lake Erie," Chapter VII, this volume.
7. W. Einsele, "Uber die Beziehungen des Eisenkreislaufs zum Phosphat-kreislauf im eutrophen See," Arch. Hydrobiol., 29, 664-686 (1936).
8. C. H. Mortimer, "The Exchange of Dissolved Substances between Mud and Water in Lakes," J. Ecol., 29, 280-329 (1941).
9. J. D. H. Williams, J. K. Syers, S. S. Shukla, R. F. Harris, and D. E. Armstrong, "Levels of Inorganic and Total Phosphorus in Lake Sediments as Related to Other Sediment Parameters," Environ. Sci. Technol., 5, 1113-1120 (1971).
10. A. L. W. Kemp, C. B. J. Gray, and A. Mudrochova, "Changes in C, N, P, and S in the Last 140 Years in Three Cores from Lakes Ontario, Erie, and Huron," Chapter VIII, this volume.
11. D. E. Carritt and S. Goodgal, "Sorption Reactions and some Ecological Implications," Deep-Sea Res., 1, 224-243 (1954).
12. J. D. H. Williams, J. K. Syers, R. F. Harris, and D. E. Armstrong, "Adsorption and Desorption of Inorganic Phosphorus by Lake Sediments in a 0.1 M NaCl System," Environ. Sci. Technol., 4, 517-519 (1970).
13. F. R. Hayes, "The Mud-Water Interface," Oceanogr. Mar. Biol. Ann. Rev., 2, 121-145 (1964).
14. S. Olsen, "Phosphate Equilibrium between Reduced Sediments and Water. Laboratory Experiments with Radioactive Phosphorus," Int. Ver. Theor. Angew. Limnol. Verh., 15, 333-341 (1964).
15. W. C. Li, D. E. Armstrong, R. F. Harris, and J. K. Syers, "Exchangeable Phosphorus in Lake Sediment," Abstracts of the 62nd Annual Meeting, American

Society of Agronomy, Tucson, Ariz., August 1970, p. 97.
16. D. B. Porcella, J. S. Kumagai, and E. J. Middlebrooks, "Biological Effects on Sediment-Water Nutrient Interchange," J. Sanit. Eng. Div., Am. Soc. Civil Eng., 96, 911-926 (1970).
17. G. P. Fitzgerald, "Aerobic Lake Muds for the Removal of Phosphorus from Lake Waters," Limnol. Oceanogr., 15, 550-555 (1970).
18. K. O. Emery and S. C. Rittenberg, "Early Diagenesis of California Basin Sediments in Relation to Origin of Oil," Am. Assoc. Petrol. Geol. Bull., 36, 735-806 (1952).
19. H. B. Moore, "Muds of the Clyde Sea Area: 1. Phosphate and Nitrogen Contents," J. Mar. Biol. Assoc. (U.K.), 16, 595-607 (1930).
20. D. A. Wentz and G. F. Lee, "Sedimentary Phosphorus in Lake Cores--Observations on Depositional Pattern in Lake Mendota," Environ. Sci. Technol., 3, 754-759 (1969).
21. S. C. Rittenberg, K. O. Emery, and W. L. Orr, "Regeneration of Nutrients in Sediments of Marine Basins," Deep-Sea Res., 3, 23-45 (1955).
22. D. A. Livingstone, "On the Sigmoid Growth Phase in the History of Linsley Pond," Am. J. Sci., 225, 364-373 (1957).
23. A. R. Gahler and W. D. Sanville, "Characterization of Lake Sediments and Evaluation of Sediment-Water Interchange Mechanisms in the Upper Klamath Lake System," presented at the 14th Conf. on Great Lakes Research, Toronto, April 1971.
24. W. Abbott, "Nutrient Studies in Hyper-Fertilized Estuarine Ecosystems: 1. Phosphorus Studies," Proc. 4th Conf. Water Pollut. Res., Prague 1969, pp. 729-739.
25. J. R. Kramer, H. E. Allen, G. W. Baulne, and N. M. Burns, "Lake Erie Time Study (LETS)," Canada Centre for Inland Waters Paper 4, Burlington, Ont., 1970.
26. E. J. Skoch and N. W. Britt, "Monthly Variation in Phosphate and Related Chemicals Found in the Sediment in the Island Area of Lake Erie, 1967-68, with Reference to Samples Collected in 1964, 1965, and 1966," Proc. 12th Conf. Great Lakes Res., Ann Arbor, Mich., May 1969, pp. 325-340.
27. G. F. Lee, "Factors Affecting the Transfer of Materials between Water and Sediments," Eutrophication Information Service Literature Review No. 1., Water Resources Center, University of Wisconsin, Madison, 1970.

28. H. M. Kanishige, "Chemical Analysis of Bottom Muds of Lake Mendota," M.S. thesis, University of Wisconsin, Madison, 1952.
29. D. J. Rochford, "Studies in Australian Estuarine Hydrology: 1. Introductory and Comparative Features," Aust. J. Mar. Freshwater Res., 2, 1-116 (1951).
30. N. M. Burns, personal communication.
31. F. J. H. Mackereth, "Some Chemical Observations on Post-Glacial Lake Sediments," Phil. Trans. Roy. Soc. London, Ser. B, 250, 165-213 (1966).
32. W. Stumm and J. J. Morgan, "Aquatic Chemistry," New York, Wiley, 1970, p. 516.
33. B. W. Bache, "Aluminum and Iron Phosphate Studies Relating to Soils: 1. Solution and Hydrolysis of Variscite and Strengite," J. Soil Sci., 14, 113-123 (1963).
34. E. T. Degens, "Geochemistry of Sediments. A Brief Survey." Englewood Cliffs, N. J., Prentice-Hall, 1965, p. 141.
35. C. Dell, personal communication.
36. J. Kjensmo, "Late and Post-Glacial Sediments in the Small Meromictic Lake Svinsjøen," Arch. Hydrobiol., 65, 125-141 (1968).
37. I. T. Rosenqvist, "Formation of Vivianite in Holocene Clay Sediments," Lithos, 3, 327-334 (1970).
38. S. S. Shukla, J. K. Syers, J. D. H. Williams, D. E. Armstrong, and R. F. Harris, "Sorption of Inorganic Phosphate by Lake Sediments," Soil Sci. Soc. Am. Proc., 35, 244-249 (1971).
39. J. D. H. Williams, J. K. Syers, D. E. Armstrong, and R. F. Harris, "Characterization of Inorganic Phosphate in Noncalcareous Lake Sediments," Soil Sci. Soc. Am. Proc., 35, 556-561 (1971).
40. J. D. H. Williams, J. K. Syers, R. F. Harris, and D. E. Armstrong, "Fractionation of Inorganic Phosphate in Calcareous Lake Sediments," Soil Sci. Soc. Am. Proc., 35, 250-255 (1971).
41. J. R. Kramer, "Equilibrium Concepts in Natural Water Systems," R. F. Gould, Ed. Advances in Chemistry Series No. 67, American Chemical Society, Washington, D.C., 1967, pp. 243-254.
42. J. R. Kramer, in "Systems Approach to Water Quality in the Great Lakes," Proceedings of the 3rd Annual Symposium on Water Resources Research, Ohio State University Water Resources Center, 1967, pp. 27-35.
43. P. D. Snow and D. S. Thompson, "Comparisons of Hydroxy-Apatite Saturations and Plankton Concentrations in Lake Erie," Proc. 11th Conf. Great Lakes Res., Milwaukee, April 1968, pp. 130-136.

44. J. C. Sutherland, J. R. Kramer, L. Nichols, and T. D. Kurtz, "Mineral-Water Equilibria, Great Lakes: Silica and Phosphorus," Proc. 9th Conf. Great Lakes Res., Chicago, March 1966, pp. 439-445.
45. J. D. H. Williams, J. K. Syers, T. W. Walker, and R. Shah, "Apatite Transformations during Soil Development," Atti. Simp. Int. Agrochim., 13, 491-501 (1969).
46. R. C. Gumerman, "Aqueous Phosphate and Lake Sediment Interaction," Proc. 13th Conf. Great Lakes Res., Buffalo, N.Y., April 1970, pp. 673-682.
47. F. R. Hayes and J. E. Phillips, "Lake Water and Sediments: IV. Radiophosphorus Equilibrium with Mud, Plants and Bacteria under Oxidized and Reduced Conditions," Limnol. Oceanogr., 3, 459-475 (1958).
48. R. L. Thomas and A. L. W. Kemp, Unpublished data, Canada Centre for Inland Waters, Burlington.
49. A. L. W. Kemp and Alena Mudrochova, "Electrodialysis: A Method for Extracting Available Nutrients in Great Lakes Sediment," Proc. 14th Conf. Great Lakes Res., Toronto, April 1971, pp. 241-251.
50. J. A. Schleicher and J. K. Kuhn, "Phosphorus Content in Unconsolidated Sediments from Southern Lake Michigan," Illinois State Geological Survey, Environ. Geol. Notes No. 39 (1970).
51. A. L. W. Kemp, "Organic Matter in the Sediments of Lakes Ontario and Erie," Proc. 12th Conf. Great Lakes Res., Ann Arbor, Mich., May 1969, pp. 237-248.
52. A. L. W. Kemp and C. F. M. Lewis, "A Preliminary Investigation of Chlorophyll Degradation Products in the Sediments of Lakes Erie and Ontario," Proc. 11th Conf. Great Lakes Res., Milwaukee, April 1968, pp. 206-229.
53. C. F. M. Lewis, "Sedimentation Studies of Unconsolidated Deposits in the Lake Erie Basin," Ph.D. thesis, University of Toronto, 1966.
54. R. L. Thomas, A. L. W. Kemp, and C. F. M. Lewis, "Distribution, Composition and Characteristics of the Surficial Sediments of Lake Ontario," J. Sediment. Petrol., 42, 66-84 (1972).
55. M. L. Jackson, "Soil Chemical Analysis," Englewood Cliffs, N.J., Prentice-Hall, 1958, p. 175.
56. J. D. H. Strickland and T. R. Parsons, "A Manual of Sea Water Analysis." Fisheries Research Board of Canada Bull. 125, Queen's Printer, Ottawa, 1965, 203 pages.
57. N. C. Mehta, J. O. Legg, C. A. I. Goring, and C. A. Black, "Determination of Organic Phosphorus in Soils: 1. Extraction Method," Soil Sci. Soc. Am. Proc., 18, 443-449 (1954).

58. S. C. Chang and M. L. Jackson, "Fractionation of Soil Phosphorus," Soil Sci., 84, 133-144 (1957).
59. C. R. Frink, "Chemical and Mineralogical Characteristics of Eutrophic Lake Sediments," Soil Sci. Soc. Am. Proc., 33, 369-372 (1969).
60. J. D. H. Williams and Tatiana Mayer, in preparation.
61. J. K. Syers, J. D. H. Williams, A. S. Campbell, and T. W. Walker, "The Significance of Apatite Inclusions in Soil Phosphorus Studies," Soil Sci. Soc. Am. Proc., 31, 752-756 (1967).
62. J. D. H. Williams, J. K. Syers, and T. W. Walker, "Fractionation of Soil Inorganic Phosphate by a Modification of Chang and Jackson's Procedure," Soil Sci. Soc. Am. Proc., 31, 736-739 (1967).
63. E. J. Duff, "Orthophosphates: Part II. The Transformations Brushite → Fluoroapatite and Monetite → Fluoroapatite in Aqueous Potassium Fluoride Solution," J. Chem. Soc., A, 33-38 (1971).
64. N. J. Barrow, "Phosphorus in Soil Organic Matter," Soils Fert., 24, 169-173 (1971).
65. W. C. Krumbein and R. M. Garrels, "Origin and Classification of Chemical Sediments in Terms of pH and Oxidation-Reduction Potentials," J. Geol., 60, 1-33 (1952).

Chapter X

A CHEMICAL MODEL FOR LAKE MICHIGAN POLLUTION: CONSIDERATIONS ON ATMOSPHERIC AND SURFACE WATER TRACE METAL INPUTS

John W. Winchester

Department of Oceanography, Florida State University
Tallahassee, Florida

THE NEED FOR A COMPREHENSIVE CHEMICAL DESCRIPTION OF LAKE MICHIGAN

There are many good reasons for keeping the waters of Lake Michigan clean. We now recognize reasons of public health, economics, and esthetics, and it is very likely that additional reasons will become apparent in future years. The time constant for water quality change in Lake Michigan is very long, and any perturbation we impose on the lake system now will take a very long time to correct through natural processes. Because the ratio of lake volume to outflow rate through the Straits of Mackinac is about 100 years (the longest of the Great Lakes except Lake Superior), water-soluble pollutants have a mean residence time of a century before they are flushed out by water raining on the lake and in the drainage basin. Pollutants that are not water soluble but are taken up by the sediments, either directly or indirectly through biological processes, may remain in the water-sediment system more or less indefinitely, depending on the extent of recycling between the uppermost sediment and the overlying water.

Unfortunately, it is not yet possible to describe in detail the occurrences and behavior of individual toxic trace elements because analytical data are almost completely lacking for important natural and industrial sources and for concentrations in the lake. Nevertheless, this detailed understanding about specific chemical substances is absolutely necessary for making intelligent decisions on water quality management in forthcoming years.

In the recent past we have seen strong evidence that the waters of Lake Michigan are changing at a rapid rate, and these changes may be related to certain changes in lake ecology which have also been observed. Beeton (1) pointed out quantitatively the degree to which Great Lakes waters have increased in concentrations of the major dissolved ions--substances that are not in themselves toxic, although they are red flags of warning about possible increases in more dangerous substances.

In 1967 we witnessed an unprecedented die-off of alewife fish in Lake Michigan. The phenomenon is still not understood; many suspect that the basis may be chemical. Smith (2) has described both this climactic event and the more gradual but by no means slow takeover during the last decade of Lake Michigan's native fish population by the alewife, an intruder from the sea. In 1969 we found that coho salmon in Lake Michigan contained enough DDT to be judged hazardous to health by the U.S. Department of Agriculture. In 1970 mercury contamination of the Great Lakes was discovered, and the extent of this condition in Lake Michigan is currently being evaluated. In the face of such revelations, we can no longer afford to be without at least an elementary chemical discription of Lake Michigan and its trace components.

The prospect of power reactor development on the shores of Lake Michigan raises some special considerations. Since we will surely experience more striking changes in the ecology of Lake Michigan in the coming years, while we are still searching for explanations and trying to develop our ability to predict future changes, we must recognize that such obvious installations as power reactors may be particularly suspect in the public eye. The more quickly we can come to interpret these changes intelligently in terms of a well-defined model for processes of pollution and environmental change, the more successful we will be in making appropriate public policy decisions.

A chemical model or description of Lake Michigan should include, probably as a first step, an evaluation of all natural and anthropogenic sources of the inorganic and the organic substances of interest. There will be natural sources of surface water runoff and of fallout directly from the atmosphere whose time variations may be predictable and perhaps small. There will also be large water and air pollution sources whose strengths, if they are not relatively constant, at least will vary according to readily obtainable indexes. Additional pollution sources may be found with more erratic time variations, but every attempt should be made to assess their importance over long time periods. Finally, there may be sources operative in the future that are not yet prominent today. These may be the pollution hazards of tomorrow and we may be called upon to control them.

In such decisions our chemical model will be vitally important. Some of the hazards may be gradual, as has been the addition of DDT to Lake Michigan during the past 20 years. Others may be sudden, such as spillage from a ship carrying a toxic cargo or the rupture of a container carried by freight train or truck on a right-of-way close to the lake shore. A useful chemical model must be designed to answer questions about possible contaminants of the future and to describe trends of the past.

For each substance considered in the model the major sinks should be determined, and by a comparison of sources and sinks a residence time and steady-state concentration in the lake may be ascertained. Some trace metals are attractive subjects because of the importance of their potential toxicity, the chance of large contamination from industrial sources, and the relative simplicity of investigation by sensitive and specific methods of analytical chemistry. Some elements exhibit strongly anionic tendencies in aqueous solution (e.g., Mo and U) and may not be strongly complexed by organic ligands in solution or attracted to clay mineral surfaces by ion exchange. For such elements we may expect the principal sink to be outflow, and the mean residence time in Lake Michigan should be about 100 years. If natural sources operate at a constant rate for at least several centuries, the concentrations in the lake should reach steady-state values. A comparison of all natural sources today with the observed concentrations in the lake can indicate whether these elements indeed behave chemically as predicted and the extent to which additional pollution sources have become significant.

Elements occurring as cations in aqueous solution may exist mainly as complexes with dissolved organic material in the water, or they may be concentrated to a high degree by living organic material. In either case a full description of their chemical behavior is difficult without considerable knowledge of the organic and biological components of the water. For this reason, chemical studies should be carried out in conjunction with biological studies emphasizing the chemical environment.

Elements that are strongly held either as dissolved complexes or as components of organic or inorganic particles may have a rate of removal from the water that is more rapid than by dilution and outflow. As a result, the steady-state aqueous concentration may be held at a low level with a time constant for recovery from perturbation much shorter than 100 years. On the other hand, the lake-sediment system considered as a whole may retain these elements as long as the lake exists as a body of water--tens of thousands of years--and there is not effective cleansing of the lake by natural processes. Most toxic metals may be in this category.

Clearly, a basic chemical description or model of Lake Michigan should be a major goal in a comprehensive ecological research program.

SURFACE WATER SOURCES OF TRACE METALS IN LAKE MICHIGAN

Apparently the most extensive study of sources of toxic trace metals to Lake Michigan was published by the Federal Water Pollution Control Administration (FWPCA) in January 1968 (3). Copper, nickel, zinc, chromium, cadmium, and lead were measured, using polarography sensitive to a concentration of 10 μg/liter in waters of most of the streams tributary to Lake Michigan. Water from Lake Michigan was also extensively sampled, but concentrations were generally below the sensitivity limit. For the streams, the results for lead were not published, and cadmium and chromium were detected only in a few cases. However, since copper, nickel, and zinc were detected in nearly all the samples of the tributary streams, a total input rate of these elements from this source could be calculated. Table 1 summarizes the data reported. In most cases each stream was sampled from 18 to 55 times over a one-year period, and the sampled streams drain about 50% of the drainage basin. Note that zinc and nickel in these waters were usually close to the sensitivity limit, but copper was easily measured.

Table 1 shows mean concentrations observed from which the loading or elemental mass flow is calculated in metric tons per year from each river. If the average of the sampled streams is the same as that for the unsampled drainage, we estimate the total loading into Lake Michigan to be 2700 tons Cu/yr, 760 tons Ni/yr, and about 500 tons Zn/yr. This leads us to ask whether these values are natural or are influenced by pollution and to wonder what relation they bear to the concentrations of these elements in the lake itself.

Turekian (5) has published a valuable compilation of the concentrations of trace elements in unpolluted streams. Table 2, column 1, presents his best estimates for elements 21 to 92 in streams from various areas; these may serve as estimates for the composition of the streams draining the Lake Michigan basin before water pollution became serious. From these we can estimate some natural properties of Lake Michigan and compare present day conditions.

Column 2 of Table 2 gives maximum permissible concentrations for public water supplies according to the FWPCA (7), and column 3 reveals that these water quality criteria are regularly above the stream averages (except iron, which is complicated by solubility considerations). The geometric mean of the permissible:stream ratio is 72 (55 excluding iron and uranium). Referring to Table 1, in no case did copper or zinc come close to the maximum permissible values, but a standard for nickel has not been established. It is interesting that the FWPCA (3) reported 20 μg Cd/liter in the St. Joseph River and 40 μg Cr/liter in the Grand River, although in all other rivers cadmium and chromium were not detected. If these numbers are valid, they are close to or in excess of the maximum permissible concentrations for these elements in public water supplies.

If the stream averages of Turekian are applicable to the natural Lake Michigan basin, we can calculate the expected natural inputs of all these elements to the lake, as shown in column 4 of Table 2. We now compare the measured inputs in column 5 for copper, nickel, and zinc and the limits of chromium and cadmium. For zinc the measured input (ca. 500 tons/yr) is very close to the expected natural level of 650 tons/yr, leading us to consider the measured zinc to be natural and not mainly due to pollution. The input of copper, on the other hand, exceeds the expected natural input by 12 times, suggesting that more than 90% of present copper input is pollution derived. For nickel the observed input exceeds the expected natural input 76 times; thus nearly 99% appears to be derived from pollution.

Although these estimates are based on rather fragmentary data, the calculations indicate the value of good estimates of natural as well as total inputs of trace elements to Lake Michigan. Our tentative conclusion--

Table 1. Chemical Results of Toxic Metals, Lake Michigan Tributaries

River	Mean flow		Number of samples	Mean concentrations,[a] μg/liter			Loading,[b] tons/yr		
	ft^3/sec	km^3/yr		Cu	Ni	Zn	Cu	Ni	Zn
Manistique[c]	845	0.755	31	80	20	20	60	15	15
Manitowoc[c]	83	0.074	41	110	30	30	8	2	2
Sheboygan[c]	132	0.118	39	90	140	110	11	17	11
Milwaukee[c]	191	0.171	32	110	40	60	19	7	10
Burns Ditch[c]	150	0.134	40	70	30	ND	9	4	<1
St. Joseph[d]	2,060	1.842	55	80	10	30	150	18	58
Kalamazoo[d]	1,140	1.020	53	70	30	ND	71	31	<10
Grand[d]	1,900	1.700	52	140	40	ND	240	68	<17
Muskegon[d]	1,731	1.549	54	110	30	ND	170	46	<15
Père Marquette[d]	570	0.510	53	120	50	30	61	26	15
Fox[e]	4,420	3.956	18	90	20	ND	360	80	<40
Oconto[e]	790	0.706	42	10	30	30	7	21	21
Peshtigo[e]	890	0.796	44	150	30	40	120	24	32

Menominee[f]	3,250	2.910	NS	--	--	--	--	--	--
Ford[e]	337	0.302	45	60	30	30	18	9	9
Escanaba[e]	1,017	0.910	44	90	30	30	82	27	27
Rapid[f]	80	0.072	NS	--	--	--	--	--	--
Whitefish[f]	227	0.203	NS	--	--	--	--	--	--
Total sampled[g]	19,813	17.728		92	37	30	1,386	395	200-283
Estimated total[h]	31,400	28.1					2,700	760	∿500

[a] NS = not sampled; ND = not detected, less than 10 μg/liter.
[b] Elemental mass flow = concentration × mean flow of river, metric tons/yr.
[c] June 1963 to April 1964.
[d] March 1963 to April 1964.
[e] June 12, 1963 to May 6, 1964.
[f] June 12, 1963 to October 24, 1963.
[g] Mean gaged discharge over all years of record = 25,501 ft^3/sec from gaged drainage area of 31,940 $mile^2$. Flow considered = 90% of total from 92% of gaged area. Total area of Lake Michigan basin = 67,900 $mile^2$ = 22,400 $mile^2$ Lake Michigan + 45,500 $mile^2$ land. Twenty major rivers drain 36,400 $mile^2$ of which 31,940 $mile^2$ are gaged (i.e., 70% of Lake Michigan basin) (4).
[h] Assuming total flow proportional to area drained and loading proportional to flow.

Table 2. Natural Stream Inputs of Trace Elements to Lake Michigan

Element[a]	Average natural streams[b] μg/liter	Maximum permissible[c] μg/liter	Ratio permissible ÷ streams	Natural inflow[d] tons/yr	Measured inflow[e] tons/yr	Natural Δ(conc.),[f] ng/(liter)(yr)
21 Sc	0.004			0.13		0.027
22 Ti	3			100		20
23 V	0.9			29		6
24 Cr	1	50	50	33	>300	6.7
25 Mn	7	50	7	230		47
26 Fe	(670)	300	(0.5)	(22,000)		(4,500)
27 Co	0.2			6.5		1.3
28 Ni	0.3			10	760	2.0
29 Cu	7	1000	140	230	2700	47
30 Zn	20	5000	250	650	∿500	130
31 Ga	0.09			3		0.6
32 Ge	*,g			*		*
33 As	2	50	25	65		13
34 Se	0.2	10	50	6.5		1.3
35 Br	20			650		130
37 Rb	1			33		6.7
38 Sr	60			2000		400
39 Y	0.7			23		4.7
40 Zr	*			*		*
41 Nb	*			*		*
42 Mo	1			33		6.5
44 Ru	*			*		*
45 Rh	*			*		*
46 Pd	*			*		*
47 Ag	0.3	50	170	10		2.0
48 Cd	*	10	?	*	<300	*
49 In	*			*		*
50 Sn	*			*		*
51 Sb	1			33		6.7
52 Te	*			*		*
53 I	7			230		47
55 Cs	0.02			0.65		0.13
56 Ba	10	1000	100	330		67
57 La	0.2			6.5		1.3
58 Ce	0.06			2.0		0.4
59 Pr	0.03			1.0		0.2

60 Nd	0.2			6.5	1.3
62 Sm	0.03			1.0	0.2
63 Eu	0.007			0.23	0.047
64 Gd	0.04			1.3	0.27
65 Tb	0.008			0.26	0.053
66 Dy	0.05			1.6	0.33
67 Ho	0.01			0.33	0.066
68 Er	0.05			1.6	0.33
69 Tm	0.009			0.29	0.060
70 Yb	0.05			1.6	0.33
71 Lu	0.008			0.26	0.053
72 Hf	*			*	*
73 Ta	*			*	*
74 W	0.03			1.0	0.2
75 Re	*			*	*
76 Os	*			*	*
77 Ir	*			*	*
78 Pt	*			*	*
79 Au	0.002			0.065	0.013
80 Hg	0.07			2.3	0.47
81 Tl	*			*	*
82 Pb	3	50	17	100	20
83 Bi	*			*	*
90 Th	0.1			3.3	0.67
92 U	0.04	5000	125,000	1.3	0.27

[a]Lake Michigan concentrations of Br = 20 μg/liter and I = 1 μg/liter have been reported by Tiffan[illegible] et al. (6) and of Cr, Ni, Cu, Zn, and Cd, each <5 μg/liter have been reported by the FWPCA (3)[illegible]

[b]Turekian (5).

[c]FWPCA (7).

[d]Expected natural inflow to Lake Michigan, metric tons/yr, column 1 × 32.6 km^3/yr aver[illegible]

[e]FWPCA (3).

[f]Expected natural inflow expressed as concentration increment, column 4 ÷ 4900 k[illegible]

[g]Asterisk indicates no data or reasonable estimates are available.

mely, that stream-carried copper and nickel are largely ollution derived in the Lake Michigan basin, but that inc may be largely natural--was not made by the FWPCA in their report (3).

We have already noted that maximum permissible concentrations for public water supplies averaged only 72 times the expected natural stream concentrations, and we have also observed that current inputs for copper and nickel are 12 and 76 times higher than the expected natural rates to Lake Michigan. Thus we should be alert to the possibility that many more toxic metals may be flowing into Lake Michigan at rates far above their natural rates and approaching or exceeding safe levels from a human public health standpoint. Moreover, aquatic life may be more seriously threatened than human life by toxic trace metal pollution, and an accurate assessment of this threat is an essential part of any study of changes in ecological balance in Lake Michigan.

We can now compute an expected natural incremental increase in concentration assuming natural stream inflow and no outflow (Table 2, column 6). These numbers are of interest mainly when compared with measured concentrations in the lake, but such data are still scarce. With more lake water data we could decide whether existing concentrations can be established naturally in times shorter than the dissolved substance residence time of 100 years and whether a principally sedimentary removal route or a largely outflow route is implied. Too few trace element data are yet available for this to be done well.

AIR POLLUTION SOURCES OF TRACE METALS IN LAKE MICHIGAN

The presence along the southwestern shore of Lake Michigan of the Greater Chicago urban and industrial area, extending from southeastern Wisconsin to northwestern Indiana, makes it necessary to ask whether direct fallout of air pollution can be an important source of lake water contamination. This possibility is maximized by (1) the generally upwind location of Chicago relative to the lake, (2) the large lake water surface exposed to the atmosphere, (3) the comparatively small direct sources of water pollution in this area since the Chicago and Calumet rivers were reversed a half-century ago to flow out of Lake Michigan, (4) the century-long mean hold-up time of water in the lake before outflow, and (5) the possibility of efficient atmospheric fallout of particulate material, even in clear air, during its long trajectory over the water surface.

In order to make a quantitative study, Winchester Nifong (8) prepared an approximate inventory of 30 elements that were expected as components of air pollution particulates from the region. It is not possible to derive source strengths directly from measurements of air pollution, since dilution volumes must be known, as well as concentrations in air. Therefore, the best estimates of total particulate emissions were combined with available data on typical elemental compositions of emissions from several major sources. The results of this inventory estimate for the entire region are shown in Table 3. It should be emphasized that these estimates are lower limits, inasmuch as some large sources of certain elements may have been omitted because of lack of data, and uncertainties may exist if the emissions in this area are in any way atypical in composition. Nevertheless, this approximate inventory is a necessary first step in assessing the water pollution potential of air pollution.

A quantitative test of the adequacy of the elemental inventory is the comparison of relative composition with chemical analyses of air pollution particulates. Winchester and Nifong (8) showed comparisons of the totals in Table 3 with data of the National Air Sampling Network (NASN), and agreement within groups of elements due mainly to one of the source types was generally satisfactory. However, when all the elements were considered together, there was evidence of a bias against elements from the iron and steel industry, which is located generally downwind of the NASN sampling locations.

A more complete survey has been made by Harrison et al. (9) and Dams et al. (10) in northwest Indiana, where air was sampled at 10 in-city stations and 15 outlying stations for 24 hr (June 11-12, 1969). The in-city stations may give a more representative picture of the composition of urban sources and a more useful comparison with the inventory calculations. Table 4 presents elemental emissions from northwest Indiana alone, calculated from data recorded by Winchester and Nifong (8), along with the geometric mean concentrations observed in the in-city stations during the 24-hr survey. The ratios of emission rates to atmospheric concentrations generally lie between 1 and 10 in the units of (tons/yr)/(ng/m^3). Low values are often recorded for elements where natural sources may contribute substantially (Na, K) or where some likely industrial sources could not be considered because of lack of data (K, Cr, Br, Ag). For the most part, the ratios for other elements cluster well within an order of magnitude of each other, lending credence to the notion

Table 3. Calculated Particulate Emission of Trace Elements from Air Pollution Sources in Chicago, Milwaukee, and Northwest Indiana, after Winchester and Nifong (8)

Element	Emission rate, short tons/yr						
	Coal	Coke	Fuel Oil	Iron and Steel	Cement	Transportation	Total
Be	5	1					6
B			4				4
Na	1,120	80	274				1,500
Mg	2,090	150	56	3,360	500		6,100
Al	40,000	2,800	955	4,950	1,000		50,000
Si	55,800	4,000	433	850	4,000		65,000
P				170			170
S[a]	[550,000]	[40,000]	[150,000]	[1,000]		[7,000]	[750,000]
Cl						700	700
K			16				16
Ca	10,600	760	74	11,600	17,600		41,000
Ti	2,390	170	6				2,600
V	209	15	450				670
Cr	80	6	22				110
Mn	67	5	6	5,000			5,100
Fe	19,500	1,400	450	73,000	930		95,000
Co	25	2	28				55
Ni	109	8	1,030				1,100
Cu	109	8	28	3,400			3,500
Zn	247	18	9	4,000			4,300
As	46	3					49
Se	5.5	0.4	15			1	22
Br						800	800
Sr			6				6
Mo	45	3	3				51
Ag		3					3
Cd	12	1					13
Sn	13	1					14
Ba	176	12	45				230
Pb	328	24	33			2,000	2,400

[a]Total emission of sulfur, gaseous and particulate.

Table 4. Calculated Particulate Emission of Trace Elements from Air Pollution Sources in Northwest Indiana Alone, Based on Winchester and Nifong (8)

Element	Emission rate, short tons/yr							Station 1-10 ng/m^3 observed	Ratio calc./observed
	Coal	Coke	Fuel Oil	Iron and Steel	Cement	Gasoline	Total Calc.		
Be	1	0.35					1.35		
B			0.8				0.8		
Na	225	53	48				326	285	1.14
Mg	420	100	10	2,775	330		3,635	1,350	2.7
Al	8,000	1,900	170	4,000	660		14,730	1,850	8.0
Si	11,200	2,700	76	673	2,600		17,250		
P				135			135		
S[a]	[40,000]	[27,000]	[30,000]	[700]		[500]	[98,000]	11,500	[8.5]
Cl						35	35		
K			3				3	1,250	0.0024
Ca	2,100	510	13	9,650	11,700		24,000	3,950	6.1
Ti	480	115	1				596	190	3.1
V	42	10	80				132	9.3	14.2
Cr	16	4	4				24	54	0.44
Mn	13.5	3.5	1	4,050			4,068	180	22.6
Fe	4,000	950	80	58,000	600		63,630	6,500	9.8
Co	5.1	1.3	5				11.4	1.5	7.6
Ni	22	5.3	188				215	<50	>4.3
Cu	22	5.3	5	2,815			2,847	380	7.5
Zn	50	12	1.6	3,250			3,314	510	6.5
As	9.3	2					11.3	4.2	2.7
Se	0.4	0.3	3			0.1	4	2.6	1.5
Br						39	39	94	0.4
Sr			1				1		
Mo	9	2	0.6				11.6		
Ag			0.6				0.6	1.9	0.3
Cd	2.5	0.7					3.2		
Sn	2.6	0.7					3.3		
Ba	35	8	8				51		
Pb	66	16	6			145	233		

[a]Brackets indicate total emission of sulfur, gaseous and particulate.

that the inventory does account for the major sources of these elements. We may then assume that the calculated inputs of these elements, in tons per year, are realistic estimates.

Returning now to Table 3, we can compare atmospheric contributions of trace elements with actual river inputs to Lake Michigan from Table 1. At present this can be done only for copper, nickel, and zinc. For copper and nickel the two source strengths are comparable, but for zinc the atmospheric source is close to an order of magnitude stronger. When compared with natural river inputs (Table 2) the atmospheric contributions are seen to be stronger for all three elements.

Source, metric tons/yr	Cu	Ni	Zn
Air	3200	1000	3900
Water	2700	760	∿500
Natural water	230	10	650

Consequently, air pollution appears to be a substantial source for these elements where we may make a direct comparison with measured river inputs; it may be considerable in comparison with estimated natural river inputs for about one-third of 30 elements evaluated by Winchester and Nifong (8).

Although it is clear that for many elements trace metal contamination of the air over or near Lake Michigan may be at least comparable to that being put into the lake by river flow, it is not easy to estimate the transfer efficiency of particulates from the air to the lake surface. Data on mean wind directions show that near surface air moves from the city to the lake well over half the time, but the deposition rate of pollution particulates has apparently never been measured in this vicinity.

If the transfer mechanism were only through scavenging by rain or snow, the efficiency would be very small, since the long-term precipitation probability in a given location over the lake is small. However, as pointed out by Winchester and Nifong (8), empirical deposition velocities from clear air have been estimated to be 0.3 to 0.4 cm/sec for ^{131}I vapors from the British Windscale nuclear reactor accident and about 0.7 cm/sec for sea salt particles of >1 μ radius over the ocean.

Therefore, an average deposition velocity of about 0.5 cm/sec for pollution particulates from clear air over Lake Michigan is not unreasonable to expect. Such a velocity implies that particles initially at 100 m altitude would reach the earth's surface in 20,000 sec (5.6 hr). A wind of 5 m/sec blowing from the west would not be able to carry the particles more than 100 km, roughly the width of the lake, before the particles reached the surface. By this crude calculation, we see that the transfer probability of particles from air to water may be substantial.

As far as Lake Michigan is concerned, a comprehensive chemical model of sources, dispersion, and sinks of trace elements must include quantified atmospheric as well as water-borne input data. The vulnerability of the lake to near-permanent damage from contaminants because of the very slow natural flushing rate makes the need for such data particularly urgent at this time.

REFERENCES

1. A. M. Beeton, "Eutrophication of the St. Lawrence Great Lakes," Limnol. Oceanogr., 10, 240-254 (1965).
2. S. H. Smith, "That Little Pest, the Alewife," Limnos, 1 (2), 13-20 (1968).
3. Federal Water Pollution Control Administration, "Water Quality Investigations, Lake Michigan Basin, Physical and Chemical Quality Conditions." U.S. Department of Interior, Chicago, 1968.
4. Federal Water Pollution Control Administration, "Water Pollution Problems of Lake Michigan and Tributaries." U.S. Department of Interior, Chicago, 1968.
5. K. K. Turekian, "Elements, Hydrosphere Distribution of," in "1969 Yearbook of Science and Technology." McGraw-Hill, New York, 1969.
6. M. A. Tiffany, J. W. Winchester, and R. H. Loucks, "Natural and Pollution Sources of Iodine, Bromine, and Chlorine in the Great Lakes," J. Water Pollut. Control Fed., 41, 1319-1329 (1969).
7. Federal Water Pollution Control Administration, "Report of the Committee on Water Quality Criteria--Surface Water Criteria for Public Water Supplies." U.S. Department of Interior, Washington, D.C., 1968.
8. J. W. Winchester, and G. D. Nifong, "Water Pollution in Lake Michigan by Trace Elements From Pollution Aerosol Fallout," Water, Air, Soil Pollut., 1, 50-64 (1971).
9. P. R. Harrison, K. A. Rahn, R. Dams, J. A. Robbins, J. W. Winchester, S. S. Brar, and D. M. Nelson, "Areawide Trace Metal Concentrations Measured by Multielement Neutron Activation Analysis--A One Day Study in

Northwest Indiana," J. Air Pollut. Control Assoc., 21, 563-570 (1971).

10. R. Dams, J. A. Robbins, K. A. Rahn, and J. W. Winchester, "Quantitative Relationships Among Trace Elements Over Industrialized N. W. Indiana," in "Nuclear Techniques in Environmental Pollution." International Atomic Energy Agency, Vienna, 1971, pp. 139-157.

Chapter XI

DETERGENT DEVELOPMENTS AND THEIR IMPACT ON WATER QUALITY

J. R. Duthie

Environmental Water Quality Research Department
Procter & Gamble Company, Cincinnati, Ohio

The ubiquitous presence of detergent products in American homes and industries, and their normal disposal in wastewater make it appropriate to consider the impact of these products on water quality. For many years, the detergent industry has made it a practice to assess carefully all safety aspects of the products, the materials used in them, and the materials being considered for use in them. The human safety record of detergent products has long been outstanding, and the industrywide changeover in surfactants to more degradable forms has often been cited as an example of responsible action in the environmental area (1,2).

More recently, the importance of the nutrient contribution of detergent products has been hotly debated in public forums. In these debates the control of phosphate from detergent sources has been overemphasized, and insufficient attention has been paid to alternate contributions of phosphate (3) and other nutrients from human wastes, from agricultural activities, from land runoff, and from other sources. Certainly more complete phosphate removal is possible by chemical precipitation in waste treatment, and this added process step generally provides added benefits of BOD reduction and suspended solids removal. Nevertheless, pressure to reduce or remove phosphates from detergents has made this the overwhelming technical objective of the industry. Without arguing the merits of such a product change, some facts are presented about the overall assessment of detergent materials, along with a discussion of new developments and their potential impact.

DETERGENT BACKGROUND

The detergent industry, made up of formulator/marketer companies and chemical/petroleum companies, produces annually about 6 billion lb of cleaning products. Products for every conceivable cleaning purpose are in widespread daily use in homes, hospitals, hotels, restaurants, and schools and they are bought by every municipality, county, and state. Detergents are fundamental to operations of many industries, textile and food processing being just two important examples. Although individual product applications result in a wide variety of final formulations, some generalizations can be made about the kinds of materials used in most detergent products (4).

Surface Active Agents

A variety of the surface active agents, or surfactants, are used, and many products contain mixtures tailored for a specific purpose or set of properties. Anionic types dominate in household products with linear alkylate sulfonate (LAS) being the best known and most widely used. Other important anionic classes include alcohol sulfates, sulfates of alcohol ethoxylates, and alkyl glycerol sulfonates. Nonionic-type surfactants are also used and now make up about 20% of the overall surfactant total. Nonionic agents are more diverse in structure, and proportionately they find greater usage in specialty products and industrial usage. Important nonionic types include the alcohol ethoxylates, the alkyl and alkanol amides, and the amine oxides and condensates.

The selection of the surfactant combination for a particular application depends on many factors, including the surface to be cleaned, the temperature, the water conditions, the use concentration, the type of soil to be removed, and the desired foaming characteristics. In general, surface active agents provide a reduction in the surface tension of water and a reduction of interfacial tension between soils and surface to be cleaned, thereby aiding in the removal, emulsification, and suspension of the soil.

Builders

The second class of compounds comprises an important portion of most detergent formulations designed for heavy cleaning. The addition of so-called builders to a product conditions the water and increases the effectiveness of the surface active agent. Unless the calcium and magnesium ions and the trace amounts of iron and manganese contained in the wash water or the soil are properly sequestered, or tied up by builders, they can adversely affect the performance of washing product.

Complex phosphates are nonprecipitating builders and have been broadly utilized in detergents. In fact, consumer acceptance of detergents for laundry largely depend on the commercial development of sodium tripolyphosphate (STP). Previously, household detergents were used only in light-duty applications such as hand dishwashing and fine fabric washing. Phosphates additionally contribute a number of other useful properties, including the following:

1. The ability to peptize and suspend certain kinds of particulate matter.
2. The ability to decrease the critical micelle concentration of surfactants and, thereby, making the surfactants more efficient.
3. The ability to buffer the washing solution at a pH that aids in fatty soil removal but is safe for normal surfaces, fibers, colors and, above all, the skin of the user.

Other Materials

Additional inorganic salts may be present for a variety of reasons. Sodium sulfate normally occurs in laundry detergents at around 15% as a by-product of the surfactant-making process. Its positive effect in the presence of truly effective detergent builders is modest, and it must be considered as essentially inert. Another inorganic additive in laundry detergents is silicate, which provides corrosion protection, basicity, and improvement in the physical properties of detergent granules. Inorganic bleaching compounds like sodium perborate or chlorinated trisodium phosphate may be present in specific applications.

Organic additives may include enzymes, fluorescent whitening agents, sanitizing agents, and perfumes. These latter materials are present in relatively low amounts and their addition depends on the specific purpose and usage of the product.

Synthetic detergents have largely replaced soaps for all cleaning product usages except personal cleanliness, where bar soaps predominate; light-duty washing, where soap flakes or granules persist; and in commercial laundry applications, where soap can be effective when accompanied by chemical water softening and other supplementary processes. Soap is discussed further in connection with formulation alternatives in future products.

QUALIFICATION OF A DETERGENT MATERIAL

The qualification by Procter & Gamble of a new material or a basically new detergent formula divides itself into five main areas: performance, compatibility, human safety, environmental effect, and feasibility. The kinds of testing that go into assessment in each area are outlined in the following sections.

Performance

It is the job of the detergent manufacturer to provide products that will perform well regardless of user interest, difficulty of the task, or the range of usage conditions encountered. Complete and comprehensive pretesting of formula changes is necessary to assure, for example, that a new laundry product will fulfill the following requirements:

1. Give good removal of lipid, food, and particulate soils.
2. Prevent graying or yellowing from redeposition of removed soil or reaction products of water impurities.
3. Assure good fabric whiteness and color maintenance.
4. Retain good softness and odor.

Reliable detergency evaluations, therefore, require repeated testing on a variety of naturally soiled clothes and household linens under many washing conditions. Reliance on a single-cycle laboratory washing test using an artificially soiled cloth is obviously inadequate.

Although laundry products for home use are the largest single class of detergent products, there are, of course, products used for a variety of cleaning tasks in hospitals, hotels, restaurants, schools, food processing plants, textile mills, and many other situations. It is important to remove germs from household linens, diapers, and so on. Yet effective cleaning for sanitation and health reasons can be even more critical in institutional applications. Appropriate assessment of performance for each application is a part of overall qualification of any new product developed for these uses.

Compatibility

Detergent products must be compatible with materials of construction of washers and dishwashers and with the growing variety of textiles, utensils, and surfaces they are to clean. The evaluation includes small-scale laboratory testing, realistic long-term washings, controlled home-use studies, and finally careful follow-up on shipping tests or test markets of all new products. The compatibility of detergents with equipment deserves special mention in the sanitation of clean-in-place installations, which are used extensively in the food and dairy industries. Here the nonprecipitating properties of complex phosphates prevent the accumulation of troublesome deposits due to reaction with hardness ions that were present originally in the water or contributed by the milk or food. This assures clean surfaces that are easily sanitized without disassembling for laborious and less reliable hand cleaning.

Human Safety

The human safety record of detergent products has been outstanding despite almost universal exposure to these products and rather casual storage around homes and institutions. An overall plan for arriving at safety judgments on detergent products has been discussed by Carter (5), and he has illustrated it by means of a block diagram showing the types of information needed and their interaction. Figure 1 presents his idealized, and oversimplified plan, indicating the types of biological and other data to be considered and how information is used to arrive at a final safety judgment. It is unlikely that any actual material will follow this pattern exactly, so this chart should be considered as a model and not as a rigid evaluation procedure.

The individual parts of this plan are covered in

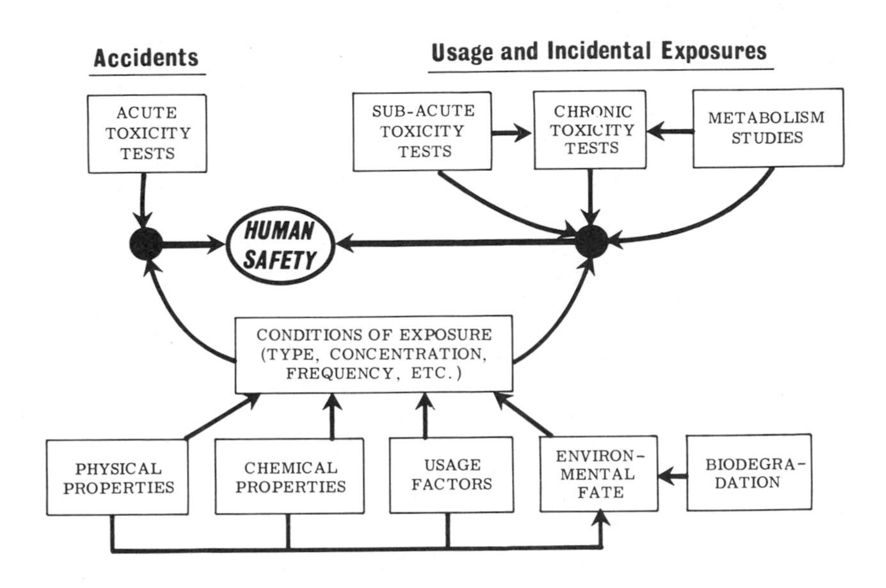

Fig. 1. Safety assessment of soap and detergent materials.

somewhat more detail, but a brief overview may be useful. At the top of the diagram appear the types of biological data to be considered. At the bottom are indicated various factors that are considered important in estimating exposure conditions. Acute toxicity tests are valuable to permit estimates of the effects of accidental exposures, as, for example, ingestion by children. The subacute and chronic studies aid in estimating hazards stemming from intentional use exposures and from such unintentional exposures as might occur for various reasons (e.g., contact with residues remaining on dishes or with traces of the material itself or its decomposition or reaction products occurring in water). More detailed consideration of the individual portions of Carter's model follows.

Physical and Chemical Properties

The initial data sought in the safety evaluation are those relating to the material's physical properties and form, its chemical nature, and the type and condition of the proposed use. Physical form and proposed use are particularly relevant to the assessment of accidental hazards because they largely determine the types of accidents that might occur and the probable frequencies of each.

The characterization and chemical properties of the material are important in planning subsequent biological tests, but toxicological judgments seldom can be made solely on the basis of chemical structure. Information regarding types of usage and usage concentrations is essential for relating practical exposures to test conditions.

Environmental Fate

The environmental fate of a new material is important in its own right, and assessment in this category is discussed later. It is important that biodegradability, chemical lability, and other factors be considered in determining human exposure parameters both for proper planning of chronic toxicity tests and in interpreting test results.

Acute Toxicity Tests. The first toxicity tests relate to the acute type of exposure. These relatively quick tests on animals serve to highlight any important negatives. Falling into three groups involving ingestion, eye instillation, and skin contact, the acute toxicity tests supply the basis for estimating the consequences of various types of exposure to high concentrations of the material. It is useful to determine the emetic dose because man's well-developed emetic reflex eliminates today's synthetic detergents promptly and rather completely from the stomach; thus it can act as a built-in safety factor against toxic consequences of accidental ingestion.

With some formulations it is important to investigate the local effect of ingested material on the lining of the esophagus and stomach. Highly alkaline materials like those of some nonphosphate detergent brands have the capability of producing corrosion, prolonged irritation, or scarring, thereby raising questions (6) about the safety of such formulations and possibly requiring precautionary labeling under the Federal Hazardous Substances Labeling Act.

Subacute Toxicity Tests. Following completion of the acute toxicity tests, subacute animal studies involving prolonged exposures of up to approximately three months are begun. At this phase of testing, emphasis is heavily focused on skin and percutaneous toxicity, and the potentials for contact sensitization and photosensitization are determined. Subacute feeding tests are designed to reveal which sites in the animal body are first to be adversely affected by the material. In addition to providing significant data regarding systemic toxicity, these tests supply information to aid in the design of chronic feeding studies aimed at establishing no-effect levels for long-term or life-span exposures.

Chronic Studies. Chronic feeding studies on animals help to assess the effects of low-level, long-term exposure and to define the maximum level at which no adverse effect is produced. Levels in a properly designed chronic feeding study should include at least one that is high enough so that a clearcut adverse effect stemming from the treatment is evident. This identifies the organ system or tissue that is most sensitive to the toxic effect of the material. There should also be a progression of lower levels, and the lowest, at least, should have no effect on the most sensitive target system.

The test material may be administered either in water or in diet, and the period of experimentation is usually extended for the lifetime (2 years) of the small laboratory rodent or for two or more years for nonrodent species. The animals are observed for growth, and all the normal parameters of health and the results of detailed pathological examinations are evaluated. The experimentally determined no-adverse-effect level is then related to the estimated maximum human intake levels through a calculated factor of safety.

Additives for general food use are generally accepted as safe if the no-effect level established in animal studies exceeds the estimated human intake a hundred times or more. This food factor provides a conservative guideline for evaluation of detergent products. The potential of ingestion or other exposure in relation to carcinogenicity, teratogenicity, and mutagenicity is also determined.

Environmental Effect

Because detergents are being used universally, because they necessarily contain materials that are effective at relatively low washing concentrations, and because they are normally discharged to wastewater systems, their potential impact on the water environment and on the reuse of our waters for drinking supply must be assessed.

It may be useful to consider for a moment the likely concentration of detergent ingredients in wastes and water. In the United States, a broadly used class of surfactant or detergent builder will appear in the influent to a treatment plant at about 1/20,000 of the concentration that appears in products (7). A surfactant then, used at 15 to 20% in a detergent product, will appear in untreated wastewater at levels of 8 to 10 mg/liter. Phosphates from detergents are used at a higher level and give a wastewater concentration of about 20 mg/liter (5 mg/liter on a P basis).

Our testing and judgments in the environmental area usually can be grouped into five general categories:

1. Biodegradability. Both completeness and rate of biodegradability are important, since degradation by soil and water microorganisms offers the most certain assurance the material will not build up in concentrations likely to cause an adverse effect. A variety of laboratory tests, including BOD and carbon dioxide evolution, furnish preliminary information on degradability and highlight any areas where special waste treatment testing may be required.

2. Treatability. The more practical treatability studies start in the laboratory with simulations of various aerobic degradation processes and the determination of the partition of the material in various physical separation processes. They include the determination of any adverse effect on treatment processes and special studies relating to household treatment systems and to anaerobic digestion. It is useful to extend beyond the laboratory to testing in pilot or full-scale systems such as activated sludge treatment plants, septic tanks, and trickling filters, in order to be sure that no important parameters have been overlooked in scaleup. Although septic tank vaults are often considered to be anaerobic, these household units do receive regular discharges of oxygen-saturated wastewater; therefore, degradation of compounds like trisodium nitrilotriacetate (NTA) can be substantial in these receptacles.

3. Environmental fate. The environmental fate of a material depends now, and hopefully will depend more in the future, on its degradation or removal in treatment situations. Since currently there is a very real possibility that a certain percentage of household wastes either bypasses treatment or has no treatment facilities available, a part of such studies is devoted to the biological degradation in natural waters and soils, and the adsorption or interaction with sediments and soils.

Finally, the levels in streams and lakes that occur from a preliminary or test scale usage of a new material are monitored.

4. Aquatic safety. Obviously, earlier portions of the assessment allow an estimate of the concentrations that might occur or persist in the environment. The first step in aquatic safety studies involves acute bioassays using a variety of species including bacteria, algae, invertebrates, and fish. If it is possible that significant concentrations of a new material may persist in the environment, chronic studies should be considered. Accumulation studies or interaction studies may also be appropriate. Such a life-cycle study by the FWQA (8) using LAS, the most widely employed surfactant of the industry, showed no indication of any chronic effect. With NTA there have been indications that the toxic effect of metals on aquatic organisms is diminished by a chelating agent (9).

5. Eutrophication assessments. Eutrophication assessments are designed to make it possible to judge whether the contributed level of the material or its degradation products will stimulate an undesirable level or an unbalanced type of growth. Programs for specific materials are relatively new, and to some degree procedures are still being developed. Materials of the detergent industry have served as the models for much of the thinking and testing that have been done in this area.

In the assessment of detergent materials, or in fact any materials, there are pitfalls to avoid in setting up a testing protocol and in making judgments. Likely environmental levels and the presence of other materials that might accompany the one being tested must be considered. It is obvious to knowledgeable researchers that simple addition of the test compound at highly exaggerated concentrations to laboratory media or to pristine waters contributes little in a realistic assessment. Certainly in the case of detergent materials it is more meaningful to consider the material as just one of the components of wastewater and, to the degree possible, other sources of nutrient input should be included in the testing.

For example, in testing detergent phosphates other sources of phosphorus must be considered. Furthermore, since the response may vary between waters or among tests of the same water taken at different times, a variety of water sources and time periods should be a part of the evaluation. Many of these safeguards are being written into the Provisional Algal Assay Procedure or PAAP test (10) being developed under the auspices of the Water

Quality Office of the Environmental Protection Agency, with cooperation from university and industry groups.

Short of completely natural test situations, microcosm studies of the type reported by Swisher and Mitchell (11), in which a test material is injected into a more complete simulation of a lake and species diversity is followed, appear to provide useful input. In situ ^{14}C uptake studies such as those of Goldman (12) similarly seem to be useful as short-term indicators. Further development of test procedures in this area is certainly required, and realistic field studies are desirable to help quantify and validate various laboratory simulations.

Those familiar with environmental test procedures recognize that the complexity and extent of environmental testing that may be appropriate, for certain types of compounds will vary. The decisions of what is necessary or desirable, of course, depend on the chemical nature of the material, the potential usage levels, and a number of other factors. Testing may be undertaken by groups other than the detergent industry. Since universities, foundations, and government groups do make a needed contribution to the program, their efforts are welcome, and there is an attempt to coordinate the testing by the industry to utilize the special research talents and capabilities that will extend those which it already has.

Feasibility

It may be in order to mention briefly a few factors that must be considered in product and production feasibility. The material or chemical manufacturer must consider alternate raw materials and processes, quality and consistency of product, capacity, equipment and plant design, costs, and contractual negotiations. The detergent maker must consider the compatibility of a new material with other formula ingredients, as well as optimum utilization within different kinds of formulas and in different detergent-making processes. The acceptability of the new material in the product in terms of performance, physical properties, and costs can also be important. Other significant factors include the stabilities of the material and the product in process, in storage, and in use. Most of these problems can be solved, given a safe effective material, but to make new, good quality products at a reasonable cost to the consumer requires substantial research, engineering, and testing--and these all take time.

DETERGENT FORMULA ALTERNATIVES

To understand some of the benefits of today's detergents and the problems of major changes, it may be helpful to consider alternatives. The criteria necessary to qualify a material for detergents should be borne in mind as the formula alternatives that are open to the industry are examined.

1. Reduce or remove phosphate. This remains a free option in most areas, but reduction is a legal requirement in Chicago and elsewhere. Most of the laws that have been passed as a first step prohibit the sale of laundry detergents with more than 8.7% phosphorus, which is equivalent to about 35% sodium tripolyphosphate (STP). During an initial period of time, the laws allow the sale of automatic dishwashing detergents and products for many critical uses in food processing. The laws may also restrict the amount of product that can be recommended for use by limiting phosphorus to 7 g per wash load. As a second step, these laws ban, at a not-too-distant date, the sale of any cleaning products containing any amount of phosphorus. The industry has opposed these laws in what is believed to be the interest of all the consuming public.

The reasons for opposition to a total ban on phosphorus are based on performance considerations. Figure 2 plots the stoichiometric relationship between water hardness salts and the STP in a product that complies with the 35% STP legal maximum. (The lower Great Lakes have a water hardness of about 100 mg/liter.) The horizontal line at the bottom shows the requirement of P just to neutralize the hardness that clothes contribute to the wash water by virtue of perspiration, soils, foods, and previous rinsing. In the situation depicted, which by no means represents the extreme, nearly one-half cup of product or 3 g of phosphorus in a normal load is required regardless of tap water hardness. The slanting line then represents the added phosphate required to neutralize the natural hardness in the tap water. Unless the amount of free calcium ion contributed by clothes and water in total is maintained at a very low level, performance suffers. The cleaning drop is not clifflike, but it is very steep. Certainly below 0.7:1 STP-to-calcium ratio we will surely encounter poor cleaning and poor whiteness maintenance.

Figure 3 imposes the 0.7:1 STP-to-calcium line and delineates the area of definitely good and bad performance. It should be noted that in an area where the water is of average hardness, at a one cup recommendation, com-

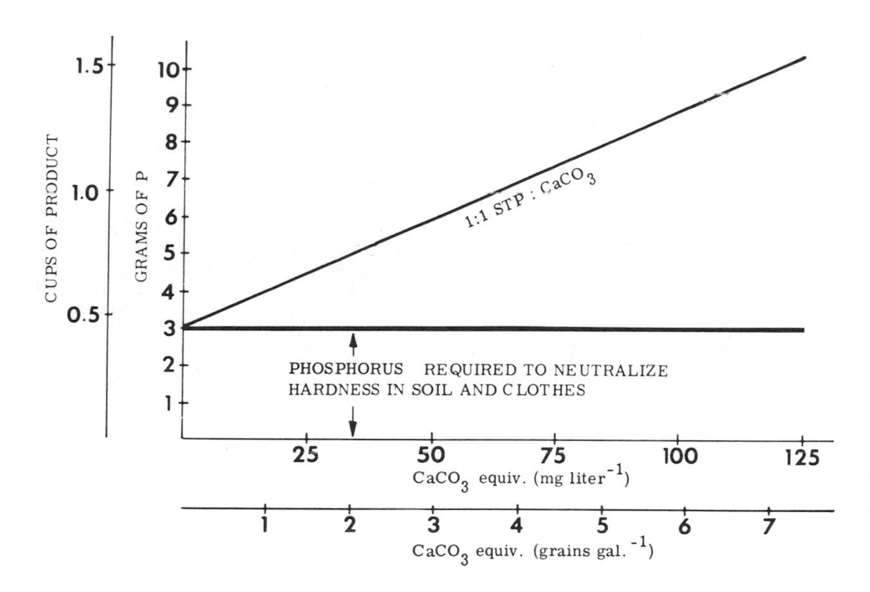

Fig. 2. Detergent phosphate requirement: stoichiometric requirement of P to neutralize hardness salts on clothes and in wash water when using 35% STP product and 17-gal capacity washer.

plying with the first phase of a law like Chicago's, will produce performance that is clearly in the poor-results area. In harder water areas or where larger volume washers are used, performance at these usage levels is, of course, even worse. Further reduction or total removal of phosphates without suitable replacements would provide extremely poor laundry performance for increasing numbers of people. Poorer cleaning of course means not just clothes that are less white, but lower standards of sanitation and health.

2. Go back to soap. This alternate has been often proposed. Soluble salts of fatty acids, known familiarly as soap, have been the traditional cleaning agents during much of recorded history. Under proper conditions they are effective in soil removal. The performance problems arise when the soap reacts with the calcium and magnesium ions, forming insoluble curd. By adequate usage of the product it is possible to suspend this insoluble precipitate during the washing process; but difficulties develop as clothes or dishes and the accompanying soap go into a

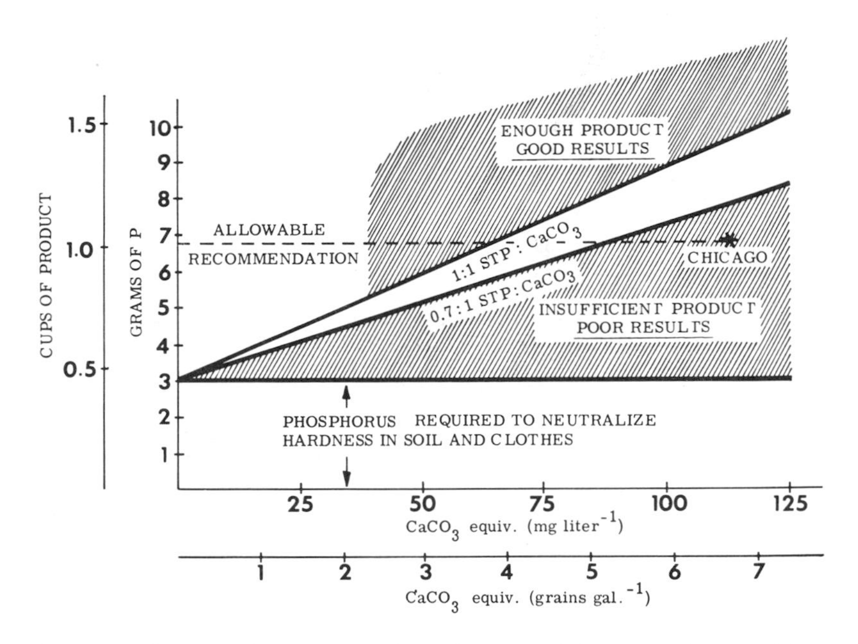

Fig. 3. Washing performance: based on stoichiometric relationship between phosphorus in normal density detergent containing 35% STP and the hardness salts in load and water of 17-gal capacity washer.

rinse, where they again meet unreacted hardness. The resulting deposit on clothes can manifest itself in many ways, and is generally more troublesome as natural water hardness increases. For example, dark colored garments washed just once in water of medium hardness using soap show such a pronounced deposit and color alteration that most consumers could be dissatisfied. Diapers, where hard water soaps and baby oils can combine to form grease balls, become completely unusable.

Suffolk County, Long Island, which lacks sewers and treatment facilities, banned the sale of all household detergents except automatic dishwasher products effective March 1, 1971, which means that soap is essentially the only washing product legally available for sale in that county. There are those who question whether such action realistically attacks the fundamental health hazards of ground water contamination by sewage (13). Of course some brands of soap have been available in this and most communities, but homemakers, particularly in harder water areas, have chosen them so infrequently that maintaining shelf display has been difficult and at times impossible. Beyond the lowered performance and added cost that characterize the use of soap--and which, we believe, make soap unacceptable to most automatic washer users--there are questions of raw material supply. At present necessary natural fat quantities are not available without disruption of human and animal food needs here and abroad (14). Additionally, there is little information about how exclusive use of soap in modern quantities would effect drainage lines and percolation fields.

3. Use NTA. The detergent industry had counted heavily on the use of NTA as at least a partial replacement for phosphates. This material is a very effective builder and can replace phosphate more than pound for pound in most household applications. Work on NTA commenced in 1961, and heavy involvement in all phases of evaluation began in 1964. More than $2 million has been spent by Procter & Gamble alone on safety and environmental testing of this material. More recently, substantial environmental testing has been undertaken by other companies and by universities and government groups.

On December 18, 1970, the Surgeon General's office requested that detergent manufacturers discontinue the use of NTA pending further tests and review of recently completed animal studies. At that time the EPA Administrator and the Surgeon General (15) made it clear that the doses of NTA and toxic metals employed in the NIEHS animal studies "were considerably higher than would ordinarily be encountered by the human population." They added, "There is no evidence at this time to indicate that anyone has been or is being harmed by the combination of NTA with metals in the environment."

Later the President of Procter & Gamble wrote "We are, of course, complying with [the Surgeon General's] request. Since that date we have had a chance to make our own evaluation of the animal studies about which the Surgeon General spoke, and we are now in the process of having these studies reviewed by independent scientific consultants. Based on the work we have done and the further scientific studies that are under way, we believe it would be premature to assume that the Surgeon General will request a permanent ban on the use of NTA in the way we had planned to use it" (16). For the present, however, NTA does not represent a viable alternate builder material.

4. Use precipitating builders. This approach, employing carbonates and silicates and incorporating them in

quite alkaline formulas, is being explored by a number of new market entries in the detergent field. A precipitating builder product may remove some soils adequately, but considering all anticipated washing conditions, poor performance may manifest itself in a number of ways. Tests on dark fabrics clearly show that a precipitating builder can cause not only fabric harshness but a dulling of the fabric color. In evaluations conducted by Consumer Bulletin (17,18), it was found that two major no-phosphate entries containing precipitating builders were poorer in cleaning than phosphate laundry products. Many no-phosphate brands are being marketed on an ecological basis, with promises mostly of what the product does not contain. Promotional statements about these products also headline biodegradability, although this can hardly be considered newsworthy, since household detergents in the United States completed conversion to "biologically soft" surfactants in 1965.

The rapidity with which some products featuring precipitating builders have appeared on the market raises some question about the thoroughness of their testing. Although longer range environmental effects may not be significant when the volume of these entries is considered, some deficiencies in overall cleaning performance, compatibility with fabrics and washers, and even human safety are suggested in preliminary testing. Safety and compatibility questions are not inextricably tied to the particular materials used as substitutes, but many of the new formulas using them are highly alkaline and therefore have carried warning labels on the package as required by the Federal Hazardous Substances Labeling Act (19). Relying on labeling for protection against accidental ingestion by children seems to be a step backward for a product class that has been stored and used without hazard in American homes for 25 years.

5. Find new substitutes. The normal course of research and product development work within the detergent industry has included a search for phosphate substitutes over the course of many years. This effort, which initially was justified on a performance basis, has been intensified because of the pressures to remove phosphate from detergents. Of more than 200 compounds screened to date, NTA was the most promising one to emerge (20). There were polycarboxylates and organic phosphonates, and classes of compounds that did show performance promise and even led to strong patent positions. Environmental testing on these two materials, however, uncovered questions about their biodegradability and their ultimate fate in the environment.

The kinds of test procedures used and judgments made recently on NTA leave some unanswered questions about clearing new builder candidates in the future. For a major detergent chemical, the long path from research to full-scale utilization, which was optimistically estimated at 7 years, has probably been extended. The minimum time just to clear for environmental and human safety alone is 3.5 years.

As the possibilities open to the industry are reviewed, it becomes painfully clear that without phosphates, short-term prospects for maintaining detergent performance at current levels are not bright. Many may consider a loss in product convenience, a somewhat lower level of household cleanliness, or poor washability on some kinds of fabrics and garments a reasonable tradeoff for an unquantified lessening of the eutrophication rate. Of somewhat greater concern must be the possibility that removal of phosphates, with their demonstrated safe, germ-killing properties, will lead to a lower standard of public health, a greater risk of infection arising out of less reliable cleaning in the food processing industries (21), or the increased risk of an ingestion hazard from highly alkaline products.

CONCLUSION

Although the detergent industry does not expect environmentalists or wastewater treatment experts to assume the industry burden of finding alternatives for phosphates, certain things can be done to create the climate in which responsible and productive changes can be made. Four specific suggestions, which could be acted on by individuals or by groups, are as follows:

1. Properly evaluate the contribution that today's detergents make to cleanliness and health standards and remember that the detergent products deliver this performance safely and economically.
2. When talking to laymen, put detergent phosphates in proper perspective. Make it clear that phosphate is an important detergent ingredient and although it is an algal nutrient, it is not a pollutant or poison, as some people have been led to believe. Help them to understand that phosphate removal from detergents, unless it is accompanied by appropriate waste treatment, has little chance of broadly arresting eutrophication. It should also be made clear that the diverse problems of pollution

certainly will not disappear coincident with phosphates being taken out of detergents.

3. Emphasize that legislation which unrealistically bans current formula ingredients before tested replacements are available can lead to unknown human and environmental risks.

4. Participate in realistic planning of waste treatment processes. Nutrient removal is certain to be a necessity in the future. A process that assumes that the input of detergent phosphates to treatment plants will cease in 1971, could be shortsighted. On the other hand, certain processes are designed to remove insoluble phosphorus compounds but omit biological treatment capable of handling soluble organic materials; these may not be suitable for detergents and other materials of the future. For major population centers, the combination of physical/chemical/biological treatment will provide maximum flexibility to remove a wide range of potential nutrients.

In asking for cooperation in these matters, it should be stressed that the detergent industry will continue to work toward the goal of phosphate removal in its products. Although the industry feels that hard evidence is still lacking to prove that unilateral reduction and removal of phosphorus from detergents will result in reduced eutrophication, it has accepted the fact that well-intentioned people believe the industry should remove phosphate from its products. Thus Procter & Gamble will remove phosphates from its products; but the company will distribute new formulas only after the new products have been tested carefully and thoroughly. We also expect to apply effort toward promoting more complete waste treatment facilities where more complete control of phosphorus and other nutrients from many sources offers an increased chance for control of eutrophication and other pollution problems.

REFERENCES

1. D. S. Black, Undersecretary of Interior, Address to Soap & Detergent Association, New York, Jan. 23, 1969 (cited by Senator Muskie, Congressional Record, March 24, 1969).
2. J. J. Gilligan, Rep. Ohio, in Congressional Record, Oct. 27, 1965.
3. F. A. Ferguson, "A Nonmyopic Approach to the Problem of Excess Algal Growths," Environ. Sci. Technol., 2,

188-193 (1968).
4. J. R. Duthie, "Laundry Detergents--What's In Them For You?" in "A Report on the 20th National Home Laundry Conference." Association of Home Appliance Manufacturers, Chicago, 1966, p. 100.
5. R. O. Carter, "Safety of Detergent Ingredients," AMA Sponsored Symposium presented before the Pharmaceutical Sciences Section, AAAS, Chicago, 1970.
6. HEW News. Food and Drug Administration Press Release, March 8, 1971.
7. W. C. Krumrei, "Trisodium Nitrilotriacetate (NTA), Environmental Safety Review--Appendix II." Statement before Subcommittee on Air and Water Pollution of the Committee on Public Works of the U.S. Senate, Pt 3, May 26-28, Washington, D.C., Government Printing Office, 1970, pp. 1131-1144.
8. Q. H. Pickering and T. O. Thatcher, "The Chronic Toxicity of Linear Alkylate Sulfonate (LAS) to Pimphales Promelas, Rafinesque," Water Pollut. Control. Fed., 42, 243-254 (1970).
9. A. B. Sprague, "Promising Anti-Pollutant: Chelating Agent NTA Protects Fish from Copper and Zinc," Nature, 220, 1345-1346 (1970).
10. Joint Industry/Government Task Force on Eutrophication, "Provisional Algal Assay Procedure," U.S. Department of the Interior, 1969.
11. R. D. Swisher and D. T. Mitchell, "NTA and the Environment," Addendum 1 to Statement of John R. Eck, Monsanto Company, before the Subcommittee on Air and Water Pollution of the Committee on Public Works of the U.S. Senate, Pt 4, Washington, D.C., Government Printing Office, 1970, pp. 1431-1439.
12. C. R. Goldman, "The Measurement of Primary Productivity and Limiting Factors in Freshwater with Carbon-14," Proc. Conf. Primary Productivity Measurements, Marine & Freshwater, University of Hawaii, August 21-September 6. U.S. Atomic Commission, Division of Technical Information, Rep. TID-7633: 103-113.
13. D. A. Okun, "Cosmetizing A Cancer," Water in the News, February, 1971.
14. E. S. Pattison, "Suggested Alternatives for Phosphates in Detergents." Appendix to Statement before the Subcommittee on Air and Water Pollution of the Committee on Public Works of the U.S. Senate, Pt 3, Washington, D.C., Government Printing Office, 1970, pp. 1101-1102.
15. W. D. Ruckelshaus and J. L. Steinfeld, Statement on NTA, Joint Press Release of the Environmental Protec-

tion Agency and Surgeon General, Dec. 18, 1970.
16. H. J. Morgens and N. McElroy, Earnings Statement of Six Months Ended December 13, 1970. Procter & Gamble Company, January 26, 1971.
17. "How Well Do The No-Phosphate Detergents Clean?" Consumer Bull., February, 1971, pp. 24-25.
18. "Non-Enzyme Laundry Detergents and Soaps," Consumer Bull., March 1971, pp. 13-16.
19. "Confiscated Detergent Will Return to Market," Wall Street Journal, March 22, 1971.
20. W. C. Krumrei, Statement before the Subcommittee on Air and Water Pollution of the Committee on Public Works of the U.S. Senate, Washington, D.C., Government Printing Office, 1970, pp. 1102-1107.
21. P. R. Elliker, "Sanitation and Environmental Factors in Detergents and Sanitizers," Environ. Health, 33, 241-250 (1970).

Chapter XII

NUTRIENT REMOVAL FROM WASTEWATER BY PHYSICAL-CHEMICAL PROCESSES

Jesse M. Cohen

Advanced Waste Treatment Research Laboratory,
Office of Research Monitoring,
Environmental Protection Agency,
Cincinnati, Ohio

The discharge of biologically treated wastewater effluents into lakes or streams causes eutrophication which is generally attributed to nutrient elements such as phosphorus or nitrogen. Data accumulated during the past 20 years indicate that increased eutrophication is caused primarily by greater discharges of phosphorus. Control of phosphorus should lead to control of eutrophication. Shapiro (1) states that no claim is made that phosphorus is the sole cause of eutrophication or that it is always the prime cause. What is claimed is that phosphorus is a "key" element because, of all major elements required by freshwater algae, phosphorus is generally present in the least amount relative to need. Limiting the availability of phosphorus, therefore, could lead to control of eutrophication. The case for removal of nitrogen, the second of the two key elements in the process, can be made on bases other than eutrophication.

Concern over eutrophication problems has provided the impetus to search for processes that reduce nutrients in discharged wastewater. Considerable effort has been directed toward the development of economic methods to achieve near complete removal of undesirable nutrients. This chapter considers the more promising advanced waste treatment methods that have been developed, or are being developed, for removal of phosphorus and nitrogen from wastewater. This discussion is restricted to the physical-chemical processes, but it is recognized that biological processes are also being developed. Only domestic or domestic-industrial wastewaters are considered, although the processes described may be equally applicable to industrial waste discharges.

PHOSPHORUS REMOVAL

Forms of Phosphorus

Phosphorus is usually present in wastewater in the form of organic phosphorus, inorganic condensed phosphates, and orthophosphates. The concentration of phosphorus in domestic wastewater is often in the range of 7 to 15 mg P/liter. Most of the organically bound phosphorus compounds are present as particulate organic matter and as bacterial cells. Very little is known about the dissolved organic phosphorus compounds, but these represent only a small fraction of the total phosphorus content. Inorganic condensed phosphates such as tripolyphosphate and pyrophosphate originate mainly in synthetic detergents. Orthophosphate is excreted in urine; it is an end-product of microbial degradation of organic phosphates and of hydrolysis of condensed phosphates. Raw sewage will contain 25 to 85% of the total phosphates as condensed phosphates, depending on the initial content and degree of hydrolysis occurring in the system. A more complete discussion of phosphorus forms can be found in a committee report of the American Water Works Association (2).

The concentration of total phosphate in raw sewage and the ratio of orthophosphate to polyphosphate vary widely during any 24-hr period. Part of this variation is caused by changes in concentration of orthophosphate, but most of it results from a large and rapid increase in polyphosphate, during the morning hours, probably the result of home laundry activities. Finstein and Hunter (3) and Laughlin (4) have observed these variations in total phosphate concentration and ratio of ortho to polyphosphate (Figure 1). To obtain effluents with low phosphorus content in any treatment process, the differing forms and the change in forms during the day must be considered. In practical terms, any process must be capable of removing both ortho and polyphosphate forms.

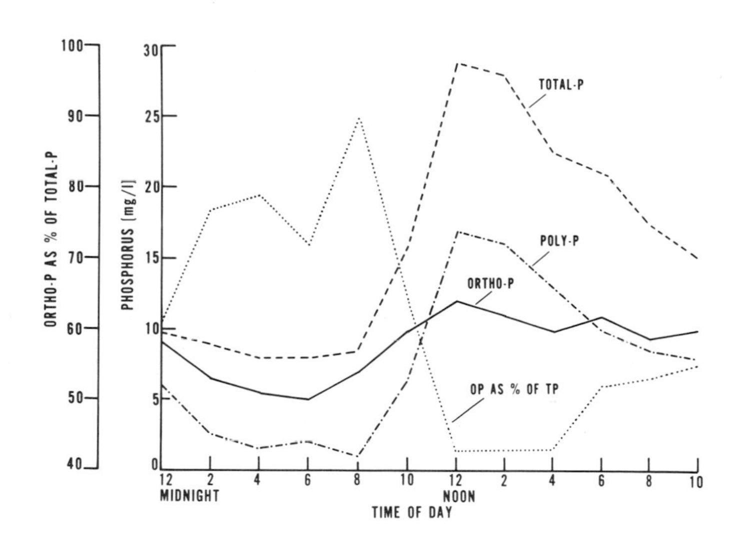

Fig. 1. Temporal variations in the concentrations of phosphate species (Ref. 3).

General Considerations

The most economical and reliable method of removing phosphates is chemical precipitation. The current technology of phosphorus removal by chemical precipitation has been described in several excellent reviews (5-9). The choice of the chemical to use and the location of its addition in a treatment system depend on several considerations, including the following:

1. The influent phosphorus concentration and the form of the phosphorus at the point of treatment.
2. The degree of phosphorus removal required and the efficiency of the various processes.
3. Characteristics of the wastewater at the point of addition of chemical.
4. Compatibility of the phosphorus removal process with other processes in the overall treatment scheme.
5. Sludge-handling facilities and ultimate disposal of the sludge.
6. Chemical costs, including transportation.
7. Effect of additional mineralization from the

precipitant anion.

These considerations have been discussed in detail by Culp (10).

A variety of chemicals can be used for precipitation of phosphate. Various salts of iron are effective (e.g., ferrous chloride or sulfate, ferric chloride or sulfate, and waste pickle liquor). Aluminum salts are also effective. Commercial alum (aluminum sulfate) is the most widely used aluminum salt, although sodium aluminate is also employed. The latter is more expensive, but it contains some excess caustic, which helps to maintain pH while contributing fewer inorganic ions than other salts.

Lime as a precipitant is very attractive because of its cheap unit cost and because lime sludge can be calcined to recover reusable lime. At the same time, the organics are also incinerated. If recarbonation is practiced with lime, no additional minerals are added, and with some wastewaters with high hardness, a reduction of minerals is obtained.

Lanthanum salts are effective precipitants, but the unit cost of the chemical requires that lanthanum be recovered for reuse; however, effective methods for recovery of lanthanum have not yet been developed.

All the precipitants cited here may require the addition of polymeric flocculant to obtain effective settling and separation of solid from liquid. Combinations of chemicals have been found effective; for example, lime plus iron or aluminum salts may be cheaper and may yield less sludge than either chemical used alone. The relative merits of chemical precipitants have been discussed by Malhotra et al. (11).

The precipitant for phosphorus removal may be added at any point in a treatment system. In any treatment system, the chemical may be added in accordance with one of three basic choices, as shown in Figure 2, although variations on these three basic schemes are possible. In scheme (a) the precipitant is added to the raw sewage and the solids are separated in a settling tank. The effluent may be further treated by biological or physical-chemical processes. In scheme (b) chemical is added to the aeration basin and sludge is removed along with the waste activated sludge. In scheme (c) the chemical is added as a final step after any biological treatment process. For good phosphate removal, a filtration step may also be necessary.

Addition of chemical directly to the raw sewage is the method of choice in physical-chemical treatment systems, and this location has substantial merit for biologi-

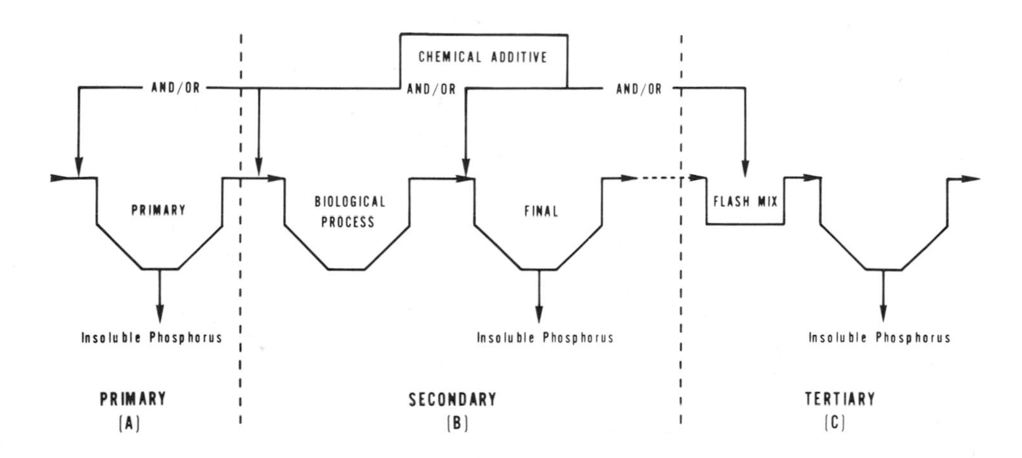

Fig. 2. Locations for chemical control of phosphorus.

cal systems. Addition at this point in the treatment not only removes phosphates but obtains clarification, thus providing suspended solids and BOD-COD removal, all at the same time. With effective clarification, 60 to 80% of the COD is removed, reducing the organic load to any subsequent process. Furthermore, precipitated phosphate not captured in the settling process will be removed by the subsequent processes, which may include activated sludge, trickling filter, or carbon adsorption. This approach can yield removals of phosphate in excess of 95% when adequate dosages of chemical are added. The process also eliminates the cost of a separate primary sedimentation tank.

Addition of the chemical to aerators in the activated sludge process involves the lowest capital cost. The only additional capital cost is for chemical storage and feeding equipment and additional sludge handling capacity. This process, however, is limited to the use of iron or aluminum salts, since pH restrictions restrict the use of lime; moreover, removals in excess of 80% cannot be obtained consistently.

For high rates of removal of phosphate in an existing biological plant, chemicals may be added as a tertiary process where there are no restrictions on kind of precipitant. This mode of addition has less appeal and is not widely practiced. The capital costs are great since, in most cases, a filter is required to achieve high removals of phosphate.

Aluminum Salts

The removal of phosphates by aluminum salts such as aluminum sulfate (alum) or sodium aluminate is technically and economically feasible. The removal of precipitated phosphates is accomplished in conventional coagulation and flocculation equipment, followed by sedimentation or filtration. Despite considerable research, the basic chemistry and the mechanism of the removal has not been thoroughly established. Lea et al. (12) have suggested that removal of phosphates involves its adsorption on precipitating aluminum hydroxides. But according to Stumm (13), Cole and Jackson (14), and Henricksen (15), the interaction of aluminum with orthophosphate results in the formation of an insoluble aluminum phosphate precipitate. The current consensus is that the interaction of aluminum with orthophosphate yields an insoluble aluminum phosphate. The excess aluminum required to achieve good removal of phosphate forms hydrolysis products that also react with a small proportion of phosphate.

The overall reaction for phosphate removal by alum can be expressed by the following equation:

$$Al_2(SO_4)_3 + HPO_4^{-2} + 3H_2O \longrightarrow$$

$$Al(OH)_3 \cdot AlPO_4\downarrow + 3SO_4^{-2} + 4H^+ \qquad (1)$$

The stoichiometry is only approximate and is dependent on pH (11,13,16). In practice, and even under optimum conditions, the amount of aluminum salts required for complete precipitation of phosphates exceeds the stoichiometric requirements. On the basis of certain solubility considerations and equilibrium data, Stumm (13) calculated the pH of minimum solubility for $AlPO_4$ to be 6.3.

In an investigation of the chemistry of precipitation of phosphates by iron and aluminum salts, Recht and Ghassemi (17) showed that minimum solubility of $AlPO_4$ occurred at pH 6. The relation between residual orthophosphate and pH when a 2:1 cation-to-orthophosphate molar ratio was used is presented in Figure 3. On either side of the optimum pH, increased residual phosphates are obtained and the settling characteristics of the precipitate become poor. Complete removal of the precipitated aluminum phosphate can yield a residual phosphate of 0.1 mg P/liter. In practice, however, complete precipitation

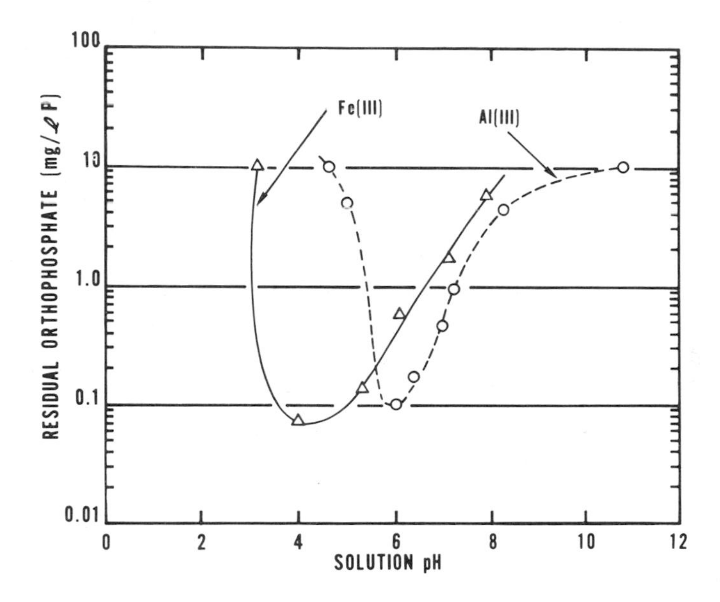

Fig. 3. Precipitation of orthophosphate with aluminum (III) and iron (III) at a 2:1 cation-to-orthophosphate molar ratio (Ref. 17).

does not occur, nor is the precipitate completely removed; thus practical residuals of phosphorus tend to range from 0.2 to 0.5 mg/liter.

To obtain removal of phosphates to the limit of the solubility of aluminum phosphate requires that an amount of aluminum ion be added in excess of the stoichiometric amount, in order to coagulate the precipitate. Hence, in practice, aluminum-to-phosphate molar ratios of 1.2 to 2.0 must be added. This relationship is illustrated in Figure 4. The data, taken from Recht and Ghassemi (17), show that at a constant pH of 6, the removal of phosphate is directly proportional to the amount of added aluminum up to a ratio of about 1.6:1 Al^{3+}/PO_4^{-3}. The existence of such a linear relationship between the amount of added aluminum and the phosphate removal obtained over so wide a range of added aluminum is offered as evidence that the reaction entails precipitation, rather than being a heterogeneous mechanism involving adsorption. From the slope of the straight-line portion of the curve, it can be calculated that approximately 1.4 moles of Al^{3+} is required for precipitation of 1 mole of orthophosphate.

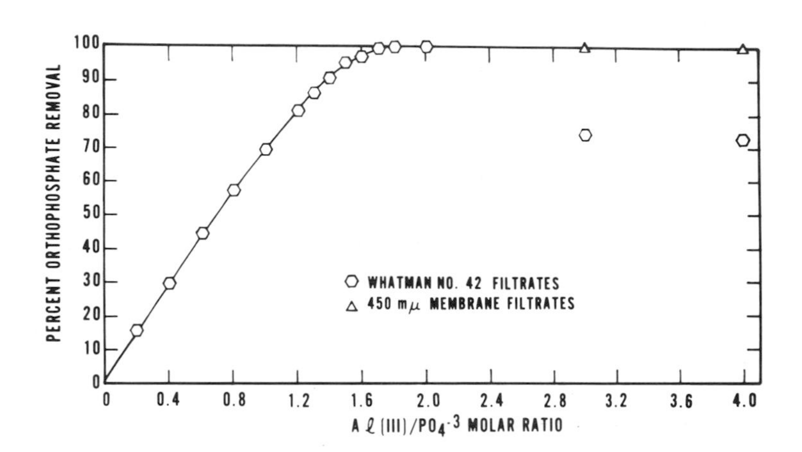

Fig. 4. Orthophosphate removal by aluminum (III) at pH 6.0 (Ref. 17).

Many investigators (13,17,18) have demonstrated that aluminum salts are inefficient precipitants for polyphosphates. To obtain removals equivalent to those for orthophosphates, an excess of precipitant must be added. The effect of various polyphosphates on alum dose is shown in Figure 5; the sewage control sample required 80 mg/liter of alum to achieve maximum removal of phosphate. Increased amounts of alum, in excess of 120 mg/liter, were required upon addition of polyphosphates equivalent to an increment of 4 mg/liter of phosphorus. Similar results were obtained by Recht and Ghassemi (17). This would suggest that, in practice, somewhat greater amounts of alum would be required when precipitating phosphate from raw sewage (since polyphosphate concentration is higher) than when alum is added as a tertiary process (where substantial hydrolysis of the polyphosphates has occurred during biological action). Extent of hydrolysis of polyphosphates during biological treatment has been described by Finstein and Hunter (3) and by Heinke and Norman (19).

Iron Salts

The chemistry and use of iron salts for phosphorus removal parallels that of aluminum salts, with the important difference that divalent as well as trivalent salts are available. The low cost of ferric salts and a history

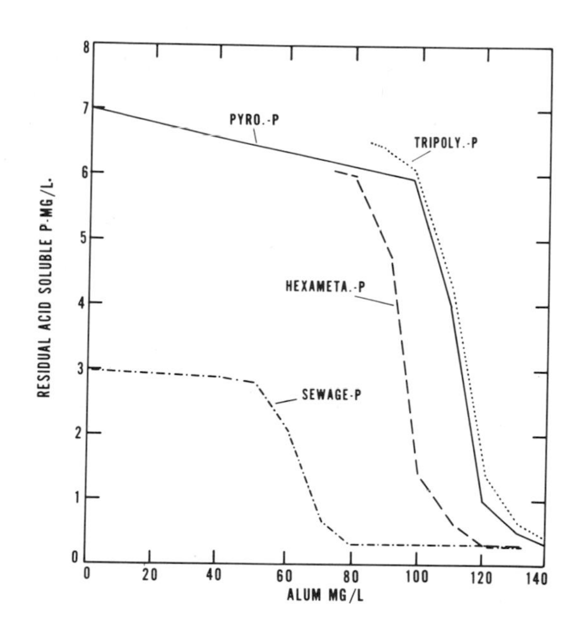

Fig. 5. Removal of phosphates by alum precipitation (Ref. 18).

of use going back to the 1930s have tended to make iron salts a common precipitant.

As with alum, the reaction of iron with phosphates is pH sensitive. Recht and Ghassemi (17) have shown that the pH of maximum precipitation of phosphate occurs at pH 3.5 to 4.5 for ferric iron (Figure 3). On either side of this pH range, decreased removals of phosphate are obtained. At a constant pH of 4.0, the removal of phosphate is directly proportional to the amount of added iron up to a molar ratio of 1.2:1 $Fe(III)/PO_4^{-3}$ (Figure 6). This linear relationship, as with aluminum, suggests that a precipitation reaction is involved. A precipitation reaction was also postulated by Gorchev and Stumm (20).

In practice, phosphate removal is not executed at the optimum pH and the amount of precipitant is generally dictated by the dosage required to obtain a floc that will settle. The need to add more metal ion than is required by stoichiometry was pointed out by Pöpel (21). Ferric salt precipitation is generally performed at a pH range of 6 to 7 and requires moler ratios of 1.2 to 2.0 $Fe(III)/PO_4^{-3}$ to obtain clarification. Ferric chloride and ferric sulfate have been used interchangeably. The choice is

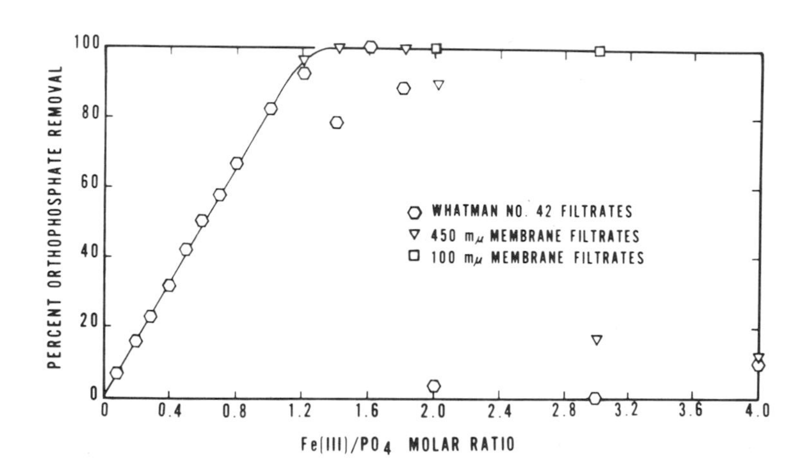

Fig. 6. Orthophosphate removal by iron (III) at pH 4.0 (Ref. 17).

usually made on the basis of cost and ease of handling.

The availability, frequently at low cost, of waste pickle liquor has generated an interest in the use of ferrous iron as a phosphate precipitant. Pickle liquors containing either ferrous sulfate or chloride have been used. The pickle solution contains some excess acidity along with 6 to 9% iron content. Most of the iron is in the ferrous form, and there is an undetermined proportion of ferric iron.

In theory, with ferrous salts a dosage of 2.70 mg Fe(II)/mg P is required for orthophosphate precipitation, in contrast to only 1.80 mg Fe(III)/mg P for ferric salts and 0.87 mg Al(III)/mg P for aluminum salts. However, this theoretical advantage of ferric and aluminum salts is partially offset by the greater tendencies of the ferric and aluminum ions to hydrolyze, thus requiring addition of excess Fe(III) or Al(III). Another advantage is the more favorable pH range for use of Fe(II).

Under contract to the Environmental Protection Agency (EPA), Ghassemi and Recht (22) have investigated the chemistry of ferrous iron precipitation of phosphates. Their data have revealed some interesting comparisons with ferric and aluminum salts.

1. Maximum orthophosphate removal with ferrous salts occurs near pH 8 in contrast to Fe(III) and Al(III) precipitation near pH 4 and 6, respectively. At pH 7.9,

94% of the phosphate and 97% of the iron are precipitated, either as the ferrous phosphate or as the hydroxide, with $Fe(II)/PO_4^{-3}$ equivalence ratio of 1.0.

2. At the respective optimum pH for orthophosphate removal and on an equivalence basis, the capacity of ferrous iron for orthophosphate removal is greater than either ferric iron or aluminum. On a concentration basis, however, both ferric iron and aluminum ion are more effective precipitants. From Figure 6, an $Fe(III)/PO_4^{-3}$ equivalence ratio of 1.2 precipitates approximately 96% of the initial orthophosphate. This represents a dosage of only 2.16 mg Fe(III)/mg P, well below the dosage of 2.70 mg Fe(II)/mg P, to obtain similar phosphate removals. A similar calculation using available data (Figure 4) shows that 1.30 mg Al(III)/mg P is required to precipitate 95% of orthophosphate under the same test conditions at an $Al(III)/PO_4^{-3}$ equivalence ratio of 1.5. The greater equivalence capacity of Fe(II) for orthophosphate removal and its higher pH of maximum effectiveness may be attributed to its less pronounced tendency to hydrolyze, as illustrated by comparing solubility equilibria for each of the metals:

$$Fe^{+2} + 2OH^- = Fe(OH)_2, \quad K_{sp} = 5 \times 10^{-15} \tag{2}$$

$$Fe^{+3} + 3OH^- = Fe(OH)_3, \quad K_{sp} = 10^{-36} \tag{3}$$

$$Al^{+3} + 3OH^- = Al(OH)_3, \quad K_{sp} = 10^{-32} \tag{4}$$

3. Rates of reaction of Fe(III) or Al(III) with orthophosphate are very fast (less than 1 sec), but the reaction rate with ferrous salts is generally slow--of the order of minutes. The results of carefully controlled laboratory tests (22) appear in Table 1, which summarizes some of the characteristics of ferrous iron precipitation. The optimum pH for both phosphate and iron removal is within a narrow pH range around 8.0. Below pH 8, phosphate is poorly removed and residual iron in the product is high. At pH values greater than 8.0, phosphate removal is poor but no residual iron is present.

In general, practical experience with ferrous salts has confirmed results from the laboratory. Use of either lime or caustic has usually been required to maintain a pH near 8.0. Thus Wukasch (23) found in full-scale tests that typical requirements for the ferrous chloride system were 10 to 25 mg/liter Fe(II) and 30 to 40 mg/liter of

Table 1. Removal of Orthophosphate by Fe(II) Precipitation[a] (22)

Initial pH	Percentage removal	
	P	Fe
6.0	7.1	8.9
7.0	34.6	38.9
7.9	94.1	96.6
8.9	68.8	ca. 100
10.0	18.3	ca. 100

[a]Initial orthophosphate concentration = 12 mg/liter of phosphorus; initial ferrous iron concentration = 32.4 mg/liter of iron.

strong base alkalinity as $CaCO_3$ (24 to 32 mg/liter of NaOH or 17 to 22 mg/liter CaO). The greater solubility of ferrous hydroxide can lead to high residuals of iron following precipitation.

In practice, dosages of Fe(II) in excess of those required for phosphate precipitation should be avoided, and pH should be maintained near 8.0 by increasing the alkalinity. It may also be necessary to add sufficient dissolved oxygen after precipitation to rapidly oxidize slightly soluble ferrous hydroxide to the very insoluble ferric hydroxide.

The final choice of metal precipitant will depend on the total cost of all added chemicals required to achieve the desired degree of treatment as well as the costs incurred for conditioning and disposal of sludges.

Coagulation of Precipitate

Good removal of phosphates in a treatment plant depends on the completeness of separation of the precipitated metal phosphate from the liquid. To achieve this, metal in excess of stoichiometric requirements is generally added to provide for coagulation of the solids that can be removed by sedimentation. Since precipitation is rarely practiced at the optimum pH for the process, a variable proportion of precipitate is extremely fine and resists sedimentation. Thus it is common practice to add an anionic polyelectrolyte to aid in the flocculation of the formed precipitates. And since sedimentation tanks rarely operate ideally, some metal phosphate will appear in the effluent. The amount, of course, will vary with design and operation of the apparatus and dosage of chemical. To obtain phosphate residuals less than 0.5 mg/liter of P requires an additional solids removal process. In practice, this can consist of a granular media filter, a carbon contractor, or a biological aeration process.

Lime

The effectiveness of lime in removing phosphorus from raw and treated wastewaters is well established (11,24-27). The chemistry of lime precipitation is entirely different from the chemistry of the hydrolysis of metals such as iron or aluminum. With the latter, metals dose is related to phosphate content and the requirement to provide enough hydrolysis products to effect coagulation of the precipitated solids. In contrast, lime requirements for phosphate removal are independent of phosphate content and are largely a function of the pH and alkalinity of the water. This dependency of lime dosage has been widely observed (28-31). Lime reacts with the alkalinity to form calcium carbonate. Excess calcium then reacts with orthophosphate to precipitate calcium hydroxyapatite. Additionally, at pH values greater than about 9.5, magnesium hydroxide will begin to precipitate until the reaction is essentially complete at pH 11. An additional reaction in the process is the adjustment of pH to remove excess calcium ion. Carbon dioxide is the chemical of choice both for pH adjustment and to provide bicarbonate ions. The series of reactions can be shown as follows:

$$Ca^{2+} + HCO_3^- + OH^- \longrightarrow CaCO_3 + H_2O \quad (5)$$

$$5Ca^{2+} + 7OH^- + 3H_2PO_4^- \longrightarrow Ca_5(OH)(PO_4)_3 + 6H_2O \quad (6)$$

$$Mg^{2+} + 2OH^- \longrightarrow Mg(OH)_2 \quad (7)$$

$$Ca^{2+} + 2OH^- + CO_2 \longrightarrow CaCO_3 + H_2O \quad (8)$$

Theoretically, the lime dose for any wastewater could be computed from stoichiometry and knowledge of the wastewater's characteristics. However, this approach is rarely used, and the usual method is to add lime to achieve some desired pH. Thus use of lime to precipitate phosphates requires a knowledge of (1) desired degree of removal of phosphate, (2) alkalinity of the wastewater, and (3) lime dosage to achieve some desired pH. Figure 7 illustrates

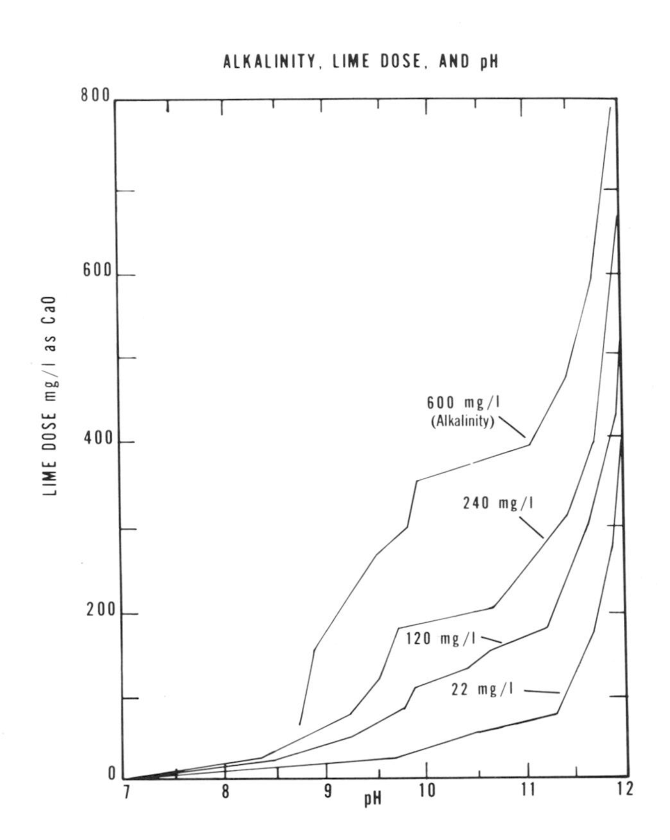

Fig. 7. Relationship between alkalinity, lime dose, and pH (Ref. 28).

the interdependency of alkalinity, pH, and lime dose (28). The lime dose to achieve a pH of, say, 11, will range from less than 100 mg/liter as CaO for a low-alkalinity water of 22 mg/liter alkalinity, to as high as 400 mg/liter for a high-alkalinity water of 600 mg/liter alkalinity.

The other variable to be determined is the pH that provides the desired removal of phosphate. Figure 8 plots the relation between pH and residual phosphate after filtration. The curve indicates that removal of phosphate is continuous and begins when the pH rises to about 9.0 and is substantially complete at pH 11.0. The choice of pH is determined by the desired residual phosphorus. Thus at pH 9.5 (the lowest pH providing any substantial

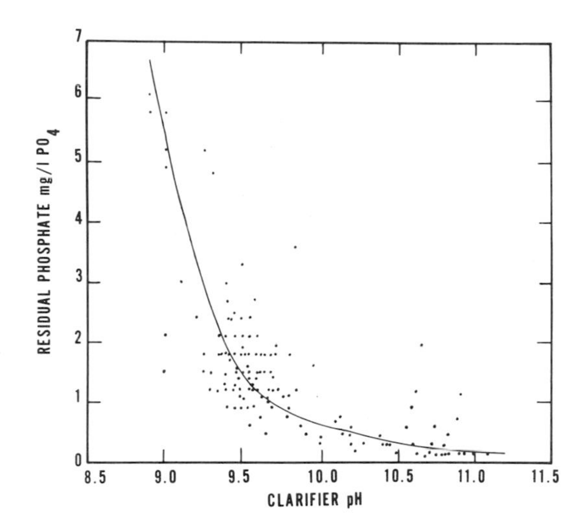

Fig. 8. Effect of pH on residual phosphate after lime clarification and filtration.

phosphorus removal), the residual phosphate is about 1.5 mg/liter as PO_4 (0.5 mg/liter P). At pH 11.0 the residual phosphate is about 0.3 mg/liter as PO_4 (0.1 mg/liter P).

As pH is increased beyond pH 9.5, the product water becomes saturated with calcium ion, resulting in high concentrations of calcium at pH values above 11. This excess calcium hardness generally must be removed to prevent postprecipitation. Two types of lime treatment are employed. Single-stage treatment, in which the wastewater is treated to some pH value and then discharged with no further treatment, is generally applicable with high alkalinity, high hardness waters and where desired residuals of phosphorus can be obtained at relatively low pH values (ca. 9.5).

At higher pH values, two-stage treatment is required. In the two-stage process, lime is added to a high pH (ca. 11.0) and the precipitated solids are separated. The product from this stage is treated with carbon dioxide to reduce the pH to approximately 9.5, which precipitates the excess calcium as calcium carbonate. Two-stage lime precipitation is clearly more costly than the single-stage process, but it is required for good removal of phosphorus and where excess hardness is undesirable. Details of two-stage lime precipitation have been recorded by Stamberg et al. (28).

One of the advantages of using lime is that sludge from the process can be calcined. This incinerates the organic matter and converts the calcium carbonate and hydroxide to reusable calcium oxide. The calcium phosphate remains as inert material and is periodically wasted to remove calcium phosphate from the system. Details of lime recovery and reuse have been published by Mulbarger et al. (30). Rand and Nemerow (32) and Slechta and Culp (24) have reported that recalcined lime can be used for phosphate precipitation.

Chemical Cost for Phosphorus Removal

The chemical cost for phosphorus removal varies widely, depending on the choice of chemical, the dose required, and the process used. Albertson and Sherwood (33) summarized chemical costs using data reported in the literature. The data (Figure 9) indicate that costs for chemicals can range from as little as \$20 to as high as \$80

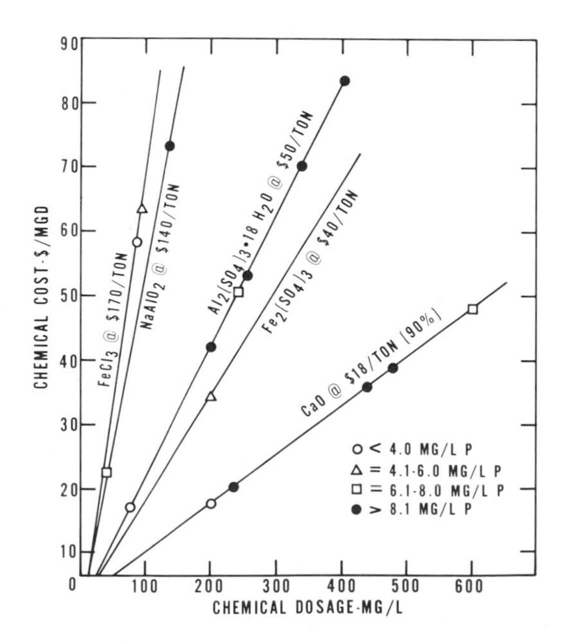

Fig. 9. Chemical costs for phosphate removal (Ref. 33).

per million gallons to achieve an 80 to 95% reduction of phosphate. To these costs must be added the expense of additional equipment, labor, and disposal of sludge.

Other Removal Processes

Lanthanum

Recht and Ghassemi (34) demonstrated that lanthanum, one of the rare earth metals, is an effective precipitant for phosphate. They proved that, with secondary effluent, lanthanum was a superior precipitant to iron and aluminum in four principal ways:

1. Lanthanum was effective over a wide pH range (pH 4.5 to 8.5) in contrast to a pH range of 5.0 to 6.5 for aluminum and pH 3.5 to 5.0 for Fe(III).
2. Lanthanum precipitated polyphosphates equally as well as orthophosphates.
3. The reaction of lanthanum with phosphate is stoichiometric and requires little excess to obtain a precipitate that will settle.
4. Almost total removal of phosphates can be obtained, leaving less than 0.1 mg/liter residual phosphorus.

Figure 10 illustrates the advantages of lanthanum over aluminum as a precipitant. At a 1:1 cation-to-orthophosphate molar ratio, nearly complete removal of phosphorus is obtained over a pH range of 5.0 to 9.5. Under the same conditions, aluminum removed barely 60% of the phosphorus. Analyses of filtrates from precipitation of phosphate by lanthanum showed that excess lanthanum was removed along with the lanthanum phosphate. Residual lanthanum was not detectable using an analytical procedure which had a limit of detectability of 0.5 mg/liter of lanthanum.

Despite the clear superiority of lanthanum as a precipitant, its higher cost compared with iron or aluminum requires that lanthanum be recovered for reuse to make the process economically feasible. Chemical methods for recovery of lanthanum are available, but they have not yet been proven, and practical use of lanthanum remains restricted. The element may find use as a "polishing" precipitant to remove last traces of phosphate after the bulk has been removed by iron, aluminum, or lime.

Activated Alumina

Anion-exchange reactions on aluminas have long been

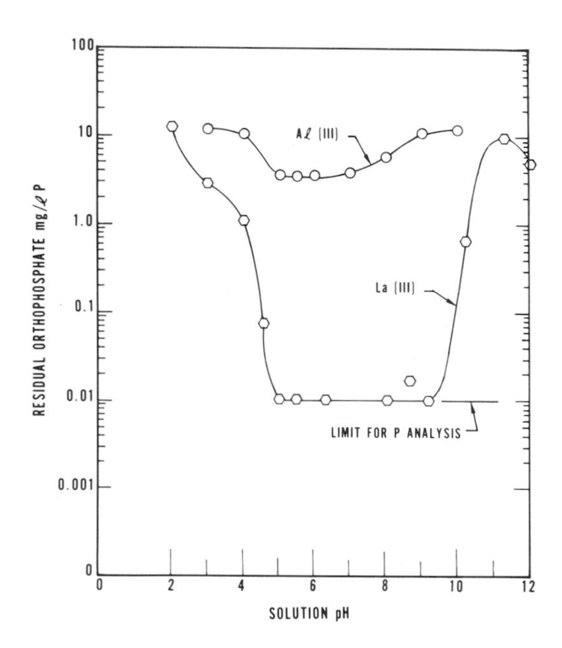

Fig. 10. Comparison of lanthanum (III) and aluminum (III) precipitation of ortho-P at 1:1 cation-to-orthophosphate molar ratio (Ref. 34).

recognized (35). The use of activated alumina for phosphate removal, however, was first proposed by Yee (36), who removed phosphate compounds from a low-level radioactive process wastewater. Development of the process for phosphorus removal from wastewaters was described by Ames and Dean (37). The removal process is achieved by using common ion-exchange techniques. The wastewater is contacted with activated alumina in a two-stage column contactor. When phosphorus appears in the effluent of the first column, the alumina is regenerated with sodium hydroxide.

The mechanism for removal of phosphate remains uncertain. From a kinetics study, Winkler and Thodos (38) concluded that removal could be described by an adsorption mechanism; although on the basis of reaction enthalpy changes, Ames and Dean (37) thought that removal probably depended on a combination of ion-exchange and chemical reactions. The difference, however, may be due to the dissimilar pretreatments of the alumina--the former using alumina treated with nitric acid and the latter using the hydroxide form.

Laboratory tests (37) on wastewater containing 10.3 mg/liter P showed that approximately 400 column volumes were required to reach 10% phosphorus breakthrough using a single column. If two columns were operated in series, more than 600 column volumes could be loaded on the first column before regeneration was required. Elution of phosphorus was accomplished with 8 column volumes of 1M sodium hydroxide followed by 20 column volumes of wash water. The 8 column volumes of regenerant containing the phosphorus, organic matter and HCO_3^- can be treated with lime. The phosphorus is removed as calcium hydroxyapatite and the HCO_3^- as calcium carbonate. Most of the organic matter is removed along with the solids, allowing reuse of the sodium hydroxide (37).

Use of the activated-alumina ion-exchange process has some advantages over precipitation. The process is selective for phosphate ion (and arsenate) and will remove both ortho and polyphosphates. The process can be controlled to produce effluents with less than 0.1 mg/liter of phosphorus. Use of alumina produces no pH change in the product, and no additional inorganic ions are added. The process appears to be unaffected by feedwater composition.

The results of the laboratory study on activated alumina were essentially confirmed in a pilot plant mounted in a trailer (39). The system consisted of a multimedia filter to pretreat the secondary effluent, three alumina columns, and facilities for lime treatment of the regenerant solution. Two alumina contactors were operated in series while the third column was being regenerated. Typical breakthrough curves are plotted in Figure 11, which represents the last in a series of phosphate loading operations. At that point, approximately 465,000 gal had been processed with 98% phosphorus removal.

Activated alumina for phosphorus removal has not yet been applied at full scale. Acceptance by the field awaits demonstration of its performance and cost. As with lanthanum, it seems likely that activated alumina may find its greatest use as a "polishing" process because of its capability of near-complete removal of phosphorus. Yee (36) estimated that about 20,000 column volumes of a waste containing about 0.3 to 1.0 mg/liter of phosphorus could be processed to 1% phosphate breakthrough in the effluent. Under these conditions, essentially complete removal of phosphorus can be obtained at reasonable cost.

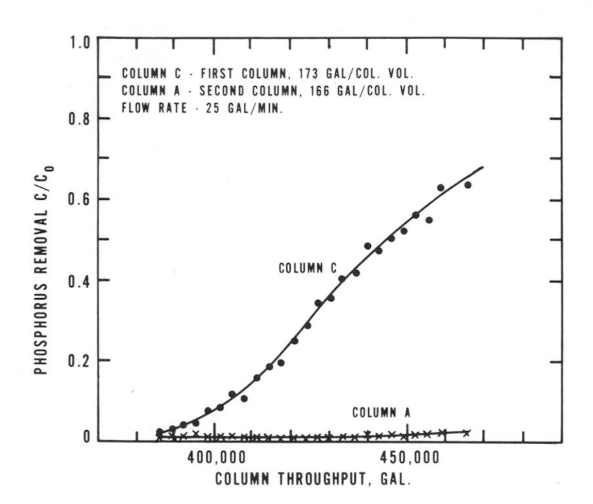

Fig. 11. Removal of phosphate by two-stage alumina columns (Ref. 39).

NITROGEN REMOVAL

Forms of Nitrogen

Approximately 90% of the total nitrogen entering a waste treatment plant is in the form of ammonia. The principal origin of ammonia is urea, which is rapidly converted to ammonia by the enzyme urease. The remaining 10% is in the form of more complex organic compounds (40). Total nitrogen concentrations in secondary effluents generally range from 12 to 42 mg/liter as nitrogen.

In addition to serving as an algal nutrient, ammonia nitrogen in effluents is undesirable because it (1) consumes dissolved oxygen during oxidation to nitrates; (2) reacts with chlorine to form chloramines, which are less effective as disinfectants than free chlorine; (3) is toxic to aquatic life; (4) is corrosive to certain metals; and (5) affects chlorine demand at downstream water plants.

Biological processes are being developed to remove nitrogen. Ammonia can be nitrified biologically to nitrate, which then is biologically denitrified to nitrogen gas. These processes are not discussed here.

Physical-chemical methods to remove nitrogen forms from wastewater are in various stages of development. The best developed processes are: ammonia stripping, natural zeolite ion exchange, and breakpoint chlorination.

Ammonia Stripping

The ammonium ion is very soluble and stable in water and is the principal nitrogen form in raw wastewaters at normal pH values. Ammonium ions exist in equilibrium with ammonia and hydrogen ions in accordance with the following equation:

$$NH_4^+ = NH_3 + H^+ \tag{9}$$

With increasing pH above 7.0, the equilibrium is displaced to the right until virtually all the nitrogen will exist in the form of molecular ammonia. Figure 12 illustrates the calculated relationship between pH and percentage of ammonia at various temperatures. The curves shift to the left as the temperature decreases. Since ammonia is volatile, it can be removed by contacting the water with air. The first requirement of a practical process is to adjust the pH. From Figure 12 we see that at any pH above 10.5, and at a temperature higher than 10°C, 90% of the ammonia is available for removal by air stripping. Early studies by Kuhn (41) of the feasibility of stripping of ammonia from wastewater revealed a major difficulty. Because of the high solubility of ammonia in water, a very large volume of air was required to strip a unit

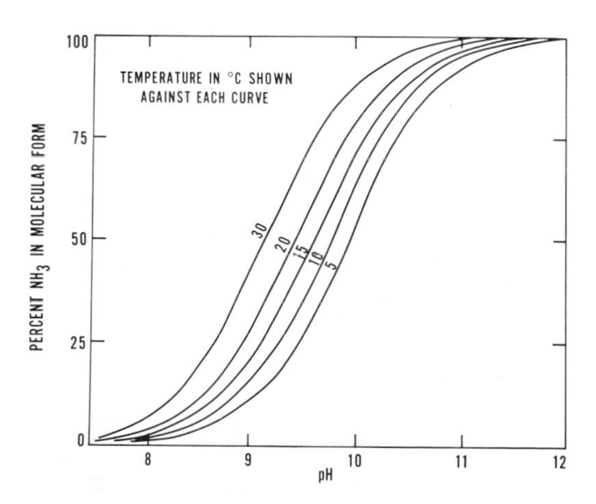

Fig. 12. Relation between pH and percentage of ammonia in aqueous solution.

volume of water: about 500 ft^3/gal to obtain 92% removal of ammonia at a pH of 11 in an experimental tower packed with Raschig rings. Bayley (42) has shown that the use of traditional types of packing such as Raschig rings results in excessive pressure losses and consequently very high power consumption.

The problem of high power cost, associated with a packed tower, was solved by investigators at the South Lake Tahoe Public Utility District and reported by Slechta and Culp (24), who used a tower packed with redwood slats such as are used for cooling water. The pressure drop across the device is very low, about 0.5 in. of water, so power costs are reduced to reasonable levels. This tower was used principally to evaluate the effect of aeration at various pH values. The results of this study appear in Figure 13.

Optimum ammonia removals at pH values above 9.0 were obtained with an aeration time of about 0.5 min or, with this particular tower, at an air/liquid loading of about 750 ft^3/gal. This study clearly showed that removals of ammonia in excess of 90% could be obtained if the pH was maintained above about 10 and if high air-to-liquid volume was applied.

Based on the results of these tests, a 25-ft-high tower filled with redwood slats was constructed. Secondary effluent was treated with caustic soda to pH 11.5,

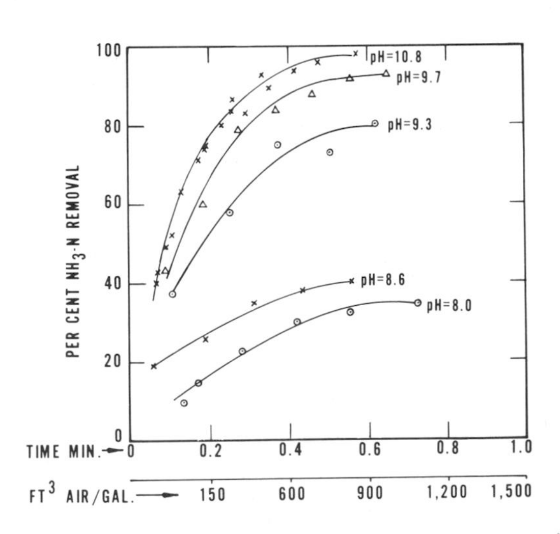

Fig. 13. Percentage of ammonia removal versus aeration time (Ref. 24).

and fed into the tower. Water and air were contacted in a countercurrent mode. Results are presented in Figure 14. For a given supply of air per gallon of wastewater, the efficiency of ammonia stripping increased with increasing packing depth. Efficiencies of the 20- and 24-ft depths were essentially the same for ammonia removals up to 90%. Higher removals up to 98% could only be obtained with the 24-ft of packing.

On the basis of this experience, a full-scale stripping tower was constructed at South Lake Tahoe. The tower was designed to remove 90% of the ammonia from 3.5 mgd (million gallons per day) lime-treated secondary effluent. Air flow is cross-flow while the water drips downward through the packing.

Initial operation of the stripping tower in the winter immediately revealed a limitation of ammonia stripping. The cold air contacting the wastewater resulted in icing on the slats. Moreover, since ammonia is more soluble in cold water than in warm water, more air was required to remove it. From theoretical considerations, difficulty with freezing may be expected when the wet-bulb temperature of the air is below 0°C. The large volumes of air used make heating uneconomical.

Another problem encountered with ammonia stripping is

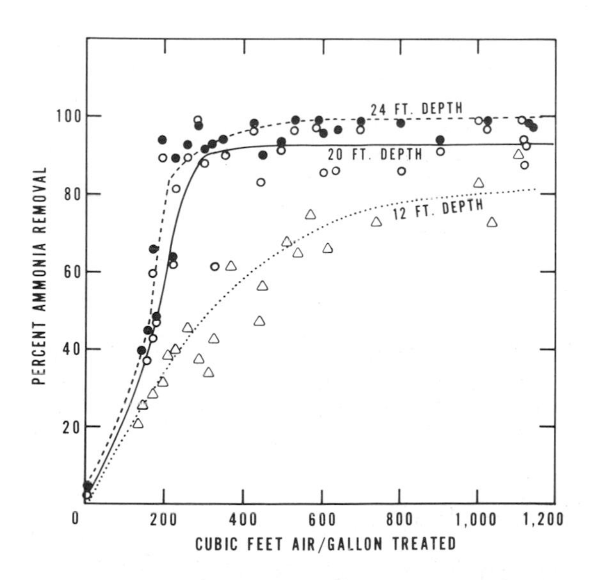

Fig. 14. Percentage of ammonia removal versus air supply for various depths of packing (Ref. 24).

the formation of scale in the tower. Because the previously treated effluent is saturated with respect to calcium carbonate, contact with large volumes of air containing carbon dioxide precipitates calcium carbonate in the tower. In the case of the tower at Tahoe, the scale could be flushed from the tower except from inaccessible areas, which could not be reached with a water jet.

O'Farrell and Frauson (43) have described similar scaling difficulties encountered with ammonia stripping in a pilot tower, except that, in contrast to the experience at Lake Tahoe, the scale was hard and adherent. The scale could only be removed from the packing grids by manual cleaning. However, a pilot tower has been operated in South Africa with scale so soft that the force of the water dripping through the tower fill has been sufficient to flush out the scale. Clearly the causes of the differences in the nature and amount of scale in various locations have not been resolved.

Cost of ammonia stripping, based on one year's operation at Lake Tahoe, has been reported by Evans and Wilson (44). The capital and operating cost, assuming continuous operation of a 7.5 mgd tower was computed to be $0.018/1000 gal. This cost does not include the expense of pH adjustment to 11.5, which is charged to phosphorus removal.

The concept of ammonia removal by air stripping is technically and economically sound. Practical application involves some restrictions. If 90% removal of ammonia nitrogen is required even in cold weather, provision must be made for increasing the air-to-water ratio, which in turn can increase costs by about 50%. In below-freezing weather the tower will be inoperable. Design of the tower must facilitate removal of accumulated solids.

A greater hindrance to the practical use of towers is the discharge of ammonia to the atmosphere. Because of the enormous dilution with air, the odor of ammonia cannot be detected. The concentration of ammonia is estimated to be 6700 $\mu g/m^3$ in the air exiting the tower. Questions can be raised, however, about the practice of transferring ammonia from a wastewater to the air. Locating a tower near a large body of water would not be recommended.

Selective Ion Exchange

Since ammonia exists in solution largely as the ammonium ion NH_4^+ at pH values less than about 9.5, removal can be obtained by conventional ion-exchange methods.

Synthetic ion-exchange resins are uneconomical because of the problems of brine disposal and, more important, because of the selectivity of ion-exchange resins for other cations, principally divalent ions, which predominate in wastewaters. The selectivity of synthetic ion-exchange resins follows the Hofmeister or lyotropic series ($Ca^{2+} > Mg^{2+} > K^+ > NH_4^+ > Na^+$) which makes these exchangers inefficient in removing ammonia in the presence of divalent ions. Recent reports (45,46) indicate that these problems may be minimized by using an ammonium-selective zeolite such as clinoptilolite. Clinoptilolite is a naturally occurring zeolite that can be used after grinding and sieving to 20 × 50 mesh. The selectivity derives from the size of the pores and from the exchangeable cation sites located throughout the three-dimensional structure of the zeolite. The selectivity is in the order $K^+ > NH_4^+ > Na^+ > Ca^{2+} > Mg^{2+}$. The combination of selectivity and an exchange capacity of about 2.0 mequiv/g make clinotpilolite a useful exchanger for ammonium ion in wastewater.

Following extensive laboratory work (45), a 100,000 gal/day demonstration pilot plant was constructed. The pilot plant included a pretreatment plant consisting of chemical flocculation and filtration apparatus and an ion-exchange regeneration facility, which contained three 750-gal ion-exchange vessels and an ammonia stripping column. Two columns were operated in series. When ammonia broke through in the first column, the third column, containing freshly regenerated clinoptilolite, was put on stream as the second column in series.

Breakthrough curves for two clinoptilolite columns with a bed volume of 500 gal are plotted in Figure 15. The flow rate was 50 gal/min of alum-coagulated secondary effluent containing an average of 16.3 mg/liter NH_3-N. The first bed was operated alone until ammonia breakthrough was detected at a throughput of 75,000 gal. At this point the second bed was placed in series with the first. Average ammonia removal was 97% for the 154,000 gal processed, and the average effluent ammonia nitrogen concentration was 0.53 mg/liter.

The loaded clinoptilolite can be regenerated by any basic solution of a pH value exceeding 11.0. Regenerants tested were solutions saturated with lime, to raise the pH to about 12, and calcium chloride alone or supplemented with sodium chloride. The presence of a significant amount of Na^+ in the lime regenerant solution permits more rapid exchange for NH_4^+ than in the case where only Ca^{2+} is available for exchange. Figure 16 presents NH_4^+ elution curves for three different regenerant

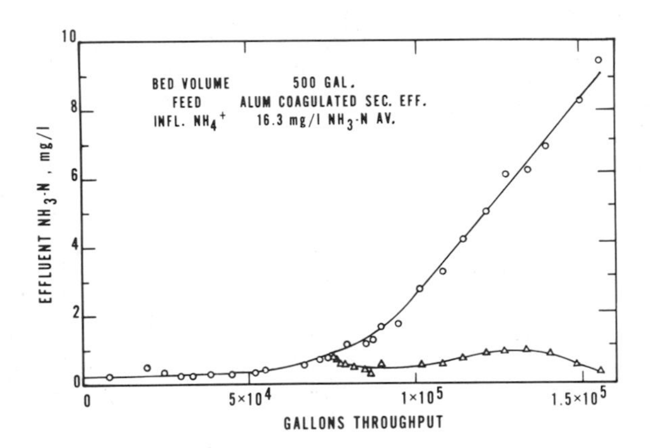

Fig. 15. Ammonia breakthrough curves for clinoptilolite columns (Ref. 45).

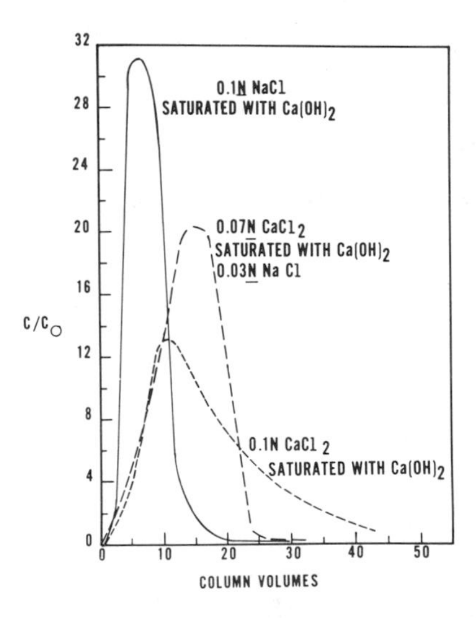

Fig. 16. NH_4^+ elution curves for three regenerant solutions.

solutions (45). Regeneration with a mixture of calcium chloride and sodium chloride in a saturated lime solution represents the most economic approach. The composition of this mixture is near that which results from continuous recycle of a regenerant solution (0.1 N NaCl + $CaCl_2$) after air stripping to remove ammonia. Following regeneration, the columns are rinsed with water to remove remaining regenerant and to reduce the pH to about 9.0.

In more extensive studies on regeneration conditions conducted at the University of California (47), regeneration at pH 12.5 was found to be more effective than at pH 11.5 or 12.0. Higher pH values, however, led to dissolution of the clinoptilolite. Tests with an ammonia-removal pilot plant showed average ammonia removal of 95.2%, with an average effluent ammonia content of 0.53 mg/liter NH_3-N.

After ammonia has been removed by air stripping, the ammonia-laden regenerant solution can be recycled. The problems of ammonia dispersal to the atmosphere are similar to those encountered in direct air stripping of ammonia. But, since the ammonia content is 20 to 50 times as high and is contained in a much smaller volume of liquid, it is feasible to heat the air used for stripping and thus avoid the problem of freezing in cold weather. Because of the high concentration of ammonia, its odor is readily detectable in the effluent gas, although, of course, the total amount does not differ from that produced by direct stripping. Because under certain conditions there could be objections to both the odor and the discharge to the atmosphere of ammonia, other methods of removal of the gas from the regenerant solution are being sought.

Pilot studies at Battelle-Northwest (45) indicated a cost per gallon of between \$0.03 and \$0.06. An estimate prepared for a 7.5 mgd plant at Lake Tahoe showed \$0.15/1000 gal. By comparison with other fixed adsorption methods, it is reasonable to suggest a cost of about \$0.10/1000 gal for ammonia removal by zeolites. Firmer cost estimates must await more pilot operations.

Chlorine Oxidation of Ammonia

It has long been observed that breakpoint chlorination in water treatment practice has resulted in oxidation of ammonia (48). Essentially, this concept provides a physical-chemical approach to removal of ammonia from wastewater. Much of the basic work on breakpoint chlorination has been done on river waters, which nominally contain ammonia nitrogen concentrations below 1 mg/liter of nitrogen. An excellent review of the chemistry of reactions between ammonia and chlorine has appeared in a series of papers by Palin (49). It is only during the past two or three years that research has been undertaken on wastewaters with concentrations on the order of 10 to 20 mg/liter of ammonia nitrogen. Whereas the objective of chlorination in water treatment is to disinfect and control odors, the objective in wastewater is to oxidize ammonia.

In the presence of aqueous ammonia, hypochlorous acid (aqueous solution of chlorine) will react to form a variety of chloramine compounds according to the following equations:

$$Cl_2 + H_2O = HOCl + H^+ + Cl^- \quad \text{hypochlorous acid} \tag{10}$$

$$NH_4^+ + HOCl \longrightarrow NH_2Cl + H_2O + H^+ \quad \text{monochloramine} \tag{11}$$

$$NH_2Cl + HOCl \longrightarrow NHCl_2 + H_2O \quad \text{dichloramine} \tag{12}$$

$$NHCl_2 + HOCl \longrightarrow NCl_3 + H_2O \quad \text{trichloramine} \tag{13}$$

Although equations 11 to 13 represent the stoichiometry of the chlorine-ammonia reactions, the relative concentration of each of the species is determined by the ratio of initial molar concentrations of chlorine and ammonia, which determines the rates of reaction; an even more important determinant of relative concentration is the pH of the system. Reactions with organic material and other nitrogen compounds also occur but are of relatively minor importance to chlorine consumption.

Several reactions have been suggested to explain the decrease in chlorine residual when increasing amounts of chlorine are added to a solution containing ammonia (i.e., the breakpoint curve). Palin (49) suggests that two reactions predominate, at least in the concentrations found in water-treatment practice. These equations are

$$NH_2Cl + NHCl_2 \longrightarrow N_2 + 3H^+ + 3Cl^- \tag{14}$$

$$2NH_2Cl + HOCl \longrightarrow N_2 + 3H^+ + 3Cl^- + H_2O \tag{15}$$

According to these equations, the stoichiometric ratio of chlorine to ammonia nitrogen at the breakpoint would be 7.6 to 1 by weight. Observed ratios approach 8.9 to 1 as a result of side reactions. In practice, breakpoint chlorination of wastewater does not quite follow the simple paths represented by equations 14 and 15. The reactions become very much more complicated and much additional research is needed to identify the multiple interactions and reaction products of chlorine and ammonia. Nevertheless, recent work by Pressley et al. (50), Lawrence et al. (51), and Cassel et al. (52) indicates that oxidation of ammonia by chlorine is technically feasible. Undesirable side reactions can be controlled so that the principal end product of the overall reaction is the desired nitrogen gas, resulting in almost total removal of ammonia nitrogen.

A laboratory study of the breakpoint process for ammonia removal was conducted at the pilot plant of the Environmental Protection Agency, in Washington, D.C., and has been described by Pressley et al. (50). Addition of chlorine to buffered solutions of ammonia at various pH values resulted in the typical breakpoint curves of Figure 17. At a pH of 7.0 the breakpoint occurred at nearly the stoichiometric ratio of approximately 8:1 weight ratio of $Cl:NH_3-N$. At pH values outside this range, increased amounts of chlorine were required for breakpoint and total residual chlorine in the product increased.

Palin (49) has offered an explanation of the typical breakpoint curve. In the ascending branch of the curve, the chlorine is present in combination with ammonia as mono- and dichloramine. The peak of the curve represents the point where all the ammonia initially present is combined to form the chloramines. In the presence of excess chlorine, the chloramines are unstable and react with free

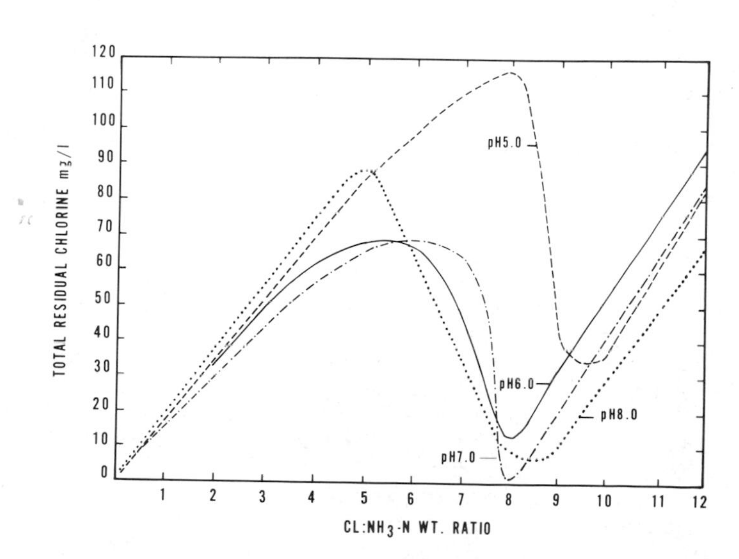

Fig. 17. Breakpoint chlorination at different pH values (Ref. 50).

chlorine; the net result is that the amount of available chlorine is decreased. This accounts for the descending leg of the curve. When all the chloramines have been converted to a form that will no longer react with free chlorine, the further addition of chlorine causes a proportionate rise in residual chlorine. This accounts for the ascending branch of the curve. The simple overall reaction of chlorine with ammonia can be expressed by the following equation:

$$2NH_3 + 3Cl_2 \longrightarrow N_2 + 6H^+ + 6Cl^- \qquad (16)$$

From the equation, it can be seen that the final products of the reaction at the breakpoint are nitrogen gas and hydrochloric acid. Also, this is the equation that is normally used to calculate the stoichiometric weight of chlorine needed to oxidize ammonia nitrogen, which is generally stated to be 7.6 parts of chlorine per part of ammonia nitrogen.

Analyses of the gas evolved from the reaction of chlorine and ammonia at the breakpoint ratio have shown that only N_2 is produced (50). The off-gas was analyzed for N_2, O_2, N_2O, NO, and NO_2. However, analyses of the solution revealed that undesirable side reactions occur, leading to the formation of nitrate and trichloramine (nitrogen trichloride). Pressley (50) has demonstrated that the formation of these end products is a function of pH (Figure 18). At pH values greater than 6.0, nitrate formation increases and can attain a concentration as high as 2.0 mg/liter of NO_3-N. At pH values less than 7.0, nitrogen trichloride is produced and can attain concentrations in excess of 2.0 mg/liter of NCl_3-N.

Nitrogen chloride is a volatile oil with a pungent odor resembling chlorine. Its vapor attacks the eyes and mucous membranes, and is most undesirable in a product water. Its formation, however, can be controlled by maintaining the pH in excess of 7.0. The concentration is negligible--0.05 mg/liter of NCl_3-N, at pH 8.0. Nitrogen trichloride is unstable and hydrolyzes on standing to ammonia and hypochlorous acid.

The reactions of chlorine with ammonia are rapid and essentially unaffected by temperature (53). Breakpoint tests at pH 6.0 in the temperature range of 5 to 40°C have not revealed significant changes in reaction products after two hours contact time.

The overall reaction, equation 16, indicates that substantial amounts of hydrochloric acid are produced. Stoichiometrically, 14.3 mg/liter of $CaCO_3$ alkalinity are required to neutralize the acid produced by the oxi-

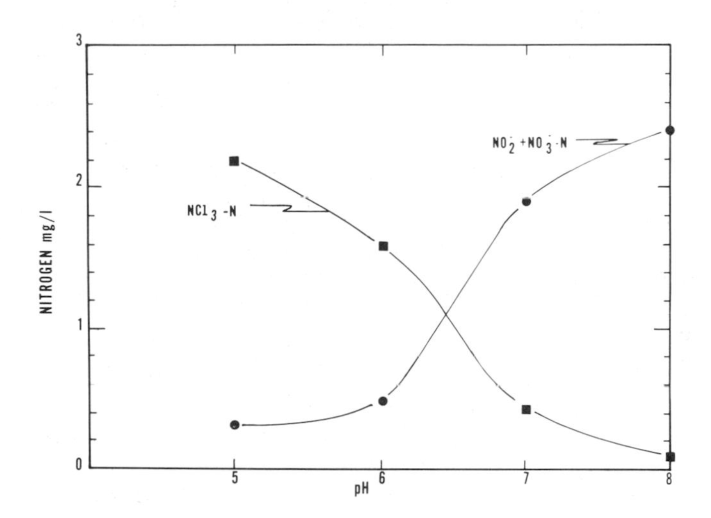

Fig. 18. Effect of pH on NO_2^- - NO_3^- and NCl_3 formation (Ref. 50).

dation of 1 mg/liter NH_3-N to N_2. Thus a wastewater containing 15 mg/liter NH_3-N will consume an alkalinity of about 215 mg/liter. To avoid the formation of undesirable nitrogen trichloride and to maintain a product water in the range of pH 6 to 7, additional alkalinity will have to be added if sufficient alkalinity is not already present.

At precisely the breakpoint, the treated solution should contain no chlorine or chloramines, and ammonia nitrogen should be absent. In practical terms, however, precise pacing of chlorine addition with ammonia nitrogen content will not be possible. To avoid discharging a product water containing free chlorine, chloramines, or chlorinated organic compounds, the treatment system generally includes some type of dechlorination process.

Magee (54), in his excellent review of dechlorination on activated carbon, showed that free chlorine is effectively reduced on the carbon. Moreover, the "life" of the carbon is extremely long. Activated carbon dechlorination plants accepting chlorinated river waters have been in operation for 7 to 18 years with no appreciable decline in activity. Recent work (51,55) has revealed that free chlorine is totally reduced by activated carbon. The evidence on removal of chloramines is less clear, and it is obvious that more research is needed to determine the fate of chloramines when applied to activated carbon.

Overall, ammonia nitrogen removal by breakpoint chlorination is technically feasible. In the range of pH 6.5 to 7.5, 95 to 99% of the ammonia can be oxidized to nitrogen gas. Capital costs for the process would be small, and the principal expenses are the cost for chlorine and a small additional cost for lime for alkalinity. Thus, for an effluent containing 15 mg/liter NH_3-N, approximately 120 mg/liter of chlorine would be required. At $0.04/lb, the cost for chlorine would be $0.038/1000 gal of treated water. Since breakpoint chlorination is a most effective disinfection process, part of the cost of ammonia removal can be credited to disinfection.

Two principal disadvantages can be cited for breakpoint chlorination. First, chlorides are added to the product water; each milligram of chlorine added is reduced to chloride ion and, for pH-controlled situations, an equivalent quantity of cation is added. For an initial ammonia nitrogen concentration of 15 mg/liter, the increment in total dissolved solids (caustic added for pH control), could be 240 mg/liter. Where additional total dissolved solids in the product water are objectionable, this process should not be considered. The second disadvantage is that large amounts of liquid chlorine must be transported and handled. A solution to this dilemma, however, is on-site electrolytic production of sodium hypochlorite. Electrochemical production of sodium hypochlorite has been practiced for sewage chlorination and treatment for years (56), but the process needs to be optimized for breakpoint treatment.

SUMMARY

Phosphorus, an important nutrient in eutrophication, can be removed from wastewater. Alternatives are available when choosing the precipitant to use and location in a treatment system. Removals in excess of 95% can be achieved. The cost is reasonable, and application at full scale is proceeding.

Nitrogen compounds, which are nutrients as well as objectionable pollutants, also can be removed from wastewater. Three physical-chemical processes are in various stages of development. Removals in excess of 90 to 95% appear to be obtainable. Additional pilot plant study is required before any of the processes can be applied at full-scale needs.

REFERENCES

1. J. Shapiro, "A Statement on Phosphorus," J. Water Pollut. Control Fed., 42, 772-775 (1970).
2. Committee Report, "Chemistry of Nitrogen and Phosphorus in Water," J. Am. Water Works Assoc., 62, 127-140 (1970).
3. M. S. Finstein and J. V. Hunter, "Hydrolysis of Condensed Phosphates during Aerobic Biological Sewage Treatment," Water Res., 1, 247-254 (1967).
4. J. E. Laughlin, "Modification of a Trickling Filter to Allow Chemical Precipitation," presented at Advances in Waste Treatment Research and Water Reuse Symposium, sponsored by Environmental Protection Agency, Dallas, January 1971.
5. J. J. Convery, "Treatment Techniques for Removing Phosphorus from Municipal Wastewaters," Water Pollution Control Research Series 17010, Environmental Protection Agency, 1970.
6. L. K. Cecil, "Problems and Practice of Phosphate Removal in Water Reuse," Chem. Eng. Progr. Symp. Ser., 63, 159-163 (1967).
7. J. B. Nesbitt, "Phosphorus Removal--The State of the Art," J. Water Pollut. Control Fed., 41, 701 (1969).
8. R. W. Bayley, "Nitrogen and Phosphorus Removal: Methods and Costs," J. Soc. Water Treat. Exam., 19 (3), 294-319 (1970).
9. K. Wuhrmann, "Objectives, Technology and Results of Nitrogen and Phosphorus Removal Processes," in "Advances in Water Quality Improvement," E. F. Gloyna and W. W. Eckenfelder, Eds., University of Texas Press, Austin and London, 1968, p. 21.
10. R. L. Culp, "The Status of Phosphorus Removal," Pub. Works, 100, 76-81 (1969).
11. S. K. Malhotra, G. F. Lee, and G. A. Rohlich, "Nutrient Removal from Secondary Effluent by Alum Flocculation and Lime Precipitation," Int. J. Air Water Poll. (Brit.), 8, 487-500 (1964).
12. W. H. Lea, G. A. Rohlich, and W. J. Katz, "Removal of Phosphate from Treated Sewage," Sewage Ind. Wastes, 26, 261-275 (1954).
13. W. Stumm, in "Advances in Water Pollution Research," W. W. Eckenfelder, Ed., Proceedings of the First International Conference on Water Pollution Research, Pergamon Press Ltd., London, 1964, pp. 216-230.
14. C. V. Cole and M. L. Jackson, "Colloidal Dihydroxy Dihydrogen Phosphates of Aluminum and Iron with Crystalline Character Established by Electron and X-Ray Diffraction," J. Phys. Colloid Chem., 54,

128-142 (1950).
15. A. Henricksen, "Laboratory Studies on the Removal of Phosphates from Sewage by the Coagulation Process," J. Hydrol., 24, 253-271 (1962).
16. J. B. Farrell, B. V. Salotto, R. B. Dean, and W. E. Tolliver, "Removal of Phosphate from Wastewater by Aluminum Salts with Subsequent Aluminum Recovery," Chem. Eng. Progr. Symp. Ser. 64, 90, 232-239 (1969).
17. H. L. Recht and M. Ghassemi, "Kinetics and Mechanisms of Precipitation and Nature of the Precipitate Obtained in Phosphate Removal from Wastewater Using Aluminum(III) and Iron(III) Salts," Water Pollution Control Research Series 17010 EKI 04/70, Environmental Protection Agency, 1970.
18. C. N. Sawyer, "Some New Aspects of Phosphates in Relation to Lake Fertilization," Sewage Ind. Wastes, 24, 768-776 (1952).
19. G. W. Heinke and J. D. Norman, "Hydrolysis of Condensed Phosphates in Wastewater," Proc. 24th Ind. Waste Conf., Purdue University, 135, 644-654 (1969).
20. H. Galal-Gorchev and W. J. Stumm, "The Reaction of Ferric Iron with Orthophosphate," J. Inorg. Nucl. Chem., 25, 567-574 (1963).
21. F. Pöpel, "Removal of Phosphate from Sewage by Coagulation with Ferric Iron and Aluminum," Proceedings of the Fourth International Conference on Water Pollution Research, Pergamon Press Ltd., London, 1969, pp. 643-653.
22. M. Ghassemi and H. L. Recht, "Phosphate Precipitation with Ferrous Iron," Water Pollution Control Research Series 17010 EKI 09/71, Environmental Protection Agency, 1971.
23. R. F. Wukasch, "New Phosphate Removal Process," Water Wastes Eng., 5, 58-60 (1968).
24. A. F. Slechta and G. L. Culp, "Water Reclamation Studies at the South Tahoe Public Utility District," J. Water Pollut. Control Fed., 39, 787-814 (1967).
25. J. C. Buzzell and C. N. Sawyer, "Removal of Algal Nutrients from Raw Wastewater with Lime," J. Water Pollut. Control Fed., 39, R16-R24 (1967).
26. L. A. Schmid and R. E. McKinney, "Phosphate Removal by a Lime-Biological Treatment Scheme," J. Water Pollut. Control Fed., 41, 1259-1276 (1967).
27. M. C. Mulbarger, E. Grossman, R. B. Dean, and O. L. Grant, "Lime Clarification, Recovery, Reuse, and Sludge Dewatering Characteristics," J. Water Pollut. Control Fed., 41, 2070-2085 (1969).
28. J. B. Stamberg, D. F. Bishop, H. P. Warner, and S. H. Griggs, "Lime Precipitation in Municipal Wastewaters,"

in "Water--1970," Chem. Eng. Progr. Symp. Ser. 107, 67, 310-320 (1971).
29. J. F. Ferguson, D. Jenkins, and W. Stumm, "Calcium Phosphate Precipitation in Wastewater Treatment," in "Water--1970," Chem. Eng. Progr. Symp. Ser. 107, 67, 279-287 (1971).
30. M. C. Mulbarger, E. Grossman, R. B. Dean, and O. L. Grant, "Lime Clarification, Recovery, Reuse, and Sludge Dewatering Characteristics," J. Water Pollut. Control Fed., 41, 2070-2085 (1969).
31. S. A. Black and W. Lewandowski, "Phosphorus Removal by Lime Addition to a Conventional Activated Sludge Plant," Toronto, Ont. Res. Publ. 36, Ontario Water Resources Commission, November 1969.
32. M. C. Rand and N. L. Nemerow, "Removal of Algal Nutrients from Domestic Wastewater," Rep. 9, Department of Civil Engineering, Syracuse University Research Institute, Syracuse, N.Y., 1965.
33. O. E. Albertson and R. L. Sherwood, "Phosphate Extraction Process," J. Water Pollut. Control Fed., 41, 1467-1490 (1969).
34. H. L. Recht and M. Ghassemi, "Phosphate Removal from Wastewaters Using Lanthanum Precipitation," Water Pollution Control Research Series 17010 EFX 04/70, Environmental Protection Agency, 1970.
35. G. Schwab, "Nature and Applications of Inorganic Alumina Chromatography," Discussions Faraday Soc., 7, 170-173 (1949).
36. W. C. Yee, "Selective Removal of Mixed Phosphates by Activated Alumina," J. Am. Water Works Assoc., 58, 239-247 (1966).
37. L. L. Ames and R. B. Dean, "Phosphorus Removal from Effluents in Alumina Columns," J. Water Pollut. Control Fed., 42, R161-R172 (1970).
38. B. F. Winkler and G. Thodos, "Kinetics of Orthophosphate Removal from Aqueous Solutions by Activated Alumina," J. Water Pollut. Control Fed., 43, 474-482 (1971).
39. L. L. Ames, "Mobile Pilot Plant for Removal of Phosphate from Wastewater by Adsorption on Alumina," Water Pollution Control Research Series 17010 EER 05/70, Environmental Protection Agency, 1970.
40. A. M. Hanson and T. F. Flynn, "Nitrogen Compounds in Sewage," Proc. 19th Ind. Waste Conf. Purdue University, 117, 32-44 (1964).
41. P. A. Kuhn, "Removal of Ammonium Nitrogen from Sewage Effluent," MS thesis, University of Wisconsin, Madison, 1956.
42. R. W. Bayley, "Desorption of Wastewater Gases in Air,"

Effl. Water Treat. J., 7, 78-84 and 154-161 (1967).
43. T. P. O'Farrell and F. P. Frauson, "Ammonia Stripping at Washington, D.C." presented at Nutrient Removal Seminar, University of Pittsburgh, Pittsburgh, February 1970.
44. D. R. Evans and J. C. Wilson, "Actual Capital and Operating Costs for Advanced Waste Treatment," presented at the 43rd Annual Conference of the Water Pollution Control Federation, Boston, October 1970.
45. B. W. Mercer, L. L. Ames, C. J. Touhill, W. J. Van Slyke, and R. B. Dean, "Ammonia Removal from Secondary Effluents by Selective Ion Exchange," J. Water Pollut. Control Fed., 42, R95-R107 (1970).
46. Battelle Memorial Institute, Pacific Northwest Laboratories, "Ammonia Removal from Agricultural Runoff and Secondary Effluents by Selective Ion Exchange," Federal Water Pollution Control Administration Rep. TWRC-5, March 1969.
47. J. H. Koon, A. B. Mener, D. Jenkins, and W. S. Kaufman, "Chemical Processing of Primary Organic Waste Streams," Report to Federal Water Pollution Control Administration, Research & Development Grant Project 17080 DAR, Sanitary Engineering Research Laboratory, University of California, Berkeley, February 1971.
48. G. C. White, "Chlorination and Dechlorination: A Scientific and Practical Approach," J. Am. Water Works Assoc., 60, 540-561 (1968).
49. A. T. Palin, A Study of the Chloro Derivatives of Ammonia and Related Compounds, with Special Reference to Their Formation in the Chlorination of Natural and Polluted Waters," Water and Water Eng., 54, 151-159, 189-200, and 248-256 (1950).
50. T. A. Pressley, D. F. Bishop, and S. G. Roan, "Nitrogen Removal by Breakpoint Chlorination," presented at the National Meeting of American Chemical Society, Chicago, September 1970.
51. A. W. Lawrence, W. S. Howard, and K. A. Rubin, "Ammonia Nitrogen Removal from Wastewater Effluents by Chlorination," presented at the Fourth Industrial Waste Conference, University of Delaware, Newark, November 1970.
52. A. F. Cassel, T. A. Pressley, W. W. Shuk, and D. F. Bishop, "Physical-Chemical Nitrogen Removal from Municipal Wastewater," presented at the 68th National Meeting of the American Institute of Chemical Engineers, Houston, March 1971.
53. J. C. Morris, "Kinetics of Reactions Between Aqueous Chlorine and Nitrogen Compounds," in "Principles and

Applications of Water Chemistry," S. D. Faust and J. V. Hunter, Eds., Wiley, New York, 1967, pp. 23-51.
54. V. Magee, "The Application of Granular Active Carbon for Dechlorination of Water Supplies," Proc. Soc. Water Treat. Exam., 5, 17-33 (1956).
55. "Ammonia Removal in a Physical-Chemical Wastewater Treatment Process," Report prepared by Ayres, Lewis, Norris and May, Inc. for Environmental Protection Agency, July 1971.
56. H. W. Marson, "Electrolytic Methods in Modern Sewage Treatment," Effluent Water Treat. J., 7, 71-73 and 75-77 (1967).

Chapter XIII

NUTRIENT REMOVAL FROM WASTEWATER BY BIOLOGICAL TREATMENT METHODS

M. W. Tenney, W. F. Echelberger, Jr.

Department of Civil Engineering, University of Notre Dame, Notre Dame, Indiana

K. J. Guter

Office of the Special Assistant to the Commanding Officer for Research and Development Activities, U.S. Army Environmental Hygiene Agency, Edgewood Arsenal, Maryland

and

J. B. Carberry

Department of Civil Engineering, University of Notre Dame, Notre Dame, Indiana

The role of biological activities in water pollution and in its control is a significant one. Pollution, for example, can induce appreciable changes in the biological relationships of a water body. These changes or imbalances are frequently regarded as indicators of water pollution. Conversely, comprehension of the fundamental microbiological and biochemical relationships associated with water pollution permits the effective utilization of biological processes in controlled reactors for the treatment of wastewaters, thereby eliminating the potential pollutional aspects of a wastewater on a receiving water.

This chapter discusses the removal from wastewaters by biological treatment methods of three dissolved nutrient impurities: organic carbon, phosphorus, and nitrogen. Although these three pollutants are often regarded as the major nutrient impurities in wastewaters, particularly in terms of their impact on receiving waters, they are by no means the only such impurities, and frequently their impact is overshadowed by that of other contaminants. Similarly, although biological treatment methods often serve to remove these three impurities from wastewaters, numerous physical and chemical treatment schemes exist which can be equally efficacious for this purpose.

FUNDAMENTAL MICROBIOLOGICAL RELATIONSHIPS

The underlying principle of biological waste treatment is to provide environmental conditions satisfactory for the controlled growth of selected microbial species that, through their metabolic activities, are capable of removing the desired pollutant(s) from a given wastewater. In this regard, the pollutant represents a food or nutrient source (substrate) to the microorganism. The following equation indicates the overall summary reaction for a microbial process.

$$\text{substrate} + \text{microbes} \xrightarrow{\text{satisfactory environmental conditions}} \text{more microbes} + \text{end products} \qquad (1)$$

Equation 1 indicates that the pollutant (substrate) can be removed from solution in one or both of the following ways:

1. By conversion to microbially produced end products.
2. By incorporation into (or onto) the cellular protoplasm of newly synthesized microbes.

In order to achieve effective treatment, the products

of equation 1 must be nonpollutional, or they must be capable of being removed by subsequent treatment processes.

The microorganisms of significance to biological waste treatment systems can be classified into two major groups (based on their metabolic activity): heterotrophs and autotrophs. Most biological treatment systems employ a community that is primarily heterotrophic or primarily autotrophic, and a predominant species within the community ultimately results; however, environmental conditions found in wastewater treatment are seldom such that these microorganisms will be found to be mutually exclusive. In fact, some proposed systems suggest the mutual culturing of both classes (1).

Both heterotrophic and autotrophic microorganisms have approximately the same elemental composition of carbon, hydrogen, oxygen, nitrogen, phosphorus, and sulfur; a widely used formulation for their cellular protoplasm is $C_{106}H_{180}O_{45}N_{16}P_1$. Various trace metals are needed as well.

The principal difference between the two classes of microorganisms is that heterotrophic organisms utilize organic material as their source of carbon, whereas autotrophic organisms utilize inorganic carbon. Heterotrophic activity can occur either in the presence of oxygen by aerobic species or in the absence of oxygen by anaerobic species. Facultative heterotrophic species are capable of metabolizing organic material under either aerobic or anaerobic conditions; aerobic metabolism is preferred, however, as greater energy yields are obtainable under these conditions.

The overall reactions for the three possible microbiologically induced carbon transformations are shown in Table 1.

1. Stabilization or oxidation of organic material (equation 1 in Table 1). In this transformation, under aerobic conditions, a fraction of the organic material is oxidized by the heterotrophic microorganisms to carbon dioxide and water in order to obtain sufficient energy to synthesize the remainder of the organic material to new cellular protoplasm.

2. Decomposition or reduction of organic material (equation 2 in Table 1). Two different classes of heterotrophic activity accomplish this transformation under anaerobic conditions: (a) initially facultative species convert the organic material to volatile acids (monocarboxylic acids), and ultimately (b) anaerobic methane-forming organisms reduce the volatile acids to carbon dioxide and methane.

3. Production of organic material (equation 3 in

Table 1. Stoichiometry of Microbially Induced Carbon Transformations.

I. HETEROTROPHIC ACTIVITIES

A. Under Aerobic Conditions:

$$\alpha C_aH_bO_cN_dP_e + \beta O_2 = \gamma C_{106}H_{180}O_{45}N_{16}P_1 + \delta CO_2 + \epsilon H_2O + \xi NO_3^- + \rho PO_4^{-3} \quad \ldots\ldots\ldots 1$$

B. Under Anaerobic Conditions:

$$\alpha C_aH_bO_cN_dP_e = \beta R\text{-}COOH = \gamma CO_2 + \delta CH_4 + \epsilon NH_4^+ + \xi H_2S + \rho PO_4^{-3} + \gamma H_2O + \phi C_{106}H_{180}O_{45}N_{16}P_1 \quad \ldots\ldots 2$$

II. AUTOTROPHIC ACTIVITY

$$106\,CO_2 + 90\,H_2O + 16\,NO_3^- + 1\,PO_4^{-3} + \text{light energy} = C_{106}H_{180}O_{45}N_{16}P_1 + 154\tfrac{1}{2}\,O_2 \quad \ldots\ldots\ldots\ldots 3$$

Table 1). In this transformation (photosynthetic activity), autotrophic microorganisms capture solar energy and utilize inorganic carbon sources in order to synthesize new cellular protoplasm and produce oxygen.

These three biochemical reactions form the basis for all microbiologically induced carbon transformations and as such are the foundation for the design of any biological treatment system.

It should be noted, however, that under treatment conditions for wastewaters in microbiological reactors, numerous other phenomena may contribute to additional or apparent luxurious removals of nutrient impurities. In selected cases, such phenomena as adsorption, absorption, precipitation, and ion exchange have been shown to account for nutrient removals in excess of their stoichiometric limit and have been reported to occur at or near the surface of the microorganisms or in the bulk of the solution.

REMOVAL OF CARBONACEOUS POLLUTANTS

The removal of carbonaceous pollutants from wastewaters by biological treatment methods can be accomplished under either aerobic or anaerobic conditions. Common aerobic treatment processes include activated sludge systems, trickling filters, oxidation ponds, and aerobic lagoons; common anaerobic treatment processes include anaerobic digestors and anaerobic lagoons. Although each process has its own operational characteristics, capabilities, and associated modifications, the basic differences between aerobic and anaerobic treatment systems involve the type of microbial species employed and the end products obtained (see Table 1, equations 1 and 2). Anaerobic treatment systems do not require costly oxygen-transfer systems and produce less biomass per unit of substrate processed; the rate of processing of organic material (substrate), however, is characteristically much greater in aerobic systems.

The two basic design considerations of any biological treatment system are:

1. Providing environmental conditions satisfactory for the removal from the wastewater of pollutants due to the metabolic activities of the microorganisms.
2. Insuring that suitable provision is made for the removal of the microorganisms from the treated wastewater after they have fulfilled their metabolic role.

The first design requirements should be viewed in terms of the reactor kinetics, and the second should be considered in connection with the flocculation of the system.

The principal components of a biological treatment system are a reactor where the substrate processing occurs and a separation device (e.g., settling tank, centrifuge, or filter), in which the microorganisms are removed from the treated wastewater. The reactor can be designed in one of two fashions, intermediate cases being taken to approach one of the two modes. In the plug flow reactor (Figure 1), the wastewater is introduced at the "head end" of a tubular or long narrow reactor and this elemental volume of liquid then essentially flows intact through the reactor. In the completely mixed reactor (Figure 2), the wastewater is introduced so that the contents of the reactor are completely mixed (e.g., by one or more of the following means: geometric configuration, mixing, and evenly distributing the influent wastewater in the reactor so that the contents are at the same concentration at any location).

Typical relationships indicating the variation in concentration with time (or distance down the reactor) of residual substrate, biomass, and cell flocculation for a plug flow reactor are given in Figure 3. Conversely, the characteristics of a completely mixed reactor are such

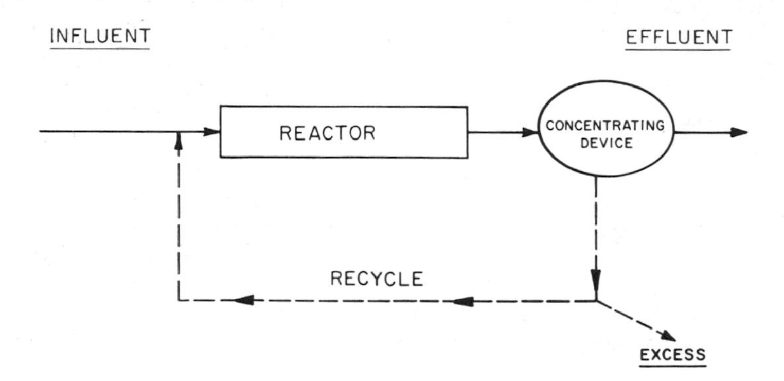

Fig. 1. Typical plug flow biological reactor.

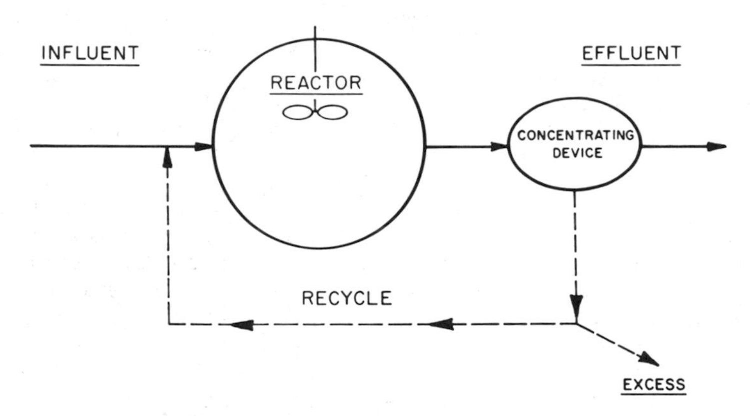

Fig. 2. Typical completely mixed biological reactor.

that operation is essentially at fixed points (steady-state concentrations) on the various curves in Figure 3.

Kinetics of the Biological Removal of Organic Material

Numerous papers have described the kinetics of substrate removal by biological systems. Excellent reviews of the technical literature on this subject as it relates

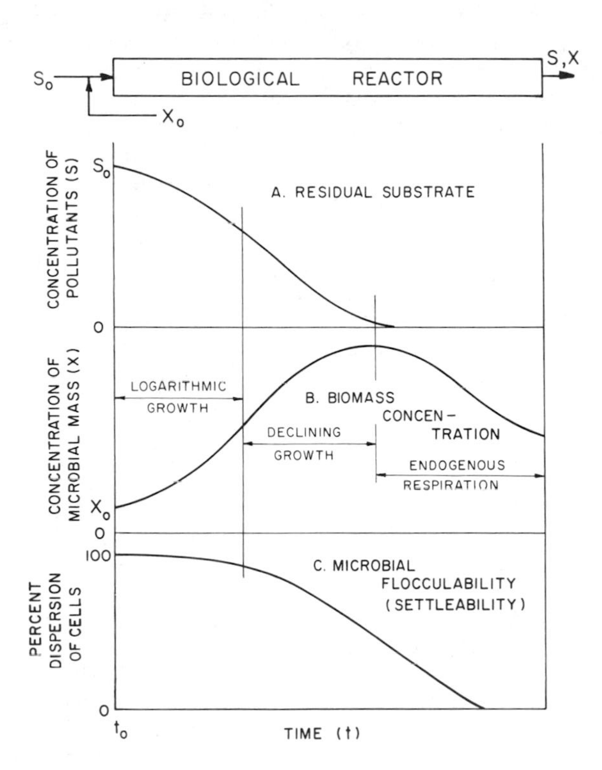

Fig. 3. Typical relationships with time (or distance) in a plug flow biological reactor.

to biological wastewater treatment have been published by Gaudy and Gaudy (2), Bungay and Bungay (3), and Pretorius (4). For the most part, the basic premise in each development follows the classical treatment of Herbert (5), which states that the growth of microorganisms with respect to time is a function of the mean cell division rate times the instantaneous concentration of microorganisms. This relationship was originally reported by Monod (6) for a pure culture of microorganisms through which a single substrate was passed in a continuous flow. The mean cell division rate was said to vary with the substrate concentration and was mathematically expressed by a function similar to the Michaelis-Menten equation for enzymatic reactions.

These relationships, or closely similar ones, have been utilized to predict such parameters as substrate concentration, cell concentration, and biomass yield from biological reactors under either batch or continuous flow operation. It has been recognized, however, that in many cases the observed values of these parameters do not correspond to those predicted from mathematical expressions using Monod relationships. In particular, these kinetics predict neither the lag phase nor the endogenous phase of batch cultivation. Still other surface uptake phenomena, including those involved in such heterotrophic processes as contact stabilization and biosorption, are totally unexplained by Monod's equations of microbial growth.

Other authors have developed different expressions [e.g., Teissier's (7) exponential expression for growth rate and the two-phase proposal of Garret and Sawyer (8)] in an attempt to overcome some of the above-mentioned limitations. The significant point about each of these models, however, is that they are all primarily empirical; that is, the experimental data were fit to a functional form. None of the authors proposes a mechanism for growth because it is always assumed that the metabolic processes are quite complex, involving a variety of rate laws, and that these processes are among the rate-determining steps for growth.

No one would deny that the metabolic processes in the cell are quite complex and not completely understood, but it may be incorrect to assume that knowledge about these rates and their dependency on substrate and enzyme concentration is necessary for a mechanistic determination of growth rate relationships. Since such diverse classes of organisms as bacteria, algae, yeasts, and even molds exhibit the same general form of growth dependency, perhaps the metabolic processes do not play a major role in establishing the growth rate.

It appears that only a few facts about the metabolic activity of the cell are required, along with a knowledge of the mode of growth, in order to accurately explain the kinetics of substrate utilization. An example of such a model based on the mode of growth is one that we have recently proposed (9). In this model of cell growth, it is assumed that every cell is either growing (however slowly) or is lysed. The mode of growth assumed in the model supposes that the cell growth occurs in a two-step manner. The first step involves the accumulation and possible loss of the limiting substrate by the organism; in the second step, the limiting substrate is ingested by the organism and possibly a subsequent cell division occurs.

This process can be put into symbolic form to represent the stoichiometry (conservation of mass) as follows:

$$A + \eta R_s \underset{r_2}{\overset{r_1}{\rightleftharpoons}} B \xrightarrow{r_3} (1 + \varepsilon)A + \sigma_1(\eta - \varepsilon)R_s \qquad (2)$$

This symbolism describes a viable cell A just after cell division combining with an amount η of substrate R_s to form an ingesting cell B. The ingesting cell B can undergo one to two processes: it can release the assimilated nutrients and in the limit revert back to its original form of A and R_s, or it can ingest the substrate and produce $1 + \varepsilon$ units of A. During the ingestion and possibly cell division, some of the substrate may be returned to the medium for use again by other cells; thus the quantity $\sigma_1(\eta - \varepsilon)$ represents the amount of substrate returned to the medium.

From the symbolic form (equation 2) it can be seen that the rate of growth will depend on three rate processes r_1, r_2, and r_3, as indicated. Expressions for these rates are presented in Table 2.

Since in cell death, a certain portion of the lysed organisms are available for food, the stoichiometry of microbial death can be represented as follows:

$$A \xrightarrow{r_4} \sigma_2 R_s \qquad (3)$$

$$B \xrightarrow{r_5} \sigma_3 R_s \qquad (4)$$

From the available evidence, it appears that cell death per unit volume occurs as a first-order rate phenomenon. That is, the fraction of cells which die per unit time is constant. Expressions for r_4 and r_5 are shown in Table 2.

The foregoing expressions for cell growth and cell death can be applied to the operation of either continuous flow or batch-operated heterotrophic reactors in order to predict such parameters as substrate utilization and cellular concentrations. Both heterotrophic and autotrophic growth follow the same growth patterns and relationships; the expressions just given are thus applicable to development of mathematical expressions for any type of biological wastewater treatment system.

Table 2. Rate Expressions for Microbial Cell Growth

Rate	Mathematical expression	Comments
Rate of absorption of substrate	$r_1 = k_1 AS$	Proportional to absorbing cell concentration and mass transfer limited (either in cell wall or surrounding film); thus S represents the difference between the limiting substrate concentration in solution S_s and that maintained by the cell S_c
Rate of loss of substrate	$r_2 = k_2 B$	Proportional to ingesting cell concentration
Rate of ingestion of substrate	$r_3 = k_3 B$	Proportional to ingesting cell concentration
Rate of cell death		
for absorbing cells	$r_4 = k_4 A$	Proportional to absorbing cell concentration
for ingesting cells	$r_5 = k_5 B$	Proportional to ingesting cell concentration

Mode of Biological Flocculation

Microorganisms must be in a flocculated, rather than a dispersed state, in order to be separated effectively from treated wastewater after they have fulfilled their metabolic role. This condition is a prerequisite for effective operation of any type of separation device (e.g., gravity settling tank, upflow clarifier, centrifuge, or filter). In general terms, flocculation consists of agglomerating the otherwise discrete microorganisms into flocs or packets so structured that under quiescent conditions they will have sufficient mass to subside from suspension or, if filtered or centrifuged, the flocs will have sufficient rigidity and associated porosity to permit the effective withdrawal of water to be achieved.

A review of the literature pertaining to biological flocculation indicates that several distinct theories on the subject have been supported throughout the last century. Early investigators, for example, explained cellular agglutination in terms of physical interparticle forces (electrostatic relationships). There is little doubt today that such relationships do play a significant role in agglutination. Yet additional information was necessary to explain why microorganisms can form stable suspensions at their isoelectric point and why net negatively charged microorganisms can be effectively flocculated by a variety of nonionic and anionic polymeric materials.

Other experimenters attributed microbial agglutination to a specific species (viz. *Zoologoea ramigera*), reporting that these microorganisms were capable of producing a gelatinous material that served to cement together the mixed population of microorganisms present (in this case, in activated-sludge systems). Today the significance of the presence of extracellular substances to induce flocculation appears to be beyond question, and it serves to explain some of the limitations of the singular physical theory cited earlier. It does not seem, however, that one species alone produces extracellular substances capable of inducing or enhancing biological flocculation; recent literature has shown that many species of both heterotrophic and autotrophic microorganisms are capable of producing exocellular substances that can serve as flocculants.

Studies in our laboratory have indicated that biological flocculation relationships such as those shown in Figure 3C are typical for mixed cultures of both heterotrophic and autotrophic microorganisms (10,11). Characteristically, microorganisms are dispersed during periods of rapid growth (high ratio of substrate to microorganism mass); whereas in periods of limited growth (low ratio of substrate to microorganism mass), agglutination occurs.

Such phenomena can be explained by a consideration of the amount of exocellular polymer present with respect to the mass, or surface area, of the microorganisms. These relationships are evident from the data in Figure 4, which indicate the accumulation of exocellular polymer (Figure 4c) as a function of growth stage (Figure 4a). The data in Figure 4 were obtained from a batch study of heterotrophic bacteria utilizing nutrient broth as substrate; these data are typical of numerous other substrates for heterotrophic bacteria and are also representative of autotrophic activity.

In the study indicated, the extent of bacterial flocculation was ascertained by light-scattering techniques and is represented as culture turbidity in Figure 4b. The importance of the ratio of exocellular polymer to microorganism mass for biological flocculation is clearly illustrated by a comparison of Figures 4d and 4b. Certainly, flocculation will not occur until sufficient polymer has accumulated with respect to the microbial mass to overcome the dispersion forces inherent to the system (e.g., agitation and surface charge).

The exocellular material, which is microbially produced, is relatively refractory to heterotrophic degradation; thus it accumulates as shown. During long periods of endogenous respiration, however, heterotrophic microorganisms will untimately degrade significant portions of this material and redispersion will result. (Similar relationships will be observed in most autotrophic wastewater treatment systems because some heterotrophic microorganisms are also generally present.) The exocellular material could originate as a waste product of metabolic activity or as polymeric fractions dispersed from cell autolysis.

Because the operation of conventional biological wastewater treatment systems relies on reseeding with flocculated microorganisms, it is not surprising that the compositions of the major constituents of the exocellular materials associated with cellular agglutination are remarkably similar among different types of biological reactors. Four general categories of organic polymers have been found in all the exocellular extracts from such microbial systems: polysaccharide, protein, RNA, and DNA. Typical relationships for these constituents (corresponding to the data reported in Figure 4) are presented in Figure 5.

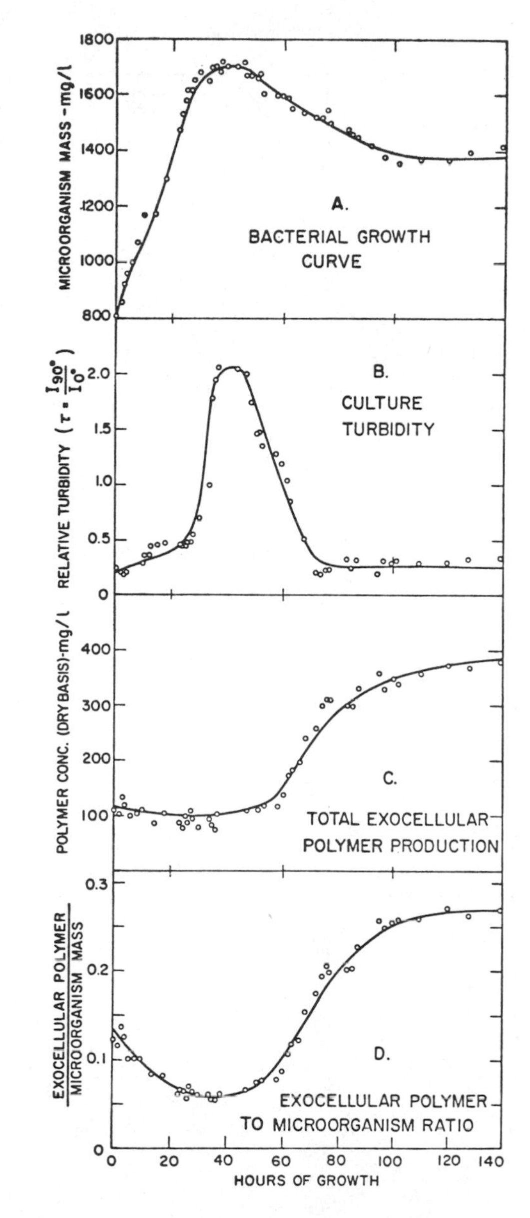

Fig. 4. Relationship between biological growth phase (Figure 4a) and biological flocculation (Figure 4b). Also shown are the accumulation of exocellular polymer (Figure 4c) and the ratio of accumulated exocellular polymer to microorganisms (Figure 4d).

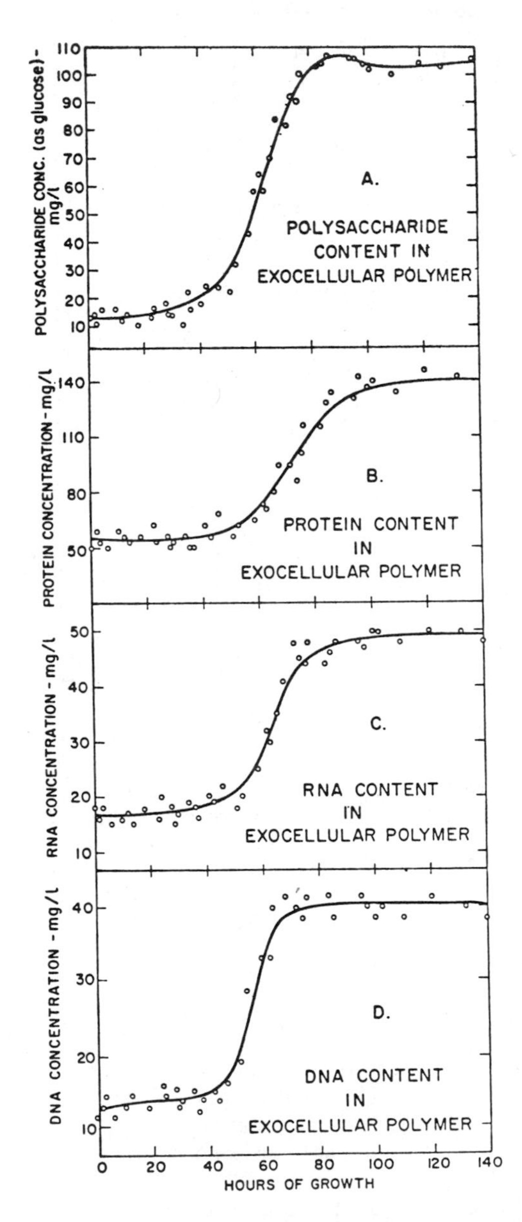

Fig. 5. Concentration of major components of exocellular polymer as a function of biological growth phase.

The phenomenon of biological flocculation can be interpreted as resulting from the flocculant action of the exocellular polymeric material. In this regard, biological autoflocculation follows relationships similar to those observed in the chemical flocculation of microorganisms (12,13). The essence of cell agglutination is based on the chemical and/or electrostatic bonding of polymeric materials to the surface of a cell, with subsequent bridging of extended polymeric segments for bonding to other cellular surfaces and/or attachment to other extended segments. In this fashion, a three-dimensional matrix or floc is formed.

In order to achieve polymer bridging between cells, either the polymeric segments must be sufficiently extended to overcome the effective distance of the net charge of repulsion due to the like electrostatic sign of the cells, or the net surface charge on the cells must be reduced. If polymeric bridging between cells is to be achieved, the resulting efficiency of biological flocculation will be governed by the extent of polymeric surface coverage of the cells. For example, insufficient polymeric coverage will result in redispersion due to insufficient bridging to withstand the forces of agitation, and excessive coverage also will cause redispersion due to the unavailability of vacant sites for extended segment bonding. Numerous environmental factors will influence biological flocculation; some of the more significant factors include pH, temperature, ionic strength, concentration of polyvalent ions, and agitation.

Such a model of biological flocculation would indicate that, in order to achieve effective separation of the microorganisms after their use in biological wastewater treatment systems, the microorganisms must enter the endogenous growth phase. This statement is true for systems in which conditioning for separation is desired solely by autoflocculative techniques. However, if the system is one in which more rapid substrate processing is desired at the sacrifice of cell autoflocculation, definite enhancement of cellular flocculation can be achieved by the addition of chemical flocculants (12).

REMOVAL OF PHOSPHATIC POLLUTANTS

Microbiological removal of phosphorus can be accomplished by heterotrophic or autotrophic microorganisms. The equations in Table 1 indicated the stoichiometry involved in heterotrophic and autotrophic metabolism of wastewater constituents.

When the composition of domestic wastewater is compared with the necessary stoichiometric relationship between carbon, nitrogen, and phosphorus for microorganisms, it appears that most wastewaters are nutritionally unbalanced because they lack sufficient carbon for heterotrophic enrichment. Consequently, only a fraction of the nitrogen and phosphorus in domestic wastewater will be biologically removed by conventional heterotrophic wastewater treatment systems.

The two most commonly utilized heterotrophic biological treatment processes, trickling-filter installations and activated-sludge aeration facilities, are reported to be capable of removing a maximum of 20 and 40% of the phosphorus concentrations, respectively, under ordinary operating conditions (14). These sludges should contain approximately 2 to 3% of their volatile suspended solids as phosphorus (15). However, a number of activated-sludge wastewater treatment plants have reported phosphate removals appreciably greater than 40% (16-19). These removals far exceed any 106:1 stoichiometric carbon-to-phosphorus ratio, and at times these treatment plants reportedly achieve 90 to 95% phosphate removal, even in phosphate-rich wastewaters.

Such excessive phosphate removal by an activated-sludge process has been termed "luxurious uptake of phosphorus," and much controversy has arisen over the mechanism by which this luxurious removal takes place. In 1965 Levin and Shapiro (20) reported that the uptake was a biological one which was dependent on oxygen transfer into the cell and relatively independent of carbon concentration or biomass concentration. In 1970 Menar and Jenkins (15) stated their opinion that such luxurious removals were due to a chemical precipitation phenomenon and that, although the extent of removal was a direct function of the aeration rate, the removal was independent of oxygen transfer. In their scheme, the aeration process allegedly served to strip the metabolically produced carbon dioxide from the sludge suspension. The resultant increase in pH would induce precipitation of some form of calcium phosphate, which is then removed with the biomass during sludge separation.

Two such divergent hypotheses are contradictory and mutually exclusive. Recent studies have been conducted in our laboratory, using South Bend, Indiana, wastewater, which is similar in composition to the previously studied wastewaters. The results of our investigations appear to support the theory that luxury uptake of phosphorus is biologically induced. The data indicate the following conclusions (21):

1. There is an optimum aeration rate for the biological removal of phosphate in a secondary aeration tank. At aeration rates above and below this optimum, there is less removal because less oxygen and/or substrate can be transferred into the cell. Data typical of this dependency are given in Figure 6. These data would indicate that an aeration rate in the range of 300 ml of air/(min)(liter) is optimum.

2. No phosphate removal occurs if the biomass is removed from the suspension and continued aeration proceeds. On the other hand, if the aeration process merely stripped the dissolved carbon dioxide and increased the

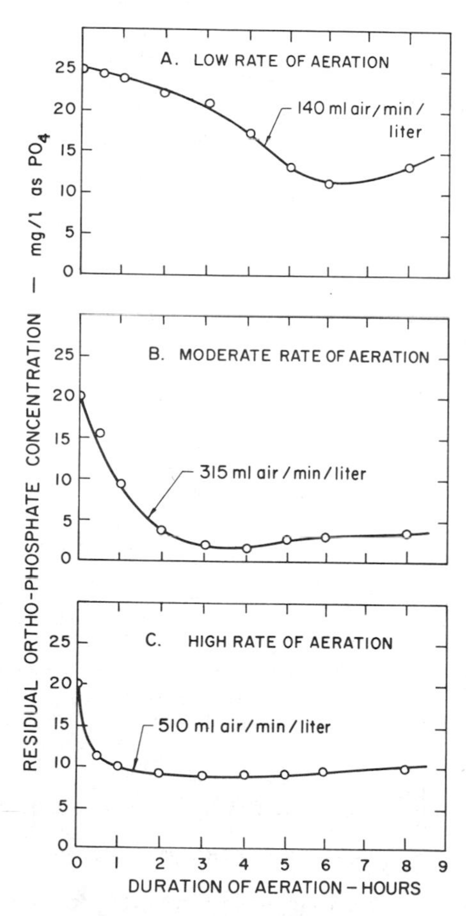

Fig. 6. The effect of aeration rate on phosphate removal in a heterotrophic biological reactor.

pH in accordance with the chemical precipitation hypothesis, then the resultant calcium phosphate precipitation should have occurred. In addition, no phosphate removal takes place if the aeration tank is purged with nitrogen or if 2,4-dinitrophenol, which decouples the biological process of oxidative phosphorylation (22), is added to the aeration tank. Typical data illustrating these findings are presented in Figure 7.

3. No concomitant removal of Ca^{2+}, Fe^{2+}, or Fe^{3+} occurs during the process of phosphate removal. In fact, solubility calculations for hydroxyapatite, $Ca_{10}(OH)_2(PO_4)_6$, the most insoluble form of calcium phosphate, indicate that the solubility product has not been exceeded. These calculations were made using the Ca^{2+} values for the moderately hard South Bend, Indiana, wastewater, averaging 100 mg/liter as $CaCO_3$ and the pH values observed during a typical experiment. Figure 8 reveals that no significant simultaneous increase in pH is measured during the time interval of maximum phosphate removal. Instead, the major phosphate removal occurs at the beginning of the treatment period, whereas the pH rises so slowly that no significant

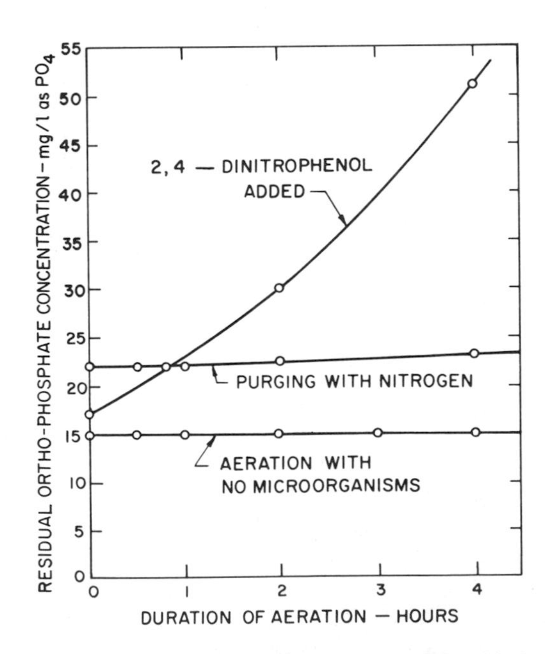

Fig. 7. The effect of various environmental conditions on residual phosphate concentration in a heterotrophic biological reactor.

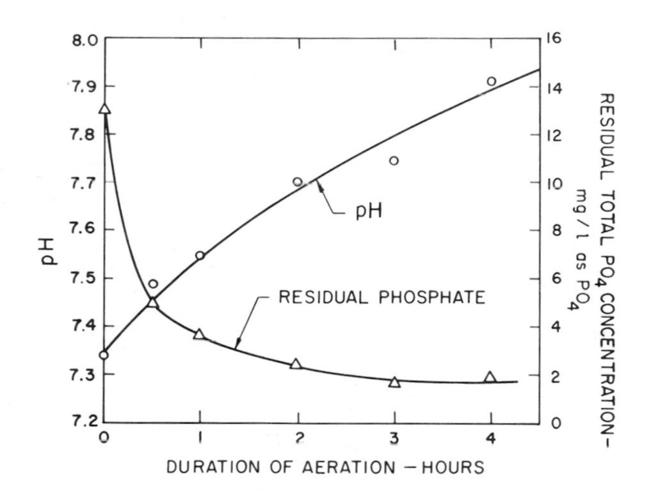

Fig. 8. Comparison of changes in pH and residual phosphate concentration in a heterotrophic biological reactor.

pH increase is observed until the entire treatment period has elapsed.

4. Figure 9 summarizes the results of detailed investigations to locate the removed phosphate within various fractions of the cell. In order to obtain these data, the cells were subjected to washing procedures similar to those used by Yall et al. (23), and the extracts were analyzed for phosphate content. These results indicate that there is only a slight increase in the phosphate concentration on the outside of the cell membrane during aeration. However, a continuous and significant increase occurs throughout the 4-hr treatment period within the soluble and insoluble fractions of the cell.

Furthermore, phosphate analyses of the biomass and the suspending medium prior to and throughout the aeration treatment suggest that the microbial cells are removing soluble phosphate from the liquid phase even though their biomass contains greater concentrations of phosphate than the suspending medium. These data are also illustrated in Figure 9.

Since the phosphate transport described in item 4 occurs against a concentration gradient, as opposed to simple diffusion according to Fick's law, energy must be

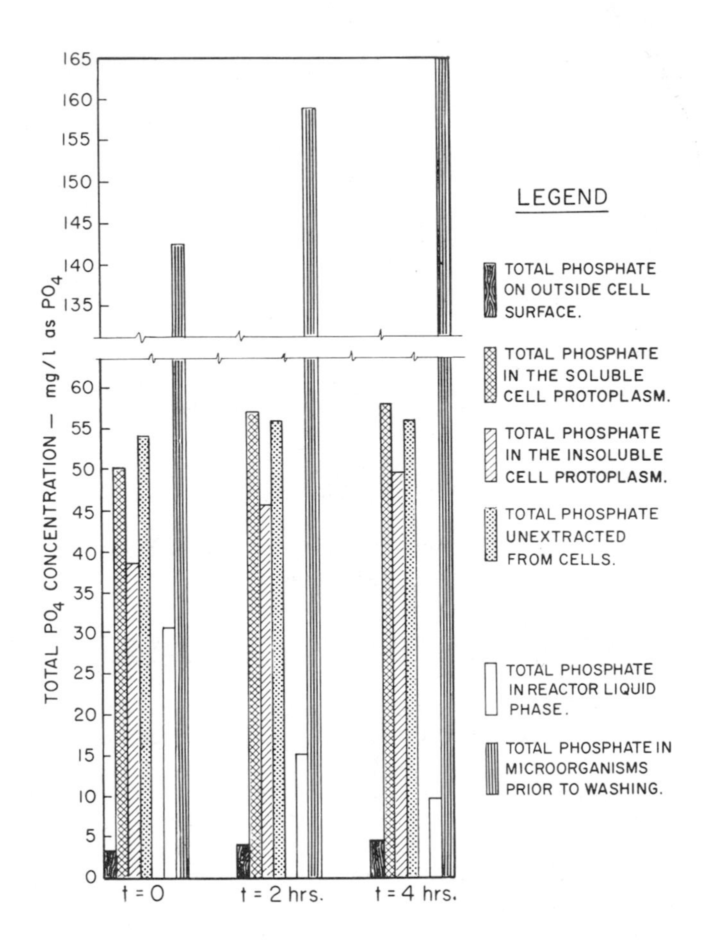

Fig. 9. Concentration of total phosphate in various cell extracts, the liquid phase, and unextracted biomass during aeration in a heterotrophic biological reactor.

expended by the cell in order to overcome the opposing diffusional effects.

This active transport system, then, appears to be the controlling mechanism for the phenomenon of luxury uptake of phosphorus by biological systems. Further work is being conducted (24) to determine whether this active transport of phosphate itself may be controlled by various cationic species such as sodium or potassium. A particular cation may simply diffuse into the microbial cell in order to equalize an existing concentration gradient for that cation. In order to neutralize any charge imbalance within the cell, the soluble phosphate is transported with the cation. The cation would be continuously pumped back out of the cell in order that the concentration gradient for that cation be preserved, and the phosphate is converted by the cell into some less highly charged form, such as a polyinorganic form or an organic ester form. If such an ion coupling exists, the observed metabolic energy expended by the cell to drive the phosphate transport system may actually be used to pump the cation out of the cell. The concentration of such cations then would determine the amount of phosphate transported into the cell.

We have not attempted to determine the form in which this excess phosphate exists within the cell. However, some authors (25) have reported the formation of volutin granules within the soluble cellular protoplasm of _Zoogloea ramigera_, a very common species in activated-sludge biomasses.

Azad and Borchardt (26) have demonstrated that autotrophic microorganisms absorb greater concentrations of phosphate than any critical amount needed for the metabolism of other essential substrates. These workers reported that pure algal cultures can take up excess phosphate until a maximum of 10% dry weight of phosphate is reached. Such cells do not grow any faster than do cells supplied by only a critical amount of phosphate. In another paper (27) the same authors suggest that algal cells will store the phosphate (e.g., in the form of meta- or polyphosphates within the cellular protoplasm) for subsequent use, should phosphate depletion occur in the medium.

The removal of phosphorus from wastewater by symbiotic autotrophic-heterotrophic (algal-bacterial) systems has been studied by several researchers (1,28-31). Systems investigated have ranged from shallow ponds requiring large surface area, to aerated reactors ("activated-algae"), similar to the activated-sludge process. McKinney et al. (1) reported that their proposed "activated-algae" system was capable of reducing the BOD, nitrogen, and phosphorus in a single process unit; they concluded that the key to the new process was related to the ability of the algae to self-flocculate for effective suspended solids removal.

Additional studies of a similar nature, reported by Humenik and Hanna (30), revealed that effective COD removal could be accomplished with a symbiotic algal-bacterial system, but no significant phosphorus uptake was observed during their testing. Hemans and Mason (31) described the pH increase resulting from algal activity in a shallow pond receiving secondary treatment plant effluent. Such a pH increase facilitated both the phosphorus removal by chemical precipitation and subsequent efficient sedimentation with the algal mass, as well as nitrogen removal by assimilation and loss to the atmosphere as ammonia gas.

REMOVAL OF NITROGENOUS POLLUTANTS

Recently much attention has been given to the removal of nitrogen from wastewaters, since most dissolved nitrogen forms are oxygen demanding and all dissolved forms are potential fertilizing elements. Nitrogen can exist in seven oxidation states and, consequently, numerous nitrogen-containing compounds are found in nature. For water pollution considerations, however, the most significant forms of nitrogen are ammonia, organic nitrogen, nitrite, and nitrate.

The primary sources of nitrogen in domestic wastewater are the end products of nitrogen metabolism in man. The largest single source of nitrogen is urea, which, along with ammonia, comprises approximately 85% of the nitrogen excreted by man. Total nitrogen values in raw domestic wastewater average approximately 25 mg/liter (as N). Some industrial wastewaters (e.g., meat packing, poultry processing, fertilizer manufacture, and animal feed lot runoff) have average total nitrogen concentrations appreciably higher than those characteristic of domestic wastewaters.

As a rule, the significant nitrogen transformations in polluted water follow the classical nitrogen cycle found in nature. Organic nitrogen compounds present in the excretory discharges of man and animals are broken down to ammonia nitrogen forms by bacterial action. Ammonia nitrogen is subsequently oxidized to nitrite nitrogen and then to nitrate nitrogen by bacterial action if sufficient oxygen is present (see equations 5 to 7). Conversely, in the absence of oxygen, nitrate and nitrite nitrogen are reduced to nitrogen gas (see equations 8 to 10).

The biological transformations inherent to the nitrogen cycle form the basis for nitrogen removal by biological treatment, although some removal results from nitrogen incorporation into cellular protoplasm. A major advantage of achieving biological nitrogen removal is that all forms of nitrogen ultimately are removed from solution in the nonpolluting form of nitrogen gas. The following two successive steps are involved in the biological removal of nitrogen:

1. Nitrification--the oxidation of ammonia forms to nitrate.
2. Denitrification--the subsequent reduction of nitrate to nitrogen gas.

Nitrification

When carried to completion, the oxidation of ammonia to nitrate is a two-step process. Initially, ammonia is oxidized to nitrite by a genus of strict aerobic autotrophic bacteria (Nitrosomonas) that utilize ammonia as their sole source of energy. The stoichiometry of this reaction is given in equation 5. The second step, the conversion of nitrite to nitrate, is accomplished by the Nitrobacter genus which is comprised of a specific group of autotrophic bacteria that utilize nitrite as their sole energy source. The stoichiometry of this reaction is presented in equation 6. The overall nitrification reaction is represented by the summary equation (equation 7).*

$$2NH_4^+ + 3O_2 = 2NO_2^- + 2H_2O + 4H^+ \tag{5}$$

$$2NO_2^- + O_2 = 2NO_3^- \tag{6}$$

$$NH_4^+ + 2O_2 = NO_3^- + 2H^+ + H_2O \tag{7}$$

Although it has been claimed (32) that a number of heterotrophic organisms are also capable of nitrification, it appears that the organisms of major importance in nitrification are the Nitrosomonas and Nitrobacter species (33, 34). These particular organisms have long generation times which are significantly influenced by pH, temperature, and the concentration of residual dissolved organic material. Wild et al. (35) have reported that the optimum pH for nitrification is 8.4 and that 90% of the optimum rate can be achieved in the pH 7.8 to 8.9 range. These investigators have further shown that the rate of nitrification continuously increases through the temperature range of 5 to 30°C and that residual dissolved oxygen levels greater than 1.0 mg/liter were adequate. Johnson and Schroepfer (36) define an organic load factor of lb

*It can be discerned from equation 7, that 1.0 mg/liter of ammonia nitrogen (as N) requires 4.6 mg/liter of dissolved oxygen for complete nitrification.

of BOD_5/(day)(lbs) of volatile suspended solids, which they report must be less than 0.3 in order to achieve nitrification.

Recent nitrification data obtained in our laboratory (37,38) resulted from the operation of a 50,000 gal/day domestic wastewater pilot plant study in which biological nitrification followed secondary treatment. The plant operating parameters and results are summarized in Table 3.

Table 3. Operating Parameters and Results of Nitrification and Denitrification of a 50,000 gal/day Pilot Plant Study

Parameters	Results
Nitrification unit	
Detention time	2.5 hr
MLVSS (mixed liquor volatile suspended solids)	2000 to 2500 mg/liter
Average sludge age	5 days
SVI (sludge volume index)	50 to 60
Nitrification loading factor	0.10 mg NH_4-N nitrified/(day)(mg MLVSS)
Average temperature	18°C
pH	7.5
Cell yield	0.05 mg MLVSS/mg NH_4-N removed
Nitrification achieved	> 90%
Denitrification unit	
Detention time	2.3 hrs
MLVSS	2000 mg/liter
Denitrification loading factor	0.06 mg NO_2-N and NO_3-N denitrified/(day)(mg MLVSS)
SVI	100
Denitrification achieved	> 90%

Denitrification

Biological dentirification is achieved under anaerobic conditions by heterotrophic microorganisms that utilize NO_3 as a hydrogen acceptor if an organic energy source is available. McCarty et al. (39) have reported that a wide variety of common facultative bacteria can accomplish denitrification (e.g., *Pseudomonas*, *Achromobacter*, *Bacillus*). The specific denitrifying capacities of these microorganisms differ, however. Some can reduce only nitrates to nitrites, some can reduce only nitrites to molecular nitrogen, and some can achieve the reduction of both nitrates and nitrites to molecular nitrogen. Thus denitrification must be considered to be a two-step process. If, for example, methanol is selected as the organic carbon source to provide energy for bacterial synthesis, then equations 8 to 10 represent the microbially induced nitrogen transformations characteristic of denitrification.

$$NO_3^- + \frac{1}{3}CH_3OH = NO_2^- + \frac{1}{3}CO_2 + \frac{2}{3}H_2O \quad (8)$$

$$NO_2^- + \frac{1}{2}CH_3OH = \frac{1}{2}N_2 + \frac{1}{2}CO_2 + \frac{1}{2}H_2O + OH^- \quad (9)$$

$$NO_3^- + \frac{5}{6}CH_3OH = \frac{1}{2}N_2 + \frac{5}{6}CO_2 + \frac{7}{6}H_2O + OH^- \quad (10)$$

It can be observed from equation 10 that 5/6 mole of methanol is required to completely reduce one mole of nitrate to molecular nitrogen. For high percentages of nitrogen removal, then, the stoichiometric amount of organic material must be present or the nitrate may be reduced only to nitrite, with no nitrogen removal occurring. External sources of carbon are generally selected rather than raw or partially treated wastewater, since the latter contains a large fraction of nonoxidized forms of nitrogen. Methanol is generally selected for this process, primarily because of its relatively low cost in comparison to similar chemicals. The requisite concentration of methanol can be calculated as follows (39):

$$O_r = 3.85(NO_3 - N) + 1.5\ (DO) \quad (11)$$

where O_r = concentration of the organic material required for denitrification (oxygen units)

$NO_3 - N$ = nitrate nitrogen concentration

DO = dissolved oxygen concentration

Data from this laboratory have been reported (37,38) for the operation of the denitrification unit in the previously described 50,000 gal/day pilot plant. The operating parameters and results obtained from this unit are also given in Table 3. No undesirable odors were observed from either the reactor or from its effluent. Between the anaerobic reactor and the final settling tank, a flash aeration chamber with a 2-min detention time was provided in order to strip the occluded nitrogen gas and enhance the ability of the sludge to settle.

SUMMARY

A review of the biological treatment methods applicable to the removal of carbonaceous, phosphatic, and nitrogenous pollutants from wastewaters has been presented. In each case, the respective biological treatment processes were described in terms of the fundamental microbiological relationships inherent to the proper functioning of the process and the parameters most influential to the process reaction(s) were elucidated.

The two primary considerations basic to the proper design and operation of any biological wastewater treatment system were cited as substrate removal and subsequent microorganism removal. A more comprehensive model of microbial growth (predicting the lag and endogenous growth phases, as well as surface phenomena) was suggested as the basis for the mathematical development of the kinetics of substrate utilization in biological wastewater treatment systems. The importance of cell flocculation as a prerequisite to the proper functioning of any microbial separation device was noted. The mechanism of cell flocculation was interpreted in terms of polymeric bonding-bridging relationships applicable both to chemically induced flocculation and to autoflocculation. In both cases the degree of flocculation is determined by the amount of polymer present with respect to the microorganism surface area. Autoflocculation of cells is induced by relatively refractory, high-molecular-weight, exocellular polymeric substances consisting mainly of polysaccharide, protein, RNA, and DNA compounds.

The organic carbon transformations obtainable by heterotrophic activity under aerobic and anaerobic conditions were indicated. The removal of phosphorus by autotrophic, heterotrophic, and combined heterotrophic-autotrophic systems was described. Luxurious removal of phosphate in biological reactors was indicated to be a biologically induced, active transport phenomenon dependent on the rate of aeration and perhaps controlled by the type and concentration of cationic species in the suspending medium. The removal of nitrogeneous impurities by biological nitrification-denitrification techniques was discussed.

ACKNOWLEDGMENTS

This investigation was supported in part by a research and development grant from the Environmental Protection Agency (17010 DTG), a National Science Foundation Fellowship awarded to Kurt J. Guter, and a research fellowship from the Environmental Protection Agency (5-Fl-WP-26, 446-02) awarded to Judith B. Carberry.

REFERENCES

1. R. E. McKinney, E. C. McGriff, R. J. Sherwood, V. N. Wahbeh, and D. W. Newport, "Ahead: Activated Algae?," Water Wastes Eng., 8, 51-52 (September 1971).
2. A. F. Gaudy, Jr., and E. T. Gaudy, "Microbiology of Waste Waters," Ann. Rev. Microbiol., 20, 319-336 (1966).
3. H. R. Bungay, III and M. L. Bungay, "Microbial Interactions in Continuous Culture," Advan. Appl. Microbiol., 10, 269-290 (1968).
4. W. A. Pretorius, "Kinetics of Anaerobic Fermentation," Water Res., 3, 545-558 (1969).
5. D. Herbert, "Continuous Culture of Microorganisms," Soc. Chem. Ind. (London), Monogr. 12, 21 (1961).
6. J. Monod, "The Growth of Bacterial Cultures," Ann. Rev. Microbiol., 3, 371-393 (1949).
7. G. Teissier, "Les Lois Quantitatives de la Croissance," Ann. Physiol. Physicochim. Biol., 12, 527-586 (1936).
8. M. T. Garrett and C. N. Sawyer, "Kinetics of Removal of Soluble B.O.D. by Activated Sludge," Proc. Ind. Waste Conf., Purdue University, 79, 51-77 (1952).
9. F. H. Verhoff, K. R. Sundaresan, and M. W. Tenney, "A Mechanism of Microbial Cell Growth," Biotechnol. Bioeng., in press.
10. J. L. Pavoni, M. W. Tenney, and W. F. Echelberger, Jr., "Fractional Composition of Bacterially Produced Exocellular Polymers and Their Relationship to Biological Flocculation," J. Water Pollut. Control Fed., 1972, in press.
11. J. L. Pavoni, M. W. Tenney, and W. F. Echelberger, Jr., "The Relationship of Algal Exocellular Polymers to Biological Flocculation," Proc. Ind. Waste Conf., Purdue University, 1971, in press.
12. M. W. Tenney and W. Stumm, "Chemical Flocculation of Microorganisms in Biological Waste Treatment," J. Water Pollut. Control Fed., 37, 1370-1388 (1965).
13. M. W. Tenney, W. F. Echelberger, Jr., R. G. Schuessler, and J. L. Pavoni, "Algal Flocculation with Syn-

thetic Organic Polyelectrolytes," J. Appl. Microbiol., 18, 965-971 (1969).
14. Environmental Protection Agency Symposium on Advanced Waste Treatment and Water Re-use, Chicago, February, 1971.
15. A. B. Menar and D. Jenkins, "Fate of Phosphorus in Waste Treatment Processes: Enhanced Removal of Phosphate by Activated Sludge," Environ. Sci. Technol., 4, 1115-1121 (1970).
16. D. Vacker, C. H. Connell, and W. N. Wells, "Phosphate Removal Through Municipal Waste-Water Treatment at San Antonio, Texas," J. Water Pollut. Control Fed., 39, 750-771 (1967).
17. J. L. Witherow, "Phosphate Removal by Activated Sludge," Proc. Ind. Waste Conf., Purdue University, 135, 1169-1194 (1969).
18. W. F. Milbury, D. McCauley, and C. H. Hawthorne, "Operation of Conventional Activated Sludge for Maximum Phosphorus Removal," J. Water Pollut. Control Fed., 43, 1890-1901 (1971).
19. R. D. Bargman, J. M. Betz, and W. F. Garber, presented at the 5th International Water Pollution Research Conference, San Francisco, 1971, in press.
20. G. V. Levin and J. Shapiro, "Metabolic Uptake of Phosphorus by Waste Water Organisms," J. Water Pollut. Control Fed., 37, 800-821 (1965).
21. J. B. Carberry, Ph.D. thesis, University of Notre Dame, 1972.
22. H. R. Mahler and E. H. Chordes, "Biological Chemistry," Harper and Row, New York, 1966, p. 528.
23. I. Yall, W. H. Boughton, R. C. Knudsen, and N. A. Sinclair, "Biological Uptake of Phosphorus by Activated Sludge," J. Appl. Microbiol., 20, 145-150 (1970).
24. F. H. Verhoff, V. J. Bierman, J. B. Carberry, and M. W. Tenney, presented at the 71st Annual Meeting of the American Institute of Chemical Engineers, Dallas, February, 1972.
25. F. A. Roinestad and I. Yall, "Volutin Granules in Zooglocaramigera," J. Appl. Microbiol., 19, 973-979 (1970).
26. H. S. Azad and J. A. Borchardt, "Variations in Phosphorus Uptake by Algae," Environ. Sci. Technol., 4, 737-743 (1970).
27. J. A. Borchardt and H. S. Azad, "Biological Extraction of Nutrients," J. Water Pollut. Control Fed., 40, 1739-1754 (1968).
28. W. J. Oswald, H. B. Gotaas, H. F. Ludwig, and V. Lynch, "Algae Symbiosis in Oxidation Ponds," Sewage

Ind. Wastes, 25, 692-705 (1953).
29. A. F. Bartsch, "Algae as a Source of Oxygen in Waste Treatment," J. Water Pollut. Control Fed., 33, 239-249 (1961).
30. F. J. Humenik and G. P. Hanna, Jr., "Algal-Bacterial Symbiosis for Removal and Conservation of Waste Water Nutrients," J. Water Pollut. Control Fed., 43, 580-594 (1971).
31. J. Hemens and M. H. Mason, "Sewage Nutrient Removal by a Shallow Algal Stream," Water Res., 2, 277-287 (1968).
32. O. K. Eylar, Jr. and E. L. Schmidt, "A Survey of Heterotrophic Micro-Organisms from Soil for Ability to Form Nitrite and Nitrate," J. Gen. Microbiol., 20, 473-481 (1959).
33. H. Less, "The Ammonia-Oxidizing Systems of *Nitrosomonas*," Biochem. J., 52, 134-139 (1952).
34. H. Lees and J. R. Simpson, "Nitrite Oxidation by *Nitrobacter*," Biochem. J., 65, 297-305 (1957).
35. H. E. Wild, Jr., C. N. Sawyer, and T. C. McMahon, "Factors Affecting Nitrification Kinetics," J. Water Pollut. Control Fed., 43, 1845-1854 (1971).
36. W. K. Johnson and G. J. Schroepfer, "Nitrogen Removal by Nitrification and Denitrification," J. Water Pollut. Control Fed., 36, 1015-1036 (1964).
37. W. F. Echelberger, Jr. and M. W. Tenney, "Wastewater Treatment for Complete Nutrient Removal," Water and Sewage Works, 116, 396-403 (1969).
38. K. J. Guter, Ph.D. thesis, University of Notre Dame, 1971.
39. P. L. McCarty, L. Beck, and P. St. Amant, "Biological Denitrification of Wastewaters by Addition of Organic Materials," Proc. Ind. Waste Conf., Purdue University, 135, 1271-1285 (1969).

Chapter XIV

ROLE OF THE FEDERAL GOVERNMENT IN CONTROLLING NUTRIENTS IN NATURAL WATERS

A. F. Bartsch

Pacific Northwest Water Laboratory, Water Quality Office, Environmental Protection Agency, Corvallis, Oregon

In previous chapters, the subject of nutrients in natural waters has been explored in considerable depth. We have been stimulated to think about nitrogen, phosphorus, carbon, and organic materials in relation to their roles in primary production and the eutrophication process. Some of the important ancillary considerations, such as analytical techniques, modeling the system, assays, and waste treatment for nutrient removal have been discussed. The following chapter and this one have a somewhat similar role--that of focusing on the question, How does government function in attaining nutrient control?

Before addressing the question directly, there are several preliminary philosophical points to be considered. For example, what are the options in controlling eutrophication, and how does nutrient control relate to them? What nutrients in natural waters should be controlled? Where do they come from, and in what amounts?

OPTIONS FOR CONTROLLING EUTROPHICATION

There are only four major approaches that can possibly

be used today to restore lakes or control eutrophication. A necessary first step in any control campaign is to limit aquatic fertility; the second approach is to improve the food chain with hopes of harvesting a valuable crop. The third method is to stimulate diseases and parasites among the unwanted plants to keep them under control, and the fourth is to hold them down with toxic chemicals.

NUTRIENTS TO CONTROL

Without question, the most promising preventive and restorative measure is to limit fertility through curbing nutrient input. Thus every nutrient source becomes important, and each one needs to be considered for potential control. But emphasis today must be on sources for which there presently are control capabilities. This criterion is the only one that is pertinent.

Although diversion of nutrient-bearing streams away from lakes has been used successfully, the major control campaign underway in the United States is treatment of sewage or other inflows to strip them of their nutrient content. We must ask, What nutrients should be removed? This is a logical question, because lake water contains some sixteen or more chemical elements that are used in aquatic plant nutrition. All are important. An absence of one or a shortage of another can prevent, impede, or regulate plant production. However, attention has centered on phosphorus and nitrogen in this regard because experience has shown that shortages of these two elements, particularly phosphorus, function as the brakes that impede or stop production in most lakes. It is also because of the concern that in many polluted lakes, phosphorus levels are increasing more rapidly than those of other nutrients. In western Lake Erie, for example, the ratio of nitrogen to phosphorus has shifted from 35.0 in 1942 to 9.2 in 1966 (1).

There has been a lengthy dialogue on the relative contributory significance of phosphorus versus nitrogen versus carbon. The issue has been examined in considerable depth by the scientific community on several recent occasions (2-6).* It is not my intention to reopen this issue, but it is necessary to set forth a rationale for efforts to control the particular nutrients we do control.

---

*The most recent was the American Society of Limnology and Oceanography Symposium on Nutrients and Eutrophication "The Limiting Nutrient Controversy," February 10-12, 1971, Hickory Corners, Mich.

Emphasis is on phosphorus for several reasons. First, phosphorus frequently is a critical nutrient in lakes. Within limits, there is a relation between size of plant crop and the quantity of phosphorus available. This also means that exhaustion of supply limits the frequency and density of algal blooms. Second, people and their activities are the major source of phosphorus input to troubled waters, and this man-caused input is potentially subject to regulation. Third, the technology to go a long way toward regulation is now available for use on many problem lakes. Fourth, algal assays show that sewage stripped of phosphorus loses most of its algal growth capability. Restoring the phosphorus restores growth capability. And fifth, similar control technology or prospects are not as fully available with respect to nitrogen, carbon, or any other vital nutrient. In short, as pointed out by Vallentyne (5), of all the required nutrients, only phosphorus is both controlling and controllable.

Each lake, of course, is unique. It is not surprising to find lakes in which production is limited temporarily or continually by shortages of nitrogen, carbon, micronutrients, or some other factor. The desirability of curtailing these elements or factors, if we can, does not nullify the stated reasons for emphasis on phosphorus control.

Programs already underway for many lakes are based on removal of phosphorus from city sewage. Examples are Lake Superior, Lake Michigan, Lake Erie, Lake Ontario, the Upper St. Lawrence River, and the Potomac River estuary. Similar control is under consideration for Lake Huron and many small lakes as well.

WHERE DO NUTRIENTS COME FROM?

Before the initial impact of people, lakes acquired nutrients from the earth, from the air, and from the remains of plants and animals. These inputs supported growing aquatic plants such as algae and macrophytes. Fertility and productivity increased with time, and in many cases reached objectionable levels. The varying pace of contributory natural factors has left some lakes relatively untouched even today. Others have attained advanced stages of deterioration, characterized by dense algal blooms, and in some cases even by lake extinction.

It has been known for at least fifty years that human activities contribute in several ways to accelerating eutrophication. In some places, at least, the rate of change in lakes under the influence of people appears to be racing along faster than the rate of human population growth (7). We wonder, therefore, what further impact on lakes there will be as population of the United States grows beyond the present 205 million persons.

Although nutrient budgets are available for a few lakes, there is only fragmentary knowledge on the relative national importance of various nutrient sources. We know, for example, that with our present high standard of living, we shed dietary nitrogen and phosphorus rather freely in city sewage. In addition, synthetic detergents may account for 50 to 70% of the phosphorus in the sewage in some cities. We have no quantitative national data, but estimated annual discharges from city sewers contain between 200 to 550 million lb of phosphorus and about two or three times as much nitrogen (8). For many problem lakes, the sewered and nonsewered drainage from cities appears to be the principal nutrient influx.

We know even less about industries as nutrient sources (except that some, depending on process and raw materials, discharge phosphorus or nitrogen in their wastes). But there is no real knowledge even today on how many factories or what quantities are involved.

We are little better off with respect to agriculture. While there is tremendous interest in this area, there are conflicting views on the potential significance of agricultural runoff. It appears that cultivated land generally yields greater quantities of nutrients than undisturbed land. Data for six agricultural watershed runoffs show phosphorus concentrations from 33 to 110 μg/liter with a mean of 58 (1). These concentrations may seem exceedingly minute; however, a sixfold dilution or some other means of reduction is required to reach target phosphorus levels believed adequate to prevent eutrophication. The quantity and rate of nutrient losses vary with soil treatment, state of irrigation, nature of the soil, kinds of crops, fertilizer practices, slope, and other factors. There is growing awareness that barnyards and animal feedlots are focal points of intense activities that augment the release of nitrogen and phosphorus. This is to be expected when we consider the following: (1) current livestock manure production is greater than 1.6 billion tons/yr (9), (2) as much as 50% is from confinement production (10), (3) a ton of fresh manure from fattening cattle contains about 14 lb of nitrogen and 4 lb of phosphorus (11).

A recent paper (12) gives estimates of proportionate nutrient contributions to United States water supplies from various sources (Table 1*). For phosphorus, domestic waste yields 22% of the total; agricultural land, 42%;

Table 1. Estimated Amounts of Nutrients Contributed from Various Sources for Water Supplies of the United States[a]

Nutrient source	Nutrient, millions of lb/yr		Percentage of total, based on mean value of ranges given	
	Nitrogen	Phosphorus	Nitrogen	Phosphorus
Domestic waste	1,100-1,600	200-500	10	22.0
Industrial waste	>1,000	---	7	--
Rural runoff				
Agricultural land	1,500-15,000	120-1,200	60	42.0
Nonagricultural land	400-1,900	150-750	8	29.0
Farm animal waste	>1,000	---	7	--
Urban runoff	110-1,100	11-170	4	6.0
Rainfall	30-590	3-9	2	0.4

[a]Source: AWWA Task Group 2610-P (1967).

nonagricultural rural runoff, 29%; and urban runoff, 6%. These figures suggest two pertinent points. First, all inputs to any lake are significant because they are additive. It is possible to visualize circumstances in which successful lake protection or restoration appears unlikely without curbing agricultural land inputs along with municipal ones. Second, if technology to control such inputs is not now available, as it is not for many nonurban sources, we must move forward to develop it without delay.

ROLE OF GOVERNMENT

During the past few years, a growing public anxiety about degradation of lakes has stimulated action at every level of government. Some mention is made in this chapter of municipal and state actions, but most attention is on federal actions and on the Environmental Protection

*This is Table 23.14 of "Effects of Agricultural Pollution on Eutrophication," Chapter 23 in "Agricultural Practices and Water Quality," Proceedings of a Conference Concerning the Role of Agriculture in Clean Water, November 1969. Water Pollution Control Research Series, DAST-26, 13040 EYX 11/69, p. 328.

Agency in particular.

Efforts to control nutrients in natural waters involve four principal interrelated aspects. They are: (1) provision of grant funds to assist in construction of community waste treatment facilities that incidentally or by design remove nutrients before discharge, (2) intramural and extramural research to generate new knowledge and to demonstrate and activate management technology needed for effective nutrient control, (3) enforcement actions to compel compliance in preventive and restorative programs, and (4) provision of enabling and regulatory legislation.

Grant Funds for Municipal Waste Treatment

The Federal Water Pollution Control Act (Section 8) provides grant funds to assist appropriate units of government in constructing necessary waste treatment works. Since the program began in 1956, almost 11,000 construction projects have been undertaken at a total cost approaching $8 billion. Undoubtedly, this action was greatly stimulated by the federal grant share, which totaled $1.9 billion.

The treatment processes used in these projects differ widely, but all plants help somewhat to decrease the influx of nutrients to receiving waters. Advanced treatment processes are available where substantial nutrient removals are required. In light of the growing movement to control eutrophication through nutrient management, we can anticipate increasing use of advanced treatment processes to remove phosphorus, nitrogen, or both. The result will be a progressive curtailment of nutrient availability in lakes and a slowdown or turnabout of eutrophication encroachment.

Research, Development, and Demonstration

Research and development efforts supported by the Environmental Protection Agency through its National Eutrophication Research Program emphasize development and upgrading of preventive and lake restorative capabilities of the nation. Research underway relates to all four of the lake restoration possibilities already mentioned, including improved advanced waste treatment processes to remove phosphorus and nitrogen from waste streams. In addition, notable progress has been made in developing better algal assay procedures for help in assessing lake problems, viewing likely responses to remedial action, and expediting control programs.

The research and development program includes intramural and extramural projects in a dollar expenditure ratio of 1 to 4.7. Currently, more than 38 projects supported by grant and contract dollars are underway in universities and other research organizations throughout the country. Although considerable progress has been made in many important areas, major gaps in knowledge related to nutrients in natural waters still remain. Among them are: (1) understanding the eutrophication process itself in greater depth, (2) a national assessment of all nutrient inputs to surface waters, (3) efficacious technology for curbing inputs of nutrients other than phosphorus in wastewaters, and (4) effective control of nutrient inputs from diffuse sources.

Two prestigious federal reports call attention to the problem of lake deterioration and recommend actions to be taken. One is the 1970 Annual Report of the National Council on Marine Resources and Engineering Development (13). It emphasizes that "to determine the feasibility of restoring the quality of some of the Nation's seriously damaged waters [meaning lakes], existing technology should be tested on small bodies of water and new methods developed to establish the most practical and economical means for proceeding with the major task." Lake restoration was identified as one of the priority programs to receive attention during fiscal year 1971.

The other federal document is the First Annual Report of the Council on Environmental Quality (6). It recommends a concerted and comprehensive attack on eutrophication and identifies three necessary actions: "One, phase phosphates out of detergents as soon as feasible; two, find better methods to control agricultural runoff; and three, remove more of the nutrients from wastes generated by towns and cities, particularly in urban centers and in critical areas such as the Great Lakes."

In the agricultural area, there is an obvious upsurge of interest in nutrient contributions (14-18). This is not surprising in light of observations that nonurban runoff often substantially exceeds municipal sources of phosphorus input. The Agricultural Research Service of the U.S. Department of Agriculture sponsors research and development programs that assess agricultural contributions and develop appropriate management practices. The Department has action programs also (19); these are directed toward research, education, technical assistance, and in some cases, loans to improve animal management and waste-handling facilities and methods.

Enforcement

The enforcement provisions of the Federal Water Pollution Control Act have been used many times in pollution control efforts. Four actions have been directly concerned with preventing or correcting eutrophication problems. These actions have followed the formal mechanisms specified in the act and have focused on pollution problems in Lake Superior, Lake Michigan, Lake Erie, and the Potomac River estuary.

Municipalities tributary to Lake Superior from Michigan, Wisconsin, and Minnesota must provide waste treatment to achieve at least 80% reduction of total phosphorus. This action is to be accomplished by January 1975 or earlier. The indicated states have recently submitted detailed time schedules for their proposed programs, which must be completed on or before the stated deadline.

In similar action to protect and improve Lake Michigan, waste treatment is called for by all tributary municipalities of Illinois, Indiana, Michigan, and Wisconsin to achieve at least 80% reduction of total phosphorus. This goal is to be substantially accomplished by December 1972.

The action on Lake Erie established a policy of maximum phosphate removal from municipal and industrial waste sources. At present, the target is a minimum of 80% reduction of total phosphate loadings from the tributary states of Michigan, Indiana, Ohio, Pennsylvania, and New York. Construction of the required facilities was to be accomplished in 1971.

Specified waste treatment facilities now discharging to the Potomac River are to achieve such removals of total phosphorus that the resulting discharge total does not exceed 740 lb/day. The facilities cited are the Pentagon, Arlington, the District of Columbia, Alexandria, and Fairfax Westgate. Except for improvements by the District of Columbia, these actions are to be completed in the summer of 1973.

Legislation

The legislative branch of the federal government plays an indispensable role in providing enabling and regulatory legislation concerning water pollution. Several recent actions are quite directly concerned with the eutrophication problem. Among them are the 1970 amendments of the Federal Water Pollution Control Act:

1. Section 5(h) provides for contracts and grants to develop and demonstrate new or improved methods for the prevention, removal, and control of natural or manmade pollution in lakes, including the undesirable effects of nutrients and vegetation. Funds to initiate action under these new responsibilities were included in a fiscal year 1971 supplemental appropriation.
2. Section 13 provides for control of sewage from vessels, including toilet-equipped pleasure craft, such as those typically found on lakes or their tributaries.
3. Section 15 authorizes mechanisms to demonstrate new methods and techniques and to develop preliminary plans for the elimination or control of pollution within all or any part of the watersheds of the Great Lakes.

A number of pertinent bills were proposed before the 92nd Congress. Among them are several House and Senate bills that would require reduction or elimination of phosphorus in synthetic detergents. Ordinances for this purpose already have been enacted by Chicago and Akron. Several states also are considering such legislation. This is a subject of wide public interest, and tremendous publicity has been given to it.

The Clean Lakes Act of 1971 (S1017) would provide for increasing the federal grant percentage for construction of treatment works located near or adjacent to any lake and which discharge into the lake or its tributaries. Working toward a similar objective, the State of Minnesota has passed legislation setting effluent standards for phosphorus in municipal discharges. If the discharge goes directly to a lake, the phosphorus concentration must not exceed 1.0 mg/liter; if it goes to a lake by way of a river, the maximum concentration is 2.0 mg/liter.

Illinois adopted an effluent standard, effective December 31, 1971, of 1.0 mg/liter of phosphorus for discharges to Lake Michigan. The extension of this standard to the entire state is being considered. Illinois' receiving water standard of 7 μg/liter of phosphorus in Lake Michigan is now in force. Many other states are aware of the need for nutrient control and are considering similar legislative actions.

At the federal level, no legislative attempt has been made to control fertilizer use as a means of preventing nutrient runoff. Instead, the Department of Agriculture depends on research efforts to attain greater fertilizer use efficiency, water control, and improved management technology. State feedlot regulations, which have prompted a growing concern throughout the nation, have been summarized by the Agricultural Research Service (19).

In summary, under governmental stimulation, controlling municipal sources of nutrient input through advanced treatment processes is a response to the demand for voluntary and regulatory actions. Progress has been made in decreasing phosphorus inputs that originate in synthetic detergents, and action continues. The massive and important task remaining is to provide technology to cope effectively with diffuse land runoffs.

REFERENCES

1. J. Verduin, in "Agricultural Practices and Water Quality," Water Pollution Control Research Series, DAST-26, 1969, pp. 63-71.
2. W. Lange, "Effect of Carbohydrates on the Symbiotic Growth of Planktonic Blue-Green Algae with Bacteria," Nature, 215, 1277-1278 (1967).
3. L. E. Kuentzel, "Bacteria, CO_2, and Algal Blooms," J. Water Pollut. Control Fed., 41, 1737-1747 (1969).
4. P. C. Kerr, D. F. Paris, and D. L. Brockway, "The Interrelation of Carbon and Phosphorus in Regulating Heterotrophic and Autotrophic Populations in Aquatic Ecosystems," Water Pollution Control Research Series, 16050 FGS, 1970.
5. J. R. Vallentyne, Canadian Research and Development, May-June, 36-43, 49 (1970).
6. "Environmental Quality," First Annual Report of the Council on Environmental Quality together with the President's Message to Congress. Government Printing Office, Washington, D.C., 1970, p. 52.
7. R. A. Vollenweider, "Scientific Fundamentals of the Eutrophication of Lakes and Flowing Waters, with Particular Reference to Nitrogen and Phosphorus as Factors in Eutrophication," OECD Tech. Rep. DAS/CSI/ 68.27, 1968.
8. "Sources of Nitrogen and Phosphorus in Water Supplies," Report of AWWA Task Group 2610-P, J. Am. Water Works Assoc., 59, 344-366 (1967).
9. C. H. Wadleigh and C. S. Britt, in "Agricultural Practices and Water Quality," Water Pollution Control Research Series, DAST-26, 1969, pp. xix-xxvii.
10. J. B. Law and H. Bernard, Paper 69-235 presented at the Meeting of American Society Agricultural Engineers, June 1969, West Lafayette, Ind.
11. T. M. McCalla, L. R. Frederick, and G. L. Palmer, in "Agricultural Practices and Water Quality," Water Pollution Control Research Series, DAST-26, 1969, pp. 241-255.

12. D. E. Armstrong and G. A. Rohlich, in "Agricultural Practices and Water Quality," Water Pollution Control Research Series, DAST-26, 1969, pp. 314-330.
13. U.S. House of Representatives, "Annual Report of National Council on Marine Resources and Engineering Development," Government Printing Office, Washington, D.C., 1970, 284 pages. (91st Congress, 2nd Session, House Document No. 91-304.)
14. C. H. Wadleigh, "Wastes in Relation to Agriculture and Forestry," United States Department of Agriculture Miscellaneous Publ. 1965, 1968.
15. J. P. Law, Jr., J. M. Davidson, and L. W. Reed, "Degradation of Water Quality in Irrigation Return Flows," Bull. Oklahoma State University Agriculture Experiment Station, 3-684, 1970.
16. "Collected Papers Regarding Nitrates in Agricultural Waste Waters," Water Pollution Control Research Series, 13030 ELY, 1969.
17. G. Stanford, C. B. England, and A. W. Taylor, "Fertilizer Use and Water Quality," United States Department of Agriculture, ARS 41-168, 1970.
18. W. F. Schwiesow, "Summary of State Regulations Pertaining to Livestock Feedlot Design and Management," United States Department of Agriculture, ARS 42-189, 1971.
19. "Control of Agriculture-Related Pollution, A Report to the President," Secretary of Agriculture and Director, Office of Science and Technology, 1969.

Chapter XV

DEVELOPMENT OF NUTRIENT CONTROL POLICIES IN CANADA

A. T. Prince

Inland Waters Branch, Department of Energy, Mines and Resources, Ottawa, Ontario, Canada

and

J. P. Bruce

Canada Centre for Inland Waters, Burlington, Ontario, Canada

The role of scientific information in the formulation of government policies is becoming increasingly important. This is especially true where policies designed to restore and protect the environment are concerned.

An interesting example of the importance of the contribution of scientific information and judgment is seen in the recent attempts by governments to control the causes of eutrophication in Canadian lakes.

Eutrophication, the process of overenrichment of water by nutrients and the subsequent oxygen depletion caused by the decay of massive growths of aquatic vegetation, is one of the most vexing environmental problems in the developed countries of the world (1). A valuable summary of the extent of eutrophication in the world's most important lakes is given in a report of a conference sponsored in 1968 by the Organization for Economic Co-Operation and Development (2).

PHOSPHATES AND THE CANADA WATER ACT

The effects of cultural eutrophication are causing many of Canada's resort lakes, particularly in Ontario, Quebec, and British Columbia, to deteriorate rapidly. The most extensive manifestations of this problem are found in Lakes Erie and Ontario. Studies of eutrophication and other pollution problems in the lower Great Lakes were carried out by Advisory Boards to the International Joint Commission over the period from 1965 to 1969, and the results were reported to the International Joint Commission (3). Recognizing that nutrients, particularly phosphates, were the major contributors to the process of eutrophication, the Advisory Boards recommended an extensive program of nutrient discharge control, the most important step being control of phosphate discharges to the lower Great Lakes.

Three means of achieving the objective of effective phosphate control were recommended:

1. Phosphates in detergents would be replaced by substances less harmful to the environment by the year 1972.
2. A phosphate removal program would be implemented at sewage treatment plants in all large cities discharging to the Great Lakes.
3. The input of agricultural phosphates from the drainage basins would be brought under control.

The report of the Advisory Boards to the International Joint Commission was the subject of a series of public hearings held early in 1970, at which representatives of government, industry, and universities, as well as private citizens, presented arguments that supported or opposed the proposal for a phosphate control program. The argument against the efficacy of the proposed control program, which was carried largely by industrial representatives and by some scientists, brought out a difference of opinion in two main areas. The first involved establishing the causes of eutrophication (i.e., the relative importance of phosphorus as compared with other nutrients); the second required decisions on which of these nutrients can be controlled most readily and what the most effective means of control would be.

Following the hearings, the Commission issued an interim report in April 1970 (4) and a final report in December 1970 (5), confirming the recommendations of the Advisory Boards.

The governments of Canada and of Ontario, after a study of the Board's recommendations, and concurrently with the formulation of the Commission's recommendations, agreed that the federal government should introduce legislation to control the use of nutrients in cleaning products. The Canada Water Act, then under consideration, was deemed the most appropriate vehicle for this purpose. An amendment to the Act containing a clause covering nutrient control was introduced. This clause did not itself specify any particular limitation on the chemical composition of detergents or cleaning products. It did, however, permit the government, by Order in Council, to regulate the amounts of nutrients in cleaning products.

The nutrient control clause evoked vigorous debate, first in the House of Commons committee which considered the Act and later by the Senate Special Committee on Science Policy. Not unexpectedly, the arguments against control of detergent formulation paralleled the arguments advanced earlier in the public hearings before the International Joint Commission.

The Canada Water Act, incorporating the clause on nutrient control, passed the Commons and Senate in the early summer of 1970, and the first regulations under the nutrient control clause became effective August 1, 1970. The detergent industry had been informed in January 1970 of the government's intentions concerning regulations under the Act. The regulations limit the amount of phosphates in detergents, excluding dishwashing detergents, to a maximum of 20% expressed as phosphorus pentoxide (P_2O_5). Prior to the regulations most detergent powders had P_2O_5 contents from about 16 to 38%. It is estimated that this initial limitation reduces the amount of detergent phosphates entering Canadian lakes and rivers by 25 to 30%.

The arguments advanced in opposition to the nutrient section in the Canada Water Act and the counterargument that led to parliamentary approval of the section are discussed in this chapter.

SCIENTIFIC ASSISTANCE TO LEGISLATORS

The highly scientific nature of the arguments and counterarguments rendered any objective appraisal by legislators in the Commons and Senate extremely difficult. To help overcome this, the scientific staff of the Canada Centre for Inland Waters, Burlington, Ontario, the Inland Waters Branch, Ottawa, and the Freshwater Institute of the Fisheries Research Board, Winnipeg, Manitoba, prepared a nine-page pamphlet on the control of eutrophication (6). The pamphlet outlines in lay terms the rationale for a program of nutrient control and in particular a program to control the use of phosphates in detergents.

The pamphlet discusses at some length the two important questions involved in any consideration of nutrient control. First, what is the relative importance of the different nutrients in algal growth? (Or to put it another way, which of the various nutrients limits algal growth most frequently?) Second, which of the key nutrients can most readily be controlled by man?

ALGAL GROWTH AND THE RELATIVE IMPORTANCE OF CARBON, NITROGEN, AND PHOSPHORUS

The argument against instituting control of phosphorus was based on a difference of opinion about the relative importance to the eutrophication process of phosphorus as opposed to other nutrients, particularly carbon. This argument emphasized the great importance of carbon and carbon dioxide in algal growth and the symbiotic relationship between bacteria stimulated by organic wastes and the excessive growth of algal blooms.

There is no question that many different substances dissolved in water are required for algal growth and that among these, carbon, nitrogen, and phosphorus are the most important. Substances such as iron, manganese, and molybdenum, as well as calcium, potassium, magnesium, sulfur, and silica are also required, but in smaller quantities. It has been demonstrated in a number of lakes that phosphorus, nitrogen, and carbon are growth-limiting. Similar growth limitation has been reported, although less frequently, for some of the other elements. The following figures indicate roughly the relative supply of each of the three principal nutrients required to sustain extensive algal growth, assuming other nutrients to be present in adequate quantity: for every ton of phosphorus in the algal biomass, there are about 40 tons of carbon and 7 tons of nitrogen.

Carbon

Carbon is available in lakes from a large number of

natural sources. Surface waters are usually saturated with carbon dioxide from the atmosphere. Bicarbonates and carbonates are present in abundance in most lakes due to natural chemical processes, and are readily converted to carbon dioxide by a well-known chemical reaction. Since the pH of waters with algal blooms frequently increases to values between 9 and 10, it is indicated that algae can use bicarbonate ions during growth. In the near-surface growing zone, carbon in the form of bicarbonate and carbon dioxide can be continually supplied from tributaries, turbulence in the main water body, and atmospheric exchanges; in most lakes, these sources alone are sufficient to support the observed biomass.

Work in an experimental lakes project by the Fisheries Research Board of Canada suggests that carbon can be limiting in very soft waters with inorganic carbon less than 6 mg C/liter (7). It has been shown, however, that even in such lakes the introduction of inorganic fertilizers high in phosphorus and nitrogen stimulates excessive algal growth (unpublished results by R. R. Langford, University of Toronto and D. W. Schindler, Freshwater Institute, Winnipeg).

In polluted lakes, sewage wastes are another source of carbon. The strength of these wastes is usually measured as BOD (biochemical oxygen demand) a measure of the oxygen required to decompose organic substances by bacterial activity under controlled laboratory conditions. It has been claimed that carbon from BOD is an important source for algae to produce algal blooms (8).

To obtain some idea of the relative importance of these sources of carbon in a natural lake system, consider Lake Erie. The annual loading of BOD_5 (BOD in 5 days at 20°C) to the whole lake is estimated at 200,000 tons with a carbon equivalent of 75,000 tons. The annual supply of inorganic carbon from tributaries to Lake Erie is in the order of 3.5 million tons to 4.2 million tons, and about half of this amount is available for photosynthesis. The additional supply of carbon dioxide from the atmosphere is difficult to estimate, but certainly this source cannot be neglected in a complete carbon budget. Hence the carbon supply from natural sources is at least 25 times as much as the carbon from sewage wastes.

Another way to consider this question is by estimating the average algal biomass which, according to studies made by the Fisheries Research Board detachment at the Canada Centre for Inland Waters, is in the order of 60,000 tons ash-free dry weight in the uppermost 5 m of Lake Erie, with a carbon equivalent of about 30,000 tons. During periods of algal blooms the phytoplankton biomass could be 4 to 5 times as much as this average. Assuming an average photosynthesis rate of 0.4 g/(m^2)(day) (in accord with direct measurements), the time needed to produce this algal mass is in the order of days; hence the carbon from sewage would be completely insufficient to support the biomass.

Nitrogen

Nitrogen usually enters lakes in the form of nitrates, ammonia, and organic compounds. These come from a variety of natural sources including natural drainage from soils and precipitation from the atmosphere. In addition, farm fertilizers, manure, and organic wastes from municipalities contribute to the nitrogen load.

It has been observed that if sufficient phosphorus is available in the water in spring before growth starts, nitrogen can become limiting later in the summer. Sawyer's work (9) of 1947 on lakes in Wisconsin suggests that inorganic nitrogen concentrations in late winter of about 200 to 300 μg/liter (ppb) are needed along with inorganic phosphorus concentrations of at least 10 to 20 μg/liter (ppb) to stimulate blooms of algae later in the growing season. Some species of blue-green algae and bacteria can use dissolved molecular nitrogen.

Nitrogen can also be contributed by fixation of molecular nitrogen by a number of blue-green algae and nitrogen-fixing bacteria. This source, however, is likely to be important only if other nitrogen forms have been used up by algae, and it is of significance primarily in lakes whose productivity is high for other reasons.

Phosphorus

Phosphorus does not occur as abundantly in nature as either carbon or nitrogen, and even though it is required by algae in smaller quantities it is generally recognized that it frequently triggers eutrophication and is the substance limiting the overall extent of algal growth. Vollenweider in 1970 (10) analyzed some data published by Thomas (11) on 46 Swiss lakes and related the decrease in carbon, nitrogen, and phosphorus in the lakes during the growing season to the initial concentrations of the same elements in the spring. The lakes ranged from clear, oligotrophic waters to highly enriched eutrophic waters. The seasonal decreases in carbon, nitrogen, and phosphorus are due largely to uptake of the nutrients by the biomass.

Significant correlation was found between spring concentration and decrease during the growing season for each nutrient (carbon, nitrogen, phosphorus) calculated separately; but the highest correlation coefficient was found for phosphorus availability and phosphorus decrease.

Furthermore a cross-correlation analysis showed high correlation between phosphorus availability and nitrogen and carbon decrease, but low and insignificant correlation between carbon or nitrogen availability and phosphorus decrease. This illustrates the dominant function of phosphorus availability in the lake metabolism. Phosphates appear to be the key substance governing the production of algae in the 46 Swiss lakes investigated.

In Lake Erie the ratio of the available carbon in bicarbonates to total phosphorus in the waters is 175:1 in the western basin and 700:1 in the other basins. If the average carbon-to-phosphorus ratio in algal cells is taken as 40:1, then there is 4 times more bicarbonate-carbon available in the western basin and 17 times more in the lake as a whole than would be required for algal growth to completely deplete the water of phosphorus.

Similar calculations show that in the western basin of Lake Erie slightly more nitrogen is present than is needed for the maximum biomass that could be generated by available phosphorus; in the other two basins there is a two-to-threefold surplus of nitrogen, assuming an average ratio of nitrogen to phosphorus in the biomass.

To some extent these findings are corroborated by considering the annual loadings of phosphorus:nitrogen: carbon which, according to the report to the International Joint Commission, are 1.06 and 6.8 g/m^2 for phosphorus and nitrogen, respectively, and about 150 g/m^2 carbon from bicarbonates (half of this being available for photosynthesis) for the lake as a whole.

Accordingly, the nitrogen and phosphorus loading ratio is in the order as required for biomass buildup, but the available carbon from bicarbonates alone is greatly in excess of need.

Vallentyne (12) has pointed out that although the concentration of bicarbonates in the lower Great Lakes has not changed appreciably in the last century, algal growth has increased immensely, indicating that bicarbonate carbon has not been an important factor.

In summary then, these mass balance calculations show that in Lake Erie as a whole, phosphorus is generally the limiting growth factor, and work on many other lakes in North America and Europe reveals that the same is true for a large number of lakes in the world. In addition, if we wish to give a general sequence of importance, it can be said that phosphorus is most frequently the limiting element, followed in order of decreasing importance by nitrogen and carbon.

CONTROLLABILITY OF NUTRIENTS

The second important question in considering a nutrient control program is, Which of the three principal nutrients can most readily be controlled by man?

By far the largest source of soluble carbon in Canadian lakes, except those in Precambrian areas, is due to solution of calcium carbonate and magnesium carbonate sediments in the drainage basins. The resulting bicarbonate ions that arise from natural chemical processes appear to be present in quantities much more than adequate to meet the demands of biomass production. The quantity of carbon in this form is not controllable by man, nor for the most part is it created by man's activities.

Nitrogen enters lakes from rain, snow, and dustfall (ca. 16,000 tons/yr to Lake Erie and 12,000 tons/yr to Lake Ontario). Leaching from natural soils and from soils that are artificially fertilized is another source. In Lakes Erie and Ontario 60 to 70% of the nitrogen comes from diffuse sources, including agriculture, which are not readily controlled.

By contrast, 70% of the phosphates entering Lake Erie and nearly 60% of those entering Lake Ontario are from directly controllable point sources such as municipalities and industries. It is generally true throughout Europe and North America that more phosphorus inputs to lakes are from point sources of municipal or industrial wastes than is the case for nitrogen. Thus not only is phosphorus a key element in algal growth but, because it is introduced primarily by man's activities, it is also the most readily controlled of the three principal elements that are essential in algal growth.

DETERGENT REFORMULATION AND SEWAGE TREATMENT

It is estimated that approximately 50% of the municipal phosphate discharge in Canada to Lakes Erie and Ontario is from detergents; in the United States the corresponding figure is 70% (3). On this basis, detergent phosphates account for about 40% of the total phosphorus loading to the lakes.

Prior to the institution of nutrient control regulations in August 1970, detergents sold in Canada had a very wide range of phosphate content. As noted earlier, laundry detergent powders contained from 16 to 38% phosphates expressed as phosphorus pentoxide or 28 to 66% expressed as sodium tripolyphosphate (STP). Liquid detergents had, and still have, less than 1% STP.

It was argued before the legislative committees which examined the draft nutrient control legislation that the most effective way of solving the problem would be to remove the phosphates and other nutrients by sewage treatment methods, and that if this were done, reformulation of detergents would not be necessary.

There is no doubt that nutrient removal at treatment plants is a very important part of any effective nutrient control program, but there are three strong reasons for requiring detergent reformulation as an essential additional measure. The first involves timing, the second is concerned with costs of treatment facilities, and the third recognizes that many phosphate sources will remain for a long time to come beyond the reach of control facilities.

Timing

In the view of many scientists, a number of Canada's lakes are fast approaching the point at which a major change in the rate of deterioration will occur unless remedial measures are introduced quickly. Lake Ontario may already be in this position. The International Joint Commission recommended that, in the case of Lakes Erie and Ontario, nutrient removal facilities at treatment plants be completed by 1975. Many financial and political problems have to be overcome, however, before this objective can be met, and it was essential that a start be made on reducing the phosphorus loading without waiting until adequate treatment facilities become available. An immediate reduction in the phosphate content in detergents was the obvious course.

Costs of Treatment Facilities

In treatment plants which use alum to remove the phosphorus, the amount of treatment chemical required is in proportion to the amount of phosphorus to be removed from the sewage. In other cases (e.g., where lime is used in the treatment process), chemical costs do not increase in proportion to the amount of phosphorus entering the system, but sludge removal costs are likely to rise with an increase in the phosphorus precipitated. It is very difficult to assess the extent of the savings, but it is clear that the operating costs of nutrient removal facilities at sewage treatment plants will be decreased by reducing the amount of phosphorus entering the plant.

Uncontrolled Sources of Nutrients

Nutrient removal by sewage treatment will eventually be a very important element in nutrient control programs; however, certain sources of nutrients will remain indefinitely beyond the reach of control facilities. These include the combined storm and sanitary sewer systems of Canada's older cities, small municipalities for which it would be uneconomical to provide sewers and treatment facilities, and cottage and resort areas where sewage and waste treatment facilities would be prohibitively expensive and where the health and recreational value of the smaller lakes depends heavily on the success achieved in curbing cultural eutrophication.

Effectiveness of Phosphate Control

There is some scientific doubt whether a phosphate control program will in itself be sufficient to overcome the problem of lakes in an advanced stage of eutrophication. Reports by Edmondson (13) on the effects of a complete sewage diversion scheme to remove all the polluting substances being discharged into Lake Washington reveal that the most significant correlation exists between the reduced phosphate in lake waters in the winter or early spring and the reduction of algal blooms in the lake. In this particular scheme at least, control of phosphates appears to be the most important factor.

Although there is little doubt that a phosphate control program should be the cornerstone of any program to control eutrophication, in some cases additional measures such as nitrogen reductions may well be necessary.

THE PROBLEM OF INTERNATIONAL CONTROL

During the discussions on the nutrient control clause of the Canada Water Act, some doubt was expressed regarding the wisdom of instituting a national detergent phosphate reduction program when Canada's contribution of nutrients to the lower Great Lakes is significantly less than that of the United States.

In the case of Lake Erie, Canada does indeed contribute very little of the phosphate input. Only 2 to 3% of the total input comes from detergent sources in Canada. To save Lake Erie from the effects of worsening eutrophication, therefore, the responsibility for action lies largely with the United States.

In Lake Ontario, on the other hand, excluding the phosphorus input from Lake Erie by way of the Niagara River, the phosphorus load from Canadian sources is about equal to that from the United States. Detergent sources in Canada contribute about 19% of the total phosphorus input to Lake Ontario and about 43% of the total Canadian phosphorus load (3). Obviously, a reduction in Canada's contribution of phosphorus would be significant to the lake as a whole. In addition to the benefits to Lake Ontario, a general program of phosphate control will benefit resort and recreational lakes in other parts of Canada.

REPLACEMENTS FOR PHOSPHATES

The substance which appeared most promising as a replacement for phosphates in detergents was sodium nitrilotriacetate (NTA), and small quantities of this substance were used in some Canadian detergents following the introduction of the August 1, 1970, regulation limiting the use of phosphorus pentoxide to 20%.

The Canadian government is vitally interested in the potential environmental hazards of any substance that may be introduced to replace phosphates, and it has participated in a coordinated three-nation study (with the United States and Sweden) to determine the potential environmental and health impact of large amounts of NTA in lakes and rivers. Canada's participation was undertaken by the Canada Centre for Inland Waters and the Freshwater Institute of the Fisheries Research Board.

Recent findings by the U.S. Public Health Service, based on preliminary results of experiments with rats and mice, indicate the possibility that NTA may cause potentiation of the teratogenic effects of mercury and cadmium. Because of this, major detergent companies in Canada and the United States have temporarily discontinued the use of NTA.

There is still no evidence of human health or biotic toxicity problems associated with the use of NTA in small quantities, and although there is no firm Canadian government policy in the matter, it is the general view that findings of the U.S. Public Health Service require verification. However, the results suggest the need for caution in the use of large quantities of NTA until further research is completed.

It should be emphasized here that the goal of detergent reformulation to reduce phosphate to very low levels does not depend on the suitability of NTA as a substitute. Experiments are being carried out on other substitutes which will provide the complexing capability needed in a detergent. Among these are sodium citrate or citric acid. It may be possible also to achieve satisfactory results by using different washing procedures. The senior author has found that, on the basis of more than a year of continuous use in a home laundry, a low-phosphate liquid detergent plus washing soda will give completely acceptable laundry performance.

SUMMARY

Cultural eutrophication is a serious problem affecting all too many lakes. Phosphorus is the most readily controllable of the elements essential for algal growth and the key element in the lower Great Lakes and in many lakes throughout the world.

On the basis of existing scientific knowledge, a program of phosphate control is the most practical way to combat eutrophication in lakes presently affected, and to prevent the deterioration of lakes where the problem is not yet acute.

An effective program to control phosphates must have three principal aims: the reduction and eventual removal of phosphates from detergents, the removal of phosphates and possibly other nutrients at waste treatment plants, and the control of phosphates from agricultural sources. In response to the recommendations of the International Joint Commission, the Government of Ontario has announced a policy of nutrient removal at the larger municipalities in the drainage basin of the lower Great Lakes. Research on improved agricultural practices to reduce nutrient discharges is under way.

The worsening situation in many of Canada's lakes, particularly the lower Great Lakes, demanded that some effective and immediate action be taken. Anticipated delays resulting from financing and other problems made it impossible to rely on the construction of nutrient removal facilities at waste treatment plants, eventually to be one of the key factors in the control program to furnish the immediate action needed. The obvious course was to institute a program of detergent reformulation, and this was done through the medium of government legislation that limits the phosphate content in detergent products. The legislation permitting the government to institute controls on nutrients in detergents is written into the Canada Water Act.

The government is taking steps also to ensure that any substances that are introduced as replacements for phosphates do not themselves present health or environmental hazards.

The formulation of government policy in the complex area of environmental restoration and preservation presents legislators with the difficult problem of appraising often conflicting interpretations of research information. Involved in these decisions are important considerations of economics, public health, and the public interest. The task of providing the information on which policy can be built, and providing it in a form relatively easy to assimilate, falls to economists and social scientists as well as to their colleagues in the natural sciences.

ACKNOWLEDGMENTS

Review and valuable comments on the paper by Dr. R. A. Vollenweider, Chief, Lakes Division, CCIW, are gratefully acknowledged by the authors.

REFERENCES

1. R. A. Vollenweider, "Scientific Fundamentals of Eutrophication of Lakes and Flowing Waters, with Special Reference to Nitrogen and Phosphorus as Factors in Eutrophication," Technical Report Organization for Economic Co-operation and Development (OECD), 1968.
2. "Eutrophication in Large Lakes and Impoundments," Proceedings of the Uppsala Symposium. Organization for Economic Co-operation and Development, Paris, 1970.
3. Report to the International Joint Commission on the Pollution of Lake Erie, Lake Ontario, and the International Section of the St. Lawrence River, 1969.
4. International Joint Commission, "Special Report on Potential Oil Pollution, Eutrophication, and Pollution from Watercraft," 3rd Interim Report on Pollu-

tion of Lake Erie, Lake Ontario, and the International Section of the St. Lawrence River, April 1970.
5. International Joint Commission, "Pollution of Lake Erie, Lake Ontario, and the International Section of the St. Lawrence River," Final report, December 1970.
6. Canada Centre for Inland Waters, Fisheries Research Board of Canada, Inland Waters Branch, "The Control of Eutrophication," Tech. Bull. 26, Inland Waters Branch, Ottawa, 1970.
7. M. Sakamoto, "Chemical Factors Involved in the Control of Phytoplankton Production in the Experimental Lakes Area, Northwestern Ontario," J. Fish. Res. Bd. Can., 28 (2), 203-213 (1971).
8. L. E. Kuentzel, "Bacteria, Carbon Dioxide and Algal Blooms," J. Water Pollut. Control Fed., 41, 1737-1747 (1969).
9. C. N. Sawyer, "Fertilization of Lakes by Agricultural and Urban Drainage," J. New Eng. Water Works Assoc., 61 (2), 109-127 (1967).
10. R. A. Vollenweider, unpublished note, Canada Centre for Inland Waters, Fisheries Research Board Detachment, 1970.
11. E. A. Thomas, "Empirische und experimentelle Untersuchungen zur Könntnis der Minimumstoffe in 46 Seen der Schweiz. und angrenzender Gebiete," Schweiz. Ver. Gas Wasserfachm, 2, 1-15 (1953).
12. J. R. Vallentyne, "Phosphorus and the Control of Eutrophication," Can. Res. Dev. Mag., May/June, 1970.
13. W. T. Edmondson, "Phosphorus, Nitrogen and Algae in Lake Washington after Diversion of Sewage," Science, 169, 690-691 (1970).

INDEX